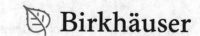

Nečas Center Series

More information about this series at https://link.springer.com/bookseries/16005

Josef Málek • Endre Süli

Editors

Modeling Biomaterials

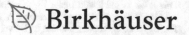 Birkhäuser

Editors
Josef Málek
Faculty of Mathematics and Physics
Charles University
Prague, Czech Republic

Endre Süli
Mathematical Institute
University of Oxford
Oxford, UK

ISSN 2523-3343   ISSN 2523-3351   (electronic)
Nečas Center Series
ISBN 978-3-030-88083-5   ISBN 978-3-030-88084-2   (eBook)
https://doi.org/10.1007/978-3-030-88084-2

Mathematics Subject Classification: 00A71, 65C20, 65C40, 65M32, 65M50, 65M60, 65Y05, 65Z05, 74A05, 74A10, 74A15, 74A20, 74A25, 74A30

This book is published under the imprint Birkhäuser, www.birkhauser-science.com by the registered company Springer Nature Switzerland AG
The registered company address is: Gewerbestrasse 11, 6330 Cham, Switzerland

Preface

The investigation of the role of mechanical and mechanochemical interactions in cellular processes and tissue development is a rapidly growing research field in life sciences and in biomedical engineering. Quantitative understanding of this important area in the study of biological systems requires the development of adequate mathematical models for the simulation of the evolution of these systems in space and time. This calls for a multidisciplinary approach since expertise in various fields is necessary.

Due to the necessity to describe the processes at different spatial and temporal scales, numerous approaches are used in the modeling of biomaterials. Methods of statistical physics may be employed at the subcellular or cellular level (e.g., in models of biomembranes or cellular interactions). When moving to the level of organs, continuum mechanics may be a more suitable approach. A combination of continuum and discrete approaches may be used, for instance, in the heart, which is typically modeled as a viscoelastic material with active stress generated by chemical interactions of actin and myosin molecules.

The objective of the present book, entitled **Modeling Biomaterials** and consisting of six independent chapters, is at least twofold. On the one hand, each chapter can be viewed as an introduction to a given field, where the reader is exposed to various approaches to modeling different materials, such as living tissues. On the other hand, each chapter has been written by active researchers with the goal of connecting the basic foundations of their field with very recent advances and open questions. The result is six chapters covering basic physical, biological, and physiological concepts as well as further research perspectives from multiple points of view.

It is our belief that the multidisciplinary nature of each chapter will stimulate the interaction of researchers from various domains of science (including applied mathematics, biophysics, mathematical biology, computational physiology, pharmacology, and medicine), for whom the book will provide a helpful starting point, particularly at the start of their research career in these interdisciplinary fields. Thanks to the wide range of approaches that arise in models of living tissues, the book may also serve as a useful reference for experienced researchers working in these fields.

While the first four chapters are universal in their approach and are not explicitly related to a specific organ, Chaps. 5 and 6 survey the recent advances in models of the cardiovascular system, including state-of-the-art clinical applications in cardiovascular medicine.

The individual chapters (lecture notes) stem from the lecture courses presented by the speakers of the international doctoral school "Modeling of Biomaterials," which took place in Kácov (Czech Republic), February 10–16, 2020. In several cases, the speakers prepared their contributions with co-workers. The book consists of the following chapters (the names of the speakers who were, at the same time, the corresponding authors for a particular chapter, are typeset in boldface):

1. **Oded Farago**: *A Beginner's Short Guide to Membrane Biophysics*
2. Georgios Misailidis, Jaroslav Ferenc, and **Charisios D. Tsiairis**: *Self-organization of Tissues Through Biochemical and Mechanical Signals*
3. Michele Righi and **Valentina Balbi**: *Foundations of Viscoelasticity and Application to Soft Tissue Mechanics*
4. **Václav Klika**: *Modelling of Biomaterials as an Application of the Theory of Mixtures*
5. Renee Miller, David Marlevi, Will Zhang, Marc Hirschvogel, Myrianthi Hadjicharalambous, Adela Capilnasiu, Maximilian Balmus, Sandra Hager, Javiera Jilberto, Mia Bonini, Anna Wittgenstein, Yunus Ahmed, and **David Nordsletten**: *Modeling Biomechanics in the Healthy and Diseased Heart*
6. **Radomír Chabiniok**, Katerina Škardová, Radek Galabov, Pavel Eichler, Maria Gusseva, Jan Janoušek, Radek Fučík, Jaroslav Tintěra, Tomáš Oberhuber, and Tarique Hussain: *Translational Cardiovascular Modeling: Tetralogy of Fallot and Modeling of Diseases*

The delivered lectures and the chapters of this book focus on mechanical and mechanochemical interactions in cellular processes and tissue development. The topics ranged from molecular dynamics simulations of lipid membranes to phenomenological continuum mechanics of tissue growth.

We wish to express our gratitude to all corresponding authors and their co-workers for their effort in preparing their chapters in a way that is accessible to nonspecialists, yet useful to the experts in the field.

The doctoral school was organized in the framework of the 4EU+ European University Alliance, which is formed by the Charles University, Heidelberg University, Sorbonne University, and the Universities of Copenhagen, Milan, and Warsaw. The school was also organized as an activity of the Nečas Center for Mathematical Modeling. The event was sponsored by the Charles University's 4EU+ project "Modelling and Analysis in Biomaterials", by the Czech Science Foundation project No. 18-12719S and by the RSJ Financial Group. We thank the members of the scientific committee of the doctoral school, namely Radomír Chabiniok, Anna Marciniak-Czochra, Michal Pavelka, Benoît Perthame, and Vít Průša, for their significant help in establishing an interesting program covering different areas of biophysics.

Last but not least, we would like to thank those who have helped with reading and reviewing the individual chapters: Christoph Allolio, Anna-Marciniak Czochra, Federica Cafario, Radomír Chabiniok, Alena Jarolímová, Moritz Mercker, Vít Průša, Renee Miller, and Ondřej Souček. We also thank Hana Bílková, Martina Kašparová, and Oldřich Ulrych for technical support concerning both the school and this book.

Josef Málek
Endre Süli

Contents

A Beginner's Short Guide to Membrane Biophysics

Oded Farago

Abstract This chapter provides a basic guide to the essentials of membrane biophysics. The review consists of three parts: The first part focuses on physico-chemical aspects of biomembranes. The origin of the hydrophobic attraction that drives the self-assembly of lipids into large structures like membranes is explained, and an overview of the thermodynamic properties of bilayer membranes is given. The second part introduces the Helfrich model for the curvature elasticity of two-dimensional manifolds. This is the most commonly used framework for studying the mechanical properties of lipid membranes. The third part of the chapter is dedicated to molecular simulations of membranes. The principles of molecular dynamics simulations are explained, and the different modeling strategies for membranes, ranging from atomistic to highly coarse-grained lattice-based models, are reviewed. These different classes of models address membrane properties and processes across a wide spectrum of lengths and time scales.

1 Introduction

A cell is the smallest unit of living organisms. The simplest organisms contain no internal organelles (prokaryotes) and have only a single cell, e.g., bacteria. Plants and animals are composed of millions of cells with complex substructures (eukaryotes). Our body, for instance, has about 200 different types of cells and an overall number of about 10^{14} cells. All cells have a cell membrane that defines the cell boundary and separates between the internal cell cytoplasm and the external environment. The membrane controls flow of materials into and out of the cell. Eukaryotic cells also contain membrane-bound organelles like the nucleus, mitochondria, and the endoplasmic reticulum. The cell's compartmentalization into organelles enables the separation of biological functions necessary for the life of organisms.

O. Farago (✉)
Department of Biomedical Engineering, Ben-Gurion University of the Negev, Be'er Sheva, Israel
e-mail: ofarago@bgu.ac.il

© The Author(s), under exclusive license to Springer Nature Switzerland AG 2021
J. Málek, E. Süli (eds.), *Modeling Biomaterials*, Nečas Center Series,
https://doi.org/10.1007/978-3-030-88084-2_1

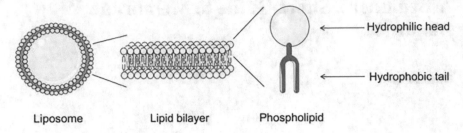

Fig. 1 The main building blocks of membranes are lipids (right). Lipids self-assemble and form bilayers composed of two sheets, where the hydrophilic head groups are facing the extracellular and intracellular aqueous environments (center). Bilayers cannot leave "open edges" and, therefore, close into spherical vesicles, also called liposomes (left). © 2017 Wiley Periodicals, Inc. Reprinted from [1] with permission

The main building blocks of membranes are lipids, which are amphiphilic molecules, i.e., molecules containing both hydrophilic ("water loving") and hydrophobic ("water hating") parts. When lipids are dispersed in water, they spontaneously form a bilayer membrane structure, as shown in Fig. 1. In this arrangement, only the hydrophilic "head groups" of the lipids come into contact with the aqueous environment, while their hydrophobic "tails" remain shielded. In order to also avoid exposure of lipid tails at the edge of the bilayer, membranes bend and close themselves, forming spherical vesicles (liposomes) whose dimensions typically vary between tens of nanometers to a few micrometers (Fig. 1). In computer simulations, planar bilayers can be stabilized and avoid closure by the application of periodic boundary conditions.

Biological membranes are composed of many different types of lipids. Phospholipids (lipids with a phosphate-containing head group) constitute the main lipid mass of eukaryotic cell membranes. The cell membrane also includes other components such as cholesterol (a lipid with a structure different from that of phospholipids), proteins, and carbohydrates (sugars) attached to some of the proteins and lipids. The compositions of the two layers (leaflets) of the membrane, the one facing the extracellular environment and the other the cell cytoplasm, are different owing to the action of specialized proteins. The compositional diversity and heterogeneity of biological membranes are essential for their functions, but it also makes the investigation of their physical properties extremely difficult. Physicists look for simplicity and universality in nature, and they do so by analyzing minimal model systems that, presumably, capture the most essential physical mechanisms governing the investigated phenomena. It is therefore not surprising that much of the advance in the field of membrane biophysics results from investigations of membranes and vesicles composed of a small number of lipid species, often only a single component and without any additional "inclusions" (which is a common biophysical term for membrane proteins). Despite their simplicity, the behavior of these biomimetic model systems is far from trivial and their studies have contributed to much of our current understanding of the thermodynamic and mechanical properties of biological membranes. From a thermodynamic perspective, the fact that lipids in water self-assemble spontaneously to form membranes and vesicles implies that these structures represent an equilibrium state of the system with a minimal free energy. Self-assembly can only occur if there is an attraction between the lipids that reduces the enthalpy of the system and overcomes the entropic tendency of the lipids to be uniformly distributed in the solvent. The attraction between the lipids originates from the "hydrophobic effect" that, in fact, arises because of the repulsion between the hydrophobic tails of the lipids and water. It plays an important role not only in the self-assembly of membranes, but also in other biological process, e.g., protein folding. In Sect. 2, we expand on the hydrophobic effect and other important aspects of lipid dispersion in water. We will also discuss some of the thermodynamics properties of biological membranes, e.g., their phase behavior.

Owing to their thinness (4–5 nm, twice the typical size of a lipid molecule) and fluidity, bilayer membranes are flexible and can quite easily bend out of plane. The ability of membranes to deform is important in many biological processes like cell division (mitosis) and vesicular transport into and out of the cell (endo- and exocytosis). Membrane deformations are often caused by forces that are applied on the membrane by other cellular components, e.g., the filaments of the cell cytoskeleton. In this context, it should also be noted that membranes exhibit natural thermal fluctuations with non-negligible amplitudes. In contrast to the softness of membranes with respect to bending deformability, their in-plane elasticity is quite limited and resembles that of a dense liquid (little compressibility, no shear rigidity). Thus, most of the focus in studies of mechanical properties of bilayer membranes has been on their bending elasticity and its relevance to biology. The most commonly used and widely accepted theoretical framework for membrane bending elasticity is that of the Helfrich effective Hamiltonian. The Helfrich model is a continuum theory describing the membrane as a two-dimensional manifold with vanishing thickness embedded in a three-dimensional space. Continuum models are very useful because they can be analyzed by different mathematical tools, which may yield important theoretical predictions. Their utility relies on the concept of separation of scales, namely the realization that knowledge of the detailed molecular structure of physical systems is often not essential for the understanding of their macroscopic properties. Instead of detailed molecular models, effective theories are constructed that describe the behavior of the system at sufficiently large length and time scales. In the continuum description, the properties of specific systems are accounted for via the coefficients appearing in the model equations. In the case of the Helfrich model, for instance, the bending energy is written as a quadratic function in the local principal curvatures of the manifold. The model features a few elastic parameters (bending modulus, Gaussian curvature modulus, spontaneous curvature) whose values characterize bilayers with different lipid compositions. The Helfrich model will be introduced and discussed in Sect. 3, which will be devoted to the mechanical properties of bilayer membranes.

Biophysics is inherently a multi-scale discipline. It deals with systems that are structurally complex with local intermolecular interactions that may trigger collective phenomena and processes across a wide range of spatial and temporal scales. Obviously, this enormous complexity cannot be addressed by continuum theories only; it requires also investigations of molecular models. In the last three decades, computer simulations have become an essential tool for biophysical research, and their importance continues to grow with the advent in commodity computing power and the improvement in computer models and methods. These models range from fully atomistic models, where water, lipids, and possibly proteins are represented with each and every atom, up to highly coarse-grained (CG) models where the lipids are represented as simple bead-spring short chains. Some CG models do not even contain an aqueous solvent explicitly and, instead, account for the hydrophobic effect via an effective attractive potential between the hydrophobic beads of the CG lipids. Different classes of models address membrane processes at different lengths and time scales. Atomistic models, for instance, allow nowadays to simulate lipid patches consisting of the order of 1000 molecules on time scales of several microseconds. CG simulations reach larger scales that depend on the level of abstraction of the simulated models and on the available computer resources. The topic of multi-scale modeling and simulations of membranes will be discussed in Sect. 4. This section will also include an explanation on the method of Molecular Dynamics (MD), where issues influencing the accuracy and efficiency of the simulations will be discussed. The discussion is important since most of the membrane simulations are nowadays conducted using open-source packages that can easily go wrong and produce "bad" results if used incorrectly.

As the title suggests, this chapter aims to serve as an introductory guide for those who have little to no knowledge in membrane biophysics. It is especially intended for graduate students in mathematics, computer science, and engineering who look for an introduction to the field in order to pursue further studies and research but have minimal background in physical chemistry. The decision to cover certain topics in this introductory review is obviously subjective; others may introduce the field from a different perspective. In particular, I decided to focus on the material science aspects of lipid membranes. The cell biology of membranes has been largely left outside the discussion, and the reader is encouraged to acquaint oneself on this topic from other sources, e.g., the book by Alberts et al. that many academic-level courses on cell biology follow [2].

2 Thermodynamics

2.1 Everything in Biology Happens in Water

Water is present both outside and inside cells, and it makes up about 60% of the weight of the human body. Everything in biology happens in water which is a very unusual solvent. Its uniqueness lies in the fact that water, H_2O, is probably among the most polar molecules in nature. It has a single oxygen atom covalently bonded to two hydrogens (see Fig. 2), but these bonds are very asymmetric and the electrons

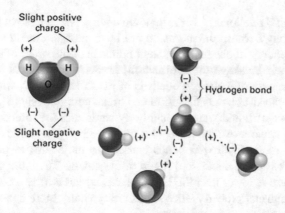

Fig. 2 A polar water molecule (left)-the oxygen and hydrogen sides are electronegative and - positive, respectively. The polarity of the water molecules allows them to engage in up to 4 hydrogen bonds with their neighbors, thus forming a dynamic hydrogen bonding network (right) that is the origin of many unusual properties of bulk water. Source: https://bodell.mtchs.org/ OnlineBio/BIOCD/text/chapter4/concept4.4.html

forming them are strongly pulled toward the oxygen. It is said that the oxygen side of water is electronegative, i.e., behaves a bit like a partially negatively charged ion, while the hydrogen is electropositive. The interaction between the electric dipoles of polar molecules is generally attractive because the dipoles can orient themselves so that the negative side of one dipole will be in close proximity with the positive side of another. Dipole–dipole attractions between molecules with hydrogen (H) atoms that are covalently bonded to electronegative atoms like oxygen (O) and nitrogen (N) are common in biology and have a special name: hydrogen bonds. In bulk water, the attraction is particularly strong because of the smallness, abundance, and high polarity of the H_2O molecules. Each water molecule can participate in up to four hydrogen bonds with its neighbors, thus forming a network of hydrogen bonds (Fig. 2).

Hydrogen bonds between H and O atoms of different water molecules are much weaker than the O-H covalent bonds of the same molecule. The strength of molecular bonding interactions is measured by the amount of energy required to break them. The distinction between weak and strong bonds is made by comparing their energy to the so-called thermal energy $k_B T$, where $k_B = 1.38 \times 10^{-23}$ J/K (Joule per degree Kelvin) is Boltzmann's constant, and T is the absolute temperature measured in Kelvin. At the physiological temperature $T = 310$ K ($= 37\,^\circ$C), the thermal energy $k_B T = 4.2 \times 10^{-21}$ J. The thermal energy, or heat, is essentially the kinetic energy of the randomly moving molecules [3]. More precisely, when a system is in thermal equilibrium, the average kinetic energy per degree of freedom is equal to $k_B T/2$. The water molecule, for instance, has six degrees of freedom (3 translational and 3 rotational).[1] If the bonding energy is much larger than $k_B T$, then the bond is strong because the atoms do not have enough kinetic energy to break it. The typical energy of chemical covalent bonds is of the order of a few hundreds of $k_B T$, and their breakage (and formation) involves a chemical reaction [4]. In contrast, the strength of hydrogen bonds is of the order of $k_B T$, which means that these are weak physical bonds that form and break continuously [5]. The hydrogen bonding network of liquid water is highly dynamic and the molecules are free to move. When water freezes, the molecules organize in a crystalline structure where the orientation of the hydrogen bonds causes the molecules to push each other further compared to the disordered liquid arrangement. Thus, the density of H_2O in the solid phase is lower than the liquid phase, which is why ice floats on water. This is a very unusual property since solids are typically more densely packed and heavier than the corresponding liquids.

The fact that ice is lighter than liquid water is only one of many unique properties of water that are attributed to their hydrogen bonding network structure [7]. Water boils at $T = 100\,°C$, which is an incredibly high boiling temperature considering its small molecular weight. For comparison, if we look at the sixth column of the periodic table, we locate sulfur (S) one row below oxygen. Generally speaking, the heavier the molecule the higher the boiling temperature of the material; therefore, we expect H_2S to have a higher boiling temperature than H_2O, but the former boils at a much lower temperature: $T = -62\,°C$. The high boiling temperature is because evaporation of liquid water requires a large amount of energy to break the hydrogen bonding network. Another unique property of liquid water is its high heat capacity (specific heat capacity at constant pressure, $C_P = 4.2 \times 10^3\ \mathrm{JK^{-1}Kg^{-1}}$). The heat capacity is the amount of energy required to raise the temperature of a unit mass of a substance by $1°$ (or the amount of energy released when it cools down by $1°$). That the heat capacity of water is high means that it takes a great deal of heat to change the temperature of water. The water in our body keeps its temperature highly regulated, and the water in the oceans prevents extreme temperature changes on Earth compared to other planets. The heat capacity of water is high because much of the heat absorbed by the water is used for the breaking of the hydrogen bonds, and only part of it serves to raise the kinetic energy of the water molecules.

2.2   The Concepts of Hydrophobicity and Hydrophilicity

Water is an excellent solvent and, in fact, more materials dissolve in water than in any other liquid (see chapter 30 in [5]). When a foreign molecule enters water, it takes a volume that was previously occupied by other water molecules and, therefore, locally modifies and even breaks the hydrogen bond network of the water.

---

[1] Water also has three vibrational degrees of freedom, but they are irrelevant at room temperature and become excited only at temperatures of several thousand degrees.

The molecule will remain in the water only if its interaction with the water is thermodynamically more favorable than the water–water interactions lost in this process; otherwise, it will not be dissolved. Substances that dissolve easily in water are "water-loving," or hydrophilic substances, while those that do not dissolve in water are "water-hating," or hydrophobic.

Ions and polar molecules are generally hydrophilic. They create electric fields that cause the water molecules around them to reorient in a manner that reduces the overall free energy of the system. In contrast, oily molecules containing hydrocarbon chains (chains of carbons and hydrogens, like the tails of the lipids in Fig. 1) are hydrophobic. This is because carbon is an electroneutral element, which means that the C-H bonds of the chain are non-polar. Therefore, they cannot replace water molecules in forming hydrogen bonds with other water molecules. The repulsion of hydrophobic materials from water is the origin of the "hydrophobic effect" that refers to the effective attraction between a-polar hydrophobic groups [8]. It arises because the free energy penalty of dissolving an oily component in water comes from the oil–water interface, where the hydrogen bonding network is damaged. To reduce the area of contact, it is better for the hydrophobic chains to segregate away from the water (despite the associated loss of translational entropy), which means that they are effectively attracted to each other.

You can see the hydrophobic effect in action when you try to pour oil into a glass of water and the oil, whose density is lower than the water, floats. In this arrangement, the surface separating the oil and water molecules is minimal. Thermodynamically, this effect can be described by a linear relationship between the free energy, G, and the hydrophobic–hydrophilic surface area, A

$$G = \gamma A, \tag{1}$$

where γ is called the surface tension and has units of energy per unit area [9]. The values of the surface tension at water–oil interfaces typically vary from 0.2 to $0.5 \, \mathrm{J/m^2}$.

2.3 Amphiphilic Molecules

Molecules with both hydrophilic and hydrophobic parts are called amphiphilic. Cleaning detergents, for instance, contain amphiphilic molecules with one hydrophobic chain that are commonly called "surface active agents," or surfactants [10]. When one tries to wash with water an oily dirt from a dish or a cloth, the high oil–water surface tension prevents the water from sticking to the dirt and removing it. Surfactants facilitate the attachment between the oil and water because they have affinity to both: Their hydrophilic parts reside in water and the hydrocarbon tails in oil. Thermodynamically, we say that they reduce the effective oil–water surface tension and even eliminate it [11]. If we add surfactants to the oil–water interface in the example above, many droplets of oil may be created and

Fig. 3 Sketch of a micelle
formed by single-tail
surfactants

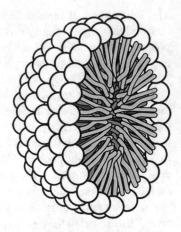

dispersed in water [12]. This is possible because the oil–water–surfactant surface tension is vanishingly small, and there is no longer a free energy penalty associated with the formation of structures with a large interfacial area.

When amphiphilic molecules are dispersed in water without oil, they need to hide their hydrophobic parts from aqueous contact. Therefore, they self-assemble into structures where only the hydrophilic parts are exposed to the water. As discussed above, this spontaneous aggregation process can be understood as if it is driven by a hydrophobic attraction between the hydrocarbon chains. Single chain amphiphilic molecules have a conical shape assemble into micelles as shown in Fig. 3 [4]. Lipids, with two tails, have more of a cylindrical shape, and so they aggregate into bilayers (Fig. 1). In contrast to micelles, there is no limit to the number of lipids that can cluster in a single bilayer. However, in order to also protect the lipids at the edge of the bilayer, the bilayer must close to form a vesicle (liposome, Fig. 1). The critical size above which this closure happens is dictated by a competition between the curvature elasticity of the vesicle and the excess energy of the open edge of the bilayer. We will return to this issue in the following Sect. 3, where we discuss the elasticity of membranes and vesicles.

2.4 Membranes are Two-Dimensional Liquids

At physiological temperatures, biological membranes are found in a liquid-crystalline state commonly referred to as the liquid-disordered phase. The term "liquid" refers to the mobility of the lipids in the membrane plane, and "disordered" refers to the state of their hydrophobic tails. Explicitly, the hydrocarbon tails of the lipids are essentially short polymer chains. When they are disordered, they switch between different molecular configurations (see Fig. 4, right). This means that the disordered state has a high entropy, which is thermodynamically desirable. The thermal motion reflecting the tendency to increase the entropy generates

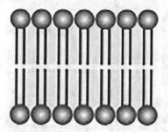 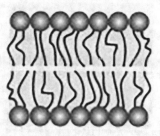

Fig. 4 A membrane undergoing a melting–freezing transition between the low-temperature "gel" phase (left), where the lipid chains are ordered and tightly packed, and the "liquid disordered" phase (right), where the chains are disordered and the lipids are free to diffuse. © 2009 Elsevier Ltd. Reprinted from [6] with permission

collision repulsion between the chains. This entropic repulsion competes with the hydrophobic effect that pulls the lipids closer to each other. If the hydrophobic attraction wins, the tails would adopt a stretched "ordered" configuration that would allow them to pack tightly. This is what happens in the "gel" phase of the membrane, where the lipid tails become highly ordered (Fig. 4, left), and the membrane "freezes," i.e., the lipids lose their mobility and, instead, arrange in a triangular fashion in the membrane plane. The transition between the gel (G) and the liquid-disordered (LD) states is essentially a melting–freezing transition that takes place at the transition temperature T_m [5]

$$T_m = \frac{H_{LD} - H_G}{S_{LD} - S_G},\tag{2}$$

where S and H denote, respectively, the entropy and enthalpy (energy) of the corresponding phases.

Biological membranes are mixtures of many different types of lipids, and their melting–freezing temperature is dictated by their lipid composition. For instance, typically each hydrocarbon chain of a lipid consists of 12–18 carbons. A greater fraction of lipids with longer tails would generally increase T_m because the hydrophobic attraction between the longer tails is stronger (i.e., H_G in Eq. 2 is lower). Another influential factor on T_m is the amount of saturated and unsaturated lipids in the mixture. Lipids are said to be saturated if all the covalent bonds connecting adjacent carbons along their tails are single covalent bonds, i.e., involve the sharing of one pair of electrons. They are unsaturated if one or more of the bonds is a double one (i.e., involving two pairs of electrons). Unsaturated lipids can be further divided into *trans* and *cis* fatty acids. The former are rare in nature but exist in industrial food and can be harmful to your health for reasons that are beyond the scope of this monograph. Lipids with *cis*-unsaturated tail or tails, on the other hand, have a very important physiological role. The *cis* double bond generates a kink in the lipid, thus making it less amendable for tight packing. In other words, the presence of unsaturated lipids increases the fluidity of the membrane and reduces

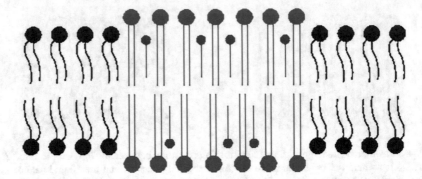

Fig. 5 Cholesterol (red) is a small amphiphilic molecule inserting itself between the lipids of the bilayer. It prefers the proximity of saturated lipids with ordered chains (blue) over unsaturated disordered ones (black), thus forming liquid-ordered membrane domains, sometimes termed as lipid "rafts"

T_m. Lipids with one saturated and one unsaturated tail are sometimes called "hybrid" lipids [13].

2.5 Biological Membranes are Heterogeneous

Another regulator of membrane fluidity is cholesterol that is a small amphiphilic molecule [14]. The smallness of the cholesterol molecules allows them to intercalate between the lipids in the bilayer (Fig. 5). When added at high enough concentration to a bilayer of saturated lipids, cholesterol induces the formation of a "liquid ordered" phase, which is a kind of intermediate state between liquid-disordered and gel phases. As the name suggests, this is (i) a liquid phase where the lipids undergo lateral diffusion (albeit slower than in the liquid-disordered phase) and (ii) a state where hydrocarbon chains are ordered and quite tightly packed, similarly to the gel phase. These features of the liquid-ordered phase reflect the dual effect of the cholesterol on the saturated lipid chains. In the liquid-disordered phase, the chains are disordered, but in the presence of cholesterol they tend to adopt more ordered configurations and, therefore, to pack more tightly. In the gel phase, on the other hand, the cholesterol disrupts the chain ordering and packing.

When cholesterol is mixed with both saturated and unsaturated lipids, it tends to associate with the former. At certain compositions, the ternary mixture would phase separate into a liquid-ordered phase enriched in saturated lipids and cholesterol and a liquid-disordered phase of mainly unsaturated lipids [15, 16]. The phase behavior of mixtures of lipids and cholesterol and, in particular, the properties of liquid-ordered domains have been in the limelight of biophysical research for many years. This is because they are perceived as simple models for lipid rafts. Lipid rafts in biological membranes are small (10–200 nm) liquid-ordered domains, enriched in saturated lipids (mainly sphingolipids), cholesterol, and often particular proteins [17, 18]. Raft domains are highly dynamic and are believed to play a key role in cellular processes such as signal transduction, cell adhesion, and membrane trafficking.

Appearance of short-lived small domains is contradictory to the notion that coexisting phases must be macroscopically separated in order to minimize the interface between them and the associated energy per unit length (line tension) [19]. The formation of raft domains implies that the line tension separating them from the embedding sea of liquid lipids is vanishingly small. Several non-exclusive mechanisms have been proposed for this observation [20, 21]. For instance, rafts may be transient fluctuations in the local composition. Since the physiological temperature is close to the melting transition temperature of many biological membranes, these fluctuations are indeed expected to be significant. Another possible reason for the reduction in the line tension between liquid-ordered domains and the liquid-disordered membrane is related to the abovementioned hybrid lipids. Such lipids tend to reside at the line interface, with the saturated tail in the raft domain and the unsaturated one in the surrounding lipid matrix. This mechanism resembles the way surfactants reduce the oil–water surface tension (Sect. 2.3). Proteins, which may attract certain lipids and repel other, may also facilitate inhomogeneous distribution of the lipids and promote raft formation. We thus conclude that the compositional complexity of biological membranes produces specialized domains on the cell surface, which are crucial for many biological processes and functions. These are typically studied using simpler biomimetic model systems containing only a small number of components that, nevertheless, exhibit a complex phase behavior.

Another class of mechanisms proposed for the stabilization of small raft domains is based on the concept of a coupling between the local membrane composition and the curvature elasticity. The basic idea is that the bending stiffness of the membrane is a function of the local composition of lipids and proteins. Under certain circumstances, an inhomogeneous distribution of the lipids (i.e., formation of different domains) may actually have a lower elastic curvature energy compared to that of a uniform membrane with the same components. This takes us to the next section in our beginner's guide to biological membranes, where we will present the elasticity theory of lipid membranes.

3 Elasticity

3.1 In-Plane Elasticity

Elasticity of isotropic two-dimensional surfaces is characterized by two elastic moduli, the compression/stretching modulus K and the shear modulus μ [22]. The latter describes the resistance of a surface to planar shear deformations, i.e., deformations that change the shape of a flat surface without changing the overall area. For liquid surfaces like a lipid bilayer in the liquid-disordered phase, the shear modulus vanishes because liquids (and gases) can flow and adopt to the change in the shape of the "container" without any elastic response. The modulus K describes the resistance of the surface to compression or stretching and is proportional to the derivative of the two-dimensional elastic (mechanical) stress, τ, with respect to the area A [23]

$$K = A_0 \frac{\partial \tau}{\partial A}, \tag{3}$$

where A_0 is the unstretched area of the surface, i.e., the area for which $\tau = 0$. Assuming a small deformation (small strain), the elastic energy can be approximated by a simple quadratic function

$$E = \frac{1}{2} K \frac{(A - A_0)^2}{A_0}, \tag{4}$$

and the tension is

$$\tau = \frac{\partial E}{\partial A} = K \frac{(A - A_0)}{A_0}. \tag{5}$$

This is Hooke's law for linear elastic materials, i.e., materials for which there is a linear relationship between the stress τ and the strain $\epsilon = (A - A_0)/A_0 = \Delta A/A_0$. Equation 3 is readily derived from Eq. 5, although it should be noted that only for linear elastic materials K is a constant. In the more general case, K is defined by Eq. 3, and it is a function of the area, $K = K(\epsilon)$.

The stress–strain relationship of lipid bilayers has the form shown in Fig. 6 [24]. The relationship is linear over most of the range $\epsilon \lesssim 0.05$, which is the strain at which lipid bilayers typically rupture. However, at very low-strain values, the

Fig. 6 The stress as a function of the strain of various lipid vesicles. © 1990, American Physical Society. Reprinted from [24] with permission

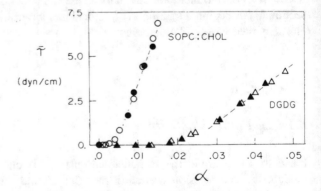

tension grows exponentially with the strain, which is a very unusual behavior as non-linear elasticity is usually associated with large deformations. The origin of this unique elastic response is the fact that bilayer membranes are not perfectly flat surfaces. They are thin sheets that also fluctuate in the normal (out-of-plane) dimension. The area measured experimentally is the area projected onto the two-dimensional membrane plane, A_p. For vesicles, this is the surface area of the sphere obtained when the fluctuations in the radius are smoothed. The membrane physical area is larger than the projected area. When stress is applied to stretch the membrane, it first flattens the thermal undulations of the membrane without changing its physical area. The response to the applied stress in this regime is of entropic nature. Reducing the amplitude of the thermal undulations means that the membrane achieves a smaller number of configurations and has a lower entropy, i.e., a higher free energy, than the unstretched state. The exponential low-tension regime in Fig. 6 characterizes this entropic elasticity. Only at higher values, after the undulations have been ironed out, the physical area of the membrane increases and linear elasticity is observed [23, 24].

The entropy associated with the membrane thermal out-of-plane undulations is directly related to the bending elasticity of membranes. Generally speaking, it is much easier (i.e., requires substantially less work) to bend a membrane than to stretch it, and it is mainly for this reason that curvature deformations play an important role in many biological processes. The reminder of this section will, therefore, be dedicated to the curvature elasticity of membranes. Before we continue, a final comment on the in-plane mechanical tension (stress) τ: There is great confusion in the literature regarding the difference between τ and the surface tension γ defined in Eq. 1 [25]. The mechanical tension describes the response of a membrane with a fixed number of lipids to an area change causing a change in the density of the lipids, i.e., a change in the average intermolecular distances between them. In contrast, the surface tension is the free energy cost to increase the surface between two fluids by a unit area. In this case, we do not stretch the interface between the fluid materials but simply create more surface area by bringing molecules from the bulk of the fluids to the boundary between them. At the interface, the molecules of the different fluids can interact with each other via interactions that are typically less favorable than the bulk interactions with molecules of the same type. This is the reason, for instance, why droplets adopt a spherical shape that has minimal surface area at a given volume. Unfortunately, in the membrane biophysics literature, it has become common to refer to τ as the "surface tension." Some of the confusion in the literature arises because a fluctuating membrane has two areas: the physical and the projected one. The correct definition of τ is not the one given in Eq. 5 but the following one [26]

$$\tau = \frac{\partial G}{\partial A_p},$$

(6)

i.e., the derivative of the free energy (rather than energy) with respect to the projected area (rather than the physical area).[2] Differentiation in Eq. 6 is performed when the number of lipids and temperature are held constant. Importantly, the free energy includes both entropic and enthalpic (energetic) contributions, which means that, in principle, Eq. 6 captures both the linear and exponential regimes in Fig. 6 for the stress–strain relationship of bilayer membranes and vesicles.

3.2   Curvature Elasticity

Curvature elasticity describes the response of a body to bending deformation and is the main factor controlling the shape of bilayer membranes and vesicles. To better understand the formalism of curvature elasticity in membranes, let us first consider a simpler example that is the bending elasticity of polymers. Polymers are long, chain-like, molecules consisting of repeating units (monomers) that are covalently bonded to each other. Polymer chains may adopt many different configurations, and if the number of monomers is sufficiently large, the framework of equilibrium statistical mechanics may be applied to describe their thermodynamic behavior, including their bending elasticity. A fundamental principle of classical equilibrium statistical mechanics is the notion that the thermal motion of the particles (the monomers in the case of polymer chains) allows the system to sample the configuration phase space and that the occurrence probability of the different states follows the Boltzmann distribution law [3, 5]

$$p_{\text{configuration}} = \frac{e^{-E_{\text{configuration}}/k_B T}}{Z},$$ (7)

where $E_{\text{configuration}}$ is the energy of the state, $k_B T$ is the thermal energy (see Sect. 2.1), and Z is the normalizing partition function

$$Z = \sum_{\text{configurations}} e^{-E_{\text{configuration}}/k_B T}.$$ (8)

The free energy of the system is related to the partition function via

$$G = -k_B T \ln Z.$$ (9)

---

[2] In contrast to A_p, the physical area A is not a thermodynamic variable that can be easily fixed. In some theoretical studies [27], the term "surface tension" is used to describe a Lagrange multiplier that fixes the *average* physical area.

From the free energy, we can calculate the relationship between the end-to-end extension of the polymer chain, R_{ee}, and the applied force, f [28]

$$f = -\frac{\partial G}{\partial R_{ee}}, \qquad (10)$$

which is analogous to Eq. 6 for lipid bilayers.[3]

At the most fundamental level, the energy of the polymer configurations is a function of the spatial coordinates of the monomers and the molecular interactions between them. Such energy functions are sometimes used in detailed molecular simulations, but they are too complicated for statistical–mechanical analytical theories. Theoretical treatments, therefore, are often based on simpler phenomenological energy functions. A simple continuum model describing the elastic behavior of polymers is the worm-like chain model [23, 29]. In this model, the polymer is depicted as a one-dimensional curve $\mathbf{R}(s) = [x(s), y(s), z(s)]$, where the coordinate s runs along the curve from $s = 0$ to $s = L$, where L is the contour (arc) length of the curve. Since there is a clear separation between the energy scales associated with the stretching (stiff) and bending (soft) elastic responses, it is possible to consider only the latter and assume that L is constant. In the case of homopolymers (polymers with one type of monomers), the configuration with minimal elastic energy is a straight line, while any curved configuration has a higher energy. Thus, the coordinates playing the role analogous to the strain in elastic materials are the local curvature

$$\mathbf{c}(s) = \frac{d^2\mathbf{R}}{ds^2}, \qquad (11)$$

which, geometrically, is the inverse of the radius of the circle that locally approximates the curve (see Fig. 7). For a fairly stiff (semi-flexible) polymer, the curvature is assumed to be small and, to a leading approximation, the elastic energy is quadratic in $\mathbf{c}(s)$, i.e., proportional to $\mathbf{c} \cdot \mathbf{c} = c^2$. The local contribution to the curvature elastic energy is $\kappa c^2(s)/2$, where κ is the bending modulus of the polymer. The total curvature elastic energy is obtained by integrating the local contributions over the contour length

$$E = \int_0^L \frac{1}{2}\kappa c^2(s)ds. \qquad (12)$$

---

[3] Equations 7–9 cannot obviously (and do not intend to) summarize the entire field of classical statistical mechanics. They are only meant to demonstrate the basic idea that the free energy of a thermodynamic system can be derived from the statistics of an ensemble of microscopic (molecular) states of that system. This concept will be used in Sect. 3.4 where the statistical–mechanical behavior of thermally fluctuating membranes will be discussed.

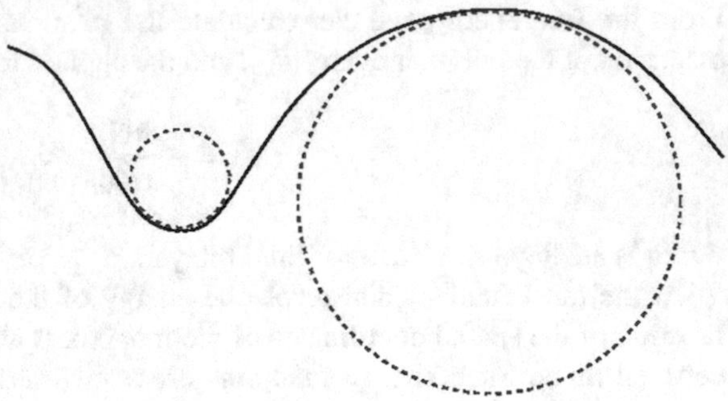

Fig. 7 In the WLC model, the polymer is presented as a one-dimensional curve in a three-dimensional space. Each point along the curve is characterized by a local curvature, which is the inverse of the radius of the circle that locally approximates the curve

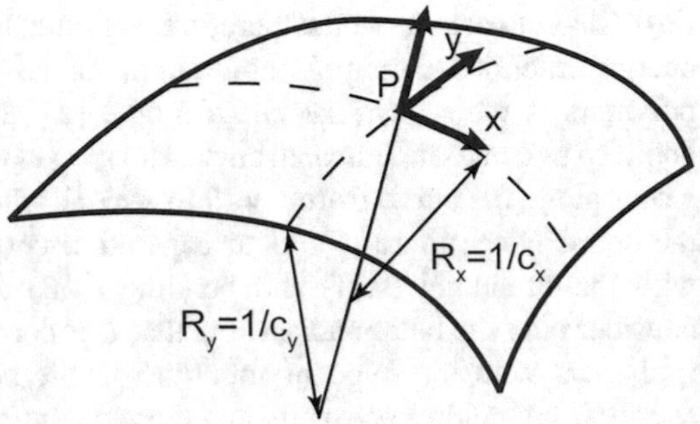

Fig. 8 In the Helfrich model, the membrane is presented as a two-dimensional manifold in a three-dimensional space. Each point along the curve is characterized by two principal curvatures. © 2014 Elsevier B. V. Reprinted from [31] with permission

3.3 Helfrich Hamiltonian for Membrane Elasticity

The curvature energy of a membrane can be expressed in a similar fashion to Eq. 12. This is known as the Helfrich model of membrane curvature elasticity [30]. In this model, the membrane is represented as a two-dimensional Riemannian manifold. At each point, the smooth surface has two principal curvatures, which are the maximum and minimum curvatures of the curves obtained from intersections between the surface and a normal plane at that point (see Fig. 8). The directions of the planes along which the principal curvatures are found are orthogonal to each other. Denoting the principal curvatures by c_1 and c_2, the local contribution to the curvature elastic energy should be expressed as a power series in c_1 and c_2, and to leading order (small curvature), it is given by a quadratic function in these variables. Instead of the principal curvatures, we can use other related invariants with respect to rigid translations and rotations of the surface, specifically, the mean curvature, $M = (c_1 + c_2)/2$, and the Gaussian curvature $K = c_1 c_2$. We also need to take into account the fact that if the bilayer is asymmetric (i.e., if the two leaflets are not identical and, for instance, contain different types of lipids at different compositions and amounts), then it may spontaneously bend. This means that the configuration with minimal elastic energy (ground state) may not be a flat one for which $M = 0$, but a state with spontaneous curvature $M = c_0/2$. Taking all together, we arrive at the famous Helfrich Hamiltonian (energy functional) for the curvature elasticity that reads

$$E = \int \left[\frac{1}{2}\kappa \, (H - c_0)^2 + \kappa_G K \right] dA, \tag{13}$$

where $H = 2M$. This expression is analogous to Eq. 12 for the curvature elasticity of linear polymers, with a few notable differences:

- The membrane has two local curvatures and, therefore, the quadratic form Eq. 13 features two elastic moduli: κ-the bending modulus (rigidity), and κ_G-the Gaussian curvature modulus.
- The membrane may have a spontaneous curvature $H = c_0$.
- The integration in Eq. 13 is carried over the curved surface. Therefore, κ and κ_G have units of energy, in contrast to the modulus in Eq. 12 that has a dimension of energy times length.

3.4 Statistical Mechanics of Membrane Fluctuations

Lipid membranes are soft materials. Typical values of bending moduli κ of lipid membranes are in the range of $10–50 \, k_B T$, which means that membranes exhibit significant shape fluctuations (see chapter 6 in [23]). These so-called undulations have relevance to many biological processes like endo- and exocytosis or, e.g., cellular adhesion [32]. In many cellular processes, membrane protrusions are induced by forces that are generated by other biological components, for instance by actin filaments of the cell cytoskeleton [33]. Lipid bilayers are also subject to strong equilibrium thermal fluctuations arising from random collisions with their environment ("heat bath" in statistical–mechanical language). One of the most common methods for measuring κ in experiments [34] and computer simulations [35] is to record and analyze the spectrum of the membrane thermal undulations. In order to understand how this "flicker analysis" works, we must look at the statistical mechanics of the Helfrich model.

Let us consider a fluctuating surface whose elastic bending energy is given by Eq. 13 that is subject to periodic boundary conditions. This is a common setup in computer simulations of lipid membranes. Furthermore, we will consider a symmetric bilayer with no spontaneous curvature and, therefore, set $c_0 = 0$ in Eq. 13. We now consider the Gauss–Bonnet theorem from differential geometry that states that the integral over the Gaussian curvature term in Eq. 13 can be written as a boundary term plus a topological constant (Euler characteristic) related to the number of "holes" in the surface (0 for a sphere, 1 for a torus, 2 for an eight-shaped pretzel, etc.). Therefore, as long as the surface of the bilayer does not undergo a topological change (as, e.g., during exocytosis when a vesicle "buds out" and detaches from the host membrane), the Gaussian curvature term in Eq. 13 has a constant value for all possible configurations and so has no influence on the physics of the system. These considerations simplify the Helfrich Hamiltonian to the following form:

$$E = \int \left[\frac{1}{2} \kappa H^2 \right] dA. \tag{14}$$

In Eq. 14, each configuration of the surface is described by the local total curvature H. Alternatively, we can use the so-called Monge parametrization and represent the surface via the height function $h(x, y)$ with respect to a reference surface of size $A_p = L^2$, where L ($-L/2 < x, y < L/2$) is the linear length of the simulation cell and A_p is the projected area of the bilayer. Note that in this parametrization, we exclude strongly fluctuating configurations with "overhangs," which cannot be described by a well-defined function $h(x, y)$. In the Monge representation,

$$H = -\nabla \cdot \left(\frac{\nabla h}{\sqrt{1 + |\nabla h|^2}} \right). \tag{15}$$

For a weakly fluctuating surface, $|\nabla h| \ll 1$, this expression is well approximated by

$$H \simeq -\nabla^2 h, \tag{16}$$

and with this form in Eq. 14, we arrive at

$$E = \int_{-L/2}^{L/2} dx \int_{-L/2}^{L/2} dy \, \frac{1}{2} \kappa \left[\nabla^2 h(x, y) \right]^2, \tag{17}$$

with the integration carried over the projected area $A_p = L^2$ rather than the surface area as in Eq. 14.[4]

Let us introduce the Fourier transform of $h(x, y)$

$$h_q = \frac{1}{L^2} \int d\mathbf{r} \, h(\mathbf{r}) e^{-i\mathbf{q} \cdot \mathbf{r}}, \tag{18}$$

where $\mathbf{r} \equiv (x, y)$ denotes a coordinate on the projected plane of the surface, and $\mathbf{q} = (2\pi/L)(n_x, n_y)$ is the wavevector of the Fourier mode (with $n_x, n_y = 0, \pm 1, \pm 2, \ldots$ being integers). Expressing $h(\mathbf{r})$ as a Fourier series

$$h(\mathbf{r}) = \frac{1}{2\pi} \sum_{n_1=-\infty}^{\infty} \sum_{n_2=-\infty}^{\infty} h_q e^{i\mathbf{q} \cdot \mathbf{r}} \equiv \frac{1}{2\pi} \sum_{\mathbf{q}} h_q e^{i\mathbf{q} \cdot \mathbf{r}} \tag{19}$$

---

[4] Carrying the integration over the projected rather than the surface area generates an error of order $|\nabla h|^2$ in the calculated Helfrich energy. The approximation in Eq. 16 involves a correction of the same order.

and using this expression in Eq. 17 give

$$E = \frac{\kappa}{2} \left(\frac{1}{2\pi}\right)^2 \sum_q \sum_p q^2 p^2 h_q h_p \int dr \, e^{i(\mathbf{q}+\mathbf{p})\cdot\mathbf{r}} = \frac{\kappa}{2} \left(\frac{L}{2\pi}\right)^2 \sum_q q^4 |h_q|^2.$$

(20)

For the second equality in Eq. 20, we use: (i) The fact that the integral in Eq. 20 is equal to $L^2 \delta_{\mathbf{q},-\mathbf{p}}$, where δ is the Kronecker delta, and (ii) that h_q and h_{-q} are complex conjugate of each other because $h(\mathbf{r})$ is a real function.

The remarkable property of Eq. 20 is the fact that the curvature elastic energy is now expressed as the sum of terms, each of which corresponds to exactly one Fourier mode, *without mode coupling*. In a statistical–mechanical language, this means that each Fourier mode is an independent degree of freedom. The contribution of a given mode to E is quadratic in $|h_q|$ and looks like the energy of a harmonic oscillator

$$E_q = \frac{1}{2} k_q |h_q|^2,$$

(21)

with an effective spring constant $k_q = \kappa (L/2\pi)^2 q^4$. From the equipartition theorem, it then follows that the average energy of the mode

$$\langle E_q \rangle = \left\langle \frac{1}{2} k_q |h_q|^2 \right\rangle = \frac{\int_{-\infty}^{\infty} d|h_q| \, (k_q |h_q|^2/2) \left[\exp\left(-k_q |h_q|^2/2k_B T\right)\right]}{\int_{-\infty}^{\infty} d|h_q| \left[\exp\left(-k_q |h_q|^2/2k_B T\right)\right]} = \frac{k_B T}{2},$$

(22)

or

$$\langle |h_q|^2 \rangle = \frac{k_B T}{k_q} = \left(\frac{2\pi}{L}\right)^2 \frac{k_B T}{\kappa q^4}.$$

(23)

From Eq. 23, we conclude that the typical amplitude of a mode with wavevector q, i.e., wavelength $\lambda = 2\pi/q$, scales proportionally to $1/q^2$: $[\langle |h_q|^2 \rangle]^{1/2} \sim 1/q^2 \sim \lambda^2$. The longer the wavelength, the larger the amplitude of the mode, which is expected since there is less curvature is involved in the modes with large λ (small q).

As the wavevector is given by $\mathbf{q} = (2\pi/L)(n_x, n_y)$, where n_x and n_y are integers, one may erroneously conclude that the membrane has infinitely many modes or degrees of freedom. This is obviously wrong since the membrane is composed of a finite number of lipids and the continuum description of the Helfrich model is applicable only on length scales that are (much) larger than the characteristic molecular size of the lipids. The question at which length scale the continuum model breaks down has been resolved by large-scale simulations of computationally simple and highly coarse-grained lipid models (see details in Sect. 4). These simulations demonstrated that Eq. 23 provides a correct description of the fluctuation spectrum down to the scales of small membrane parches consisting of about 10 lipids on each monolayer.[5] The characteristic linear size of these patches, l, is of the order of a few nanometers, and the summation in Eq. 20 should be terminated at a wavevector $q_l \propto 1/l$. An operational way to determine the cut-off l of the continuum description is as follows: In analytical calculations, it is common to replace the summation over a discrete set of q values with an integral. Furthermore, although in principle the integration should be carried over a square region in Fourier q-space, it is useful to consider a circular region. Denoting by $N_{\text{DOF}} \propto N_{\text{lipids}} \propto (L/l)^2$ the number of modes (i.e., degrees of freedom of the continuum surface), the replacement of the sum with an integral is done as follows:

$$N_{\text{DOF}} = \sum_q 1 \rightarrow \left(\frac{L}{2\pi}\right)^2 \int_{q_<}^{q_>} 2\pi q dq = \left(\frac{L}{2\pi}\right)^2 \pi \left(q_>^2 - q_<^2\right) \simeq \left(\frac{L}{2\pi}\right)^2 \pi q_>^2,$$

(24)

where the last approximation in Eq. 24 is justified since the lower bound of the integration (the smallest wavevector) $q_< = 2\pi/L$ is vanishingly small in the so-called thermodynamic limit $N_{\text{DOF}} \gg 1$. The upper bound on the integration is therefore set at

$$q_> = \frac{\sqrt{4\pi N_{\text{DOF}}}}{L} \equiv \frac{2\pi}{l}.$$

(25)

3.5 Membranes Under Tension and Confinement

As already mentioned in Sect. 3.1, the application of small tension tends to flatten the membrane fluctuations. To quantify this effect, we must return to the Helfrich Hamiltonian Eq. 14, which is written for a tensionless membrane, and add a term of the form of Eq. 4 for the stretching elastic energy. The area of the surface is equal to the projected area plus the excess area, which is "stored" in the height fluctuations, $A = A_p + \delta A$. Since the tension reduces the amplitude of the height fluctuations, we expect δA to exhibit small variations with respect to its mean value, which allows us to replace the quadratic stretching energy with a linear approximation around

$$E = \frac{1}{2}K\frac{(A-A_0)^2}{A_0} \simeq \frac{1}{2}K\frac{(A_p-A_0)^2}{A_0} + K\frac{A_p-A_0}{A_0}\delta A + O\left(\frac{\delta A}{A_p}\right)^2,$$

(26)

[5] The extrapolation of the simulated fluctuation data to the small q limit involves several technical issues that influence the determined values of the bending modulus. It is also worthwhile to mention the existence of alternative computational methods for determining κ. A comprehensive and updated review with many references on these topics can be found in [36]. A computational method to accelerate the slow relaxation of the small q (large wavelength) Fourier modes of fluctuating membranes has been proposed in [37].

where the first term is independent of δA and hence can be ignored in the fluctuation analysis. Omitting this term in Eq. 26 and using Hooke's law Eq. 5, we can also write to linear order in ΔA that the stretching energy

$$E \simeq \tau \delta A. \tag{27}$$

To linear order in $|\nabla h|^2$, the excess area is equal to

$$\delta A = \int_{-L/2}^{L/2} dx \int_{-L/2}^{L/2} dy \left[\sqrt{1 + |\nabla h|^2} - 1 \right] \simeq \int_{-L/2}^{L/2} dx \int_{-L/2}^{L/2} dy \frac{1}{2} |\nabla h|^2. \tag{28}$$

Using this expression in Eq. 27 and adding the tension term to the Helfrich Hamiltonian Eq. 17, we arrive at

$$E = \int_{-L/2}^{L/2} dx \int_{-L/2}^{L/2} dy \frac{1}{2} \left\{ \kappa \left[\nabla^2 h(x, y) \right]^2 + \tau \left[\nabla h(x, y) \right]^2 \right\}. \tag{29}$$

It is easy to check that in Fourier space, this energy expression reads

$$E = \frac{1}{2} \left(\frac{L}{2\pi} \right)^2 \sum_{\mathbf{q}} \left[\kappa q^4 + \tau q^2 \right] |h_q|^2, \tag{30}$$

and by using the equipartition theorem for this Hamiltonian, we find that

$$\left\langle |h_q|^2 \right\rangle = \frac{k_B T}{k_q} = \left(\frac{2\pi}{L} \right)^2 \frac{k_B T}{\kappa q^4 + \tau q^2}. \tag{31}$$

We thus see that a stretching tension $\tau > 0$ lowers the amplitude of all the Fourier modes w.r.t. the tensionless case $\tau = 0$. The attenuation in the amplitude of the modes is particularly significant when $\kappa q^2 \ll \tau$, which means that the modes that are mostly affected by the applied tension are the long-wavelength ones with $q \ll \sqrt{\tau/\kappa}$. Conversely, modes with wavevector $q \gg \sqrt{\tau/\kappa}$ experience a negligible reduction in their amplitudes. We also notice from Eq. 31 that the membrane may sustain a small negative (compression tension). Explicitly, according to Eq. 31, the amplitude of the longest mode with wavevector $q_< = 2\pi/L$ diverges when the tension drops below $\tau = -(4\pi^2/L^2)\kappa$. Keep in mind that this is only an estimation of the instability tension because Eq. 29 is an approximation assuming small fluctuation amplitudes. One must also keep in mind that in writing Eq. 29, we made the implicit assumption that κ is independent of τ, which may not be the case especially under strong tension.

The cell membrane does not fluctuate freely. It is connected to the cell cytoskeleton (a network of filamentous proteins, mostly actin) and is also influenced by the proximal presence of other cells (in tissues) and of the extracellular environment.

These objects limit the membrane fluctuations in a manner that is hard to evaluate. A simple way to model the fluctuations of a confined membrane is to introduce a uniform harmonic potential

$$E = \int_{-L/2}^{L/2} dx \int_{-L/2}^{L/2} dy \, \frac{1}{2} \gamma h^2. \tag{32}$$

Adding this term to Eq. 29

$$E = \int_{-L/2}^{L/2} dx \int_{-L/2}^{L/2} dy \, \frac{1}{2} \left\{ \kappa \left[\nabla^2 h(x, y) \right]^2 + \tau \left[\nabla h(x, y) \right]^2 + \gamma h(x, y)^2 \right\}, \tag{33}$$

then switching to Fourier space

$$E = \frac{1}{2} \left(\frac{L}{2\pi} \right)^2 \sum_{\mathbf{q}} \left[\kappa q^4 + \tau q^2 + \gamma \right] |h_q|^2, \tag{34}$$

and then applying the equipartition theorem gives

$$\langle |h_q|^2 \rangle = \frac{k_B T}{k_q} = \left(\frac{2\pi}{L} \right)^2 \frac{k_B T}{\kappa q^4 + \tau q^2 + \gamma}. \tag{35}$$

Back to real space, the mean square amplitude of the confined membrane is equal to

$$\langle [h(\mathbf{r})]^2 \rangle = \left(\frac{1}{2\pi} \right)^2 \sum_q \langle |h_q|^2 \rangle = \frac{k_B T}{L^2} \sum_q \frac{1}{\kappa q^4 + \tau q^2 + \gamma}. \tag{36}$$

Replacing the sum with an integral (see Eqs. 24–25)

$$\langle [h(\mathbf{r})]^2 \rangle = \frac{k_B T}{2\pi} \int_{q_<}^{q_>} \frac{q}{\kappa q^4 + \tau q^2 + \gamma} dq. \tag{37}$$

Under confinement ($\gamma > 0$), the mean square amplitude is finite, and for a strongly confined membrane ($\gamma \gg \tau/l^2$): $\langle h^2 \rangle \propto k_B T/(\gamma l^2)$. In contrast, when $\gamma = 0$, the mean square amplitude increases with the size of the membrane. For a stressed membrane: $\langle h^2 \rangle \propto (k_B T/\tau) \ln(L/l)$, and in the absence of stress $\langle h^2 \rangle \propto L^2 (k_B T/\kappa)$.

3.6   Helfrich Free Energy

In the literature on membrane elasticity, a similar expression to the Helfrich energy Eq. 13 is sometimes referred to the free energy G of membranes and vesicles. This is not just a semantic change. Consider, for instance, the question posted at the end of Sect. 2.3-why membranes close and form spherical vesicles. Thermodynamically, this implies that the free energy of the closed vesicle is lower than that of the open membrane. In principle, the free energy of both the open membrane and vesicle (macrostates) can be calculated from their partition functions that involve summations over all the relevant molecular configurations (microstates), see Eqs. 8–9. In both cases, we can use the Helfrich Hamiltonian to describe the energy of each and every configuration in the ensemble of microstates. The free energy is a function of the average profile of the microstates, which can be regarded as the profile of the macrostate. Obviously, the macrostate average profile is smoother than the profile of the individual thermal configurations. For instance, in Sect. 3.4, we considered the ensemble of configurations of a surface whose average profile is flat: $\langle h(\mathbf{r}) \rangle = 0$. Similarly, a vesicle has an average profile of a sphere of radius R. The microstate thermal configurations of the vesicle are given by $R(\Omega) = R + \delta R(\Omega)$, where Ω is the solid angle and $\langle \delta R(\Omega) \rangle = 0$. The statistical mechanics of Helfrich elasticity of nearly spherical vesicles ($\delta R \ll R$) can be conveniently calculated by representing the shape of the configurations as the sum of spherical harmonics, similarly to the way that a Fourier space representation is used in the case of a planar membrane.

The considerations that lead to the form of the phenomenological Helfrich Hamiltonian (see Sect. 3.3) are also applicable to the free energy expression. Thus, similarly to Eq. 13, we write [12]

$$G = \int \left[\frac{1}{2}\kappa \left(H - c_0 \right)^2 + \kappa_G K \right] dA, \tag{38}$$

where G is now the Helfrich free energy. The curvatures and the integration in Eq. 38 are carried over the surface corresponding to the average profile of the bilayer. As in the Helfrich Hamiltonian, the coefficients κ, κ_G, and c_0 are referred to as the bending modulus, Gaussian curvature modulus, and spontaneous curvature, respectively. However, there is a fundamental difference between the Hamiltonian coefficients and their free energy counterparts [38]. The former are material properties that depend on the local (molecular) "reality" of the lipids, i.e., the interaction energy between them and the entropy of their hydrophobic chains. The latter, on the other hand, are thermodynamic quantities obtained by averaging over the ensemble of microstates and, as such, are also influenced by the membrane thermal fluctuations at all length scales.

3.7 Edge Energy

So why does a flat membrane close itself and form a spherical vesicle? Using Eq. 38 for the curvature free energy and assuming that the bilayer membrane is symmetric ($c_0 = 0$), we get that for a flat membrane ($H = K = 0$): $G = 0$. For a spherical vesicle of radius R, we have that $H = 2/R$ and $K = 1/R^2$, and so $G = 4\pi(2\kappa + \kappa_G)$. Thus, as long as $(2\kappa + \kappa_G) > 0$ (which is indeed the case), the flat configuration is supposedly more stable. There is however an important difference between a spherical vesicle and a flat bilayer. The former is closed, while the latter is open and has edges where the lipids are exposed to the solvent. Because of the hydrophobic effect, the "edge lipids" would orient themselves in a manner shown in Fig. 9, which provides partial protection from the aqueous contact. This organization, resembling half a micelle, is not optimal since it is more suitable for surfactants with a single hydrophobic tail (see Fig. 3 and the discussion on chain packing in Sect. 2.3). Thus, some energy penalty associated with hydrophobic-water contact remains, which is proportional to the number of lipids located at the end of the flat bilayer. As this number is proportional to the perimeter of the flat bilayer, P, we can write that

$$G_{\text{edge}} = \Gamma P, \tag{39}$$

where the coefficient $\Gamma > 0$ is called the line tension of the membrane and has dimensions of energy per unit length (J/m). Adding the Helfrich bending and the edge free energies, we see that while the free energy of the vesicle is size independent, the free energy of the flat membranes grows linearly with the perimeter of the bilayer. Typical values of Γ of lipid bilayers are in the range of $10^{-2} - 10^{-1}$ J/m, which means that practically the vesicle configuration is always thermodynamically favorable over the open flat one.

Similar considerations can explain the opening of membrane pores (membrane rupture) under applied tension. When a pore is formed, the effective area of the membrane is reduced by an amount $\Delta A < 0$ which is equal to the pore area. The elastic free energy relieved by this process can be written in a form similar to Eq. 27 for the energy

$$G = \tau \Delta A. \tag{40}$$

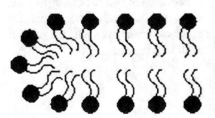

Fig. 9 The lipids residing at the edge of a bilayer sheet avoid contact with water by changing their orientation

This gain in free energy competes with the increase in the line tension energy, Eq. 39, due to the lipids residing at the perimeter of the pore. Assuming a circular hole (the shape with smallest circumference to area ratio) of radius r and adding the two free energy contributions, we get [39, 40]

$$G = 2\pi \Gamma r - \pi \tau r^2. \tag{41}$$

This free energy expression has a parabolic shape with a maximum $G_{\max} = \pi \Gamma^2/\tau$ at $r = \tau/\Gamma$. Small transient holes are constantly formed and disappear due to the motion of the lipids. This is because the free energy increases in the presence of a hole of radius $r < r_c$. Conversely, a hole of radius $r > r_c$ will continue to grow since the increase in size lowers the system free energy.[6] Thus, a stable pore will appear provided that the lipids have enough kinetic energy to allow the spontaneous formation of a pore of critical radius $r = r_c$. Statistically speaking, this is possible if the free energy required for the opening of a pore of critical radius is of the order of the thermal energy $k_B T$. We therefore expect the membrane to rupture provided that

$$G_{\max} = \frac{\pi \Gamma^2}{\tau} \simeq k_B T, \tag{42}$$

i.e., if the applied tension τ is of the order of $\Gamma^2/k_B T$. Taking the characteristic values of Γ quoted in the previous paragraph, we conclude that the rupture tension of bilayer membranes is of the order of 10^{-2} J/m^2, which is indeed consistent with the experimentally measured values [24].

4 Computer Simulations

Computer simulations are ubiquitous in many areas of science and engineering, including in biophysics where they have been essential for the development of the field. Their importance stems from the fact that biological systems are complex: They include many different components that interact with each other, which means that one can only gain limited insight into the behavior of these systems by using simple analytical models. Contemporary research in membrane biophysics, for instance, involves the investigations of increasingly complex membranes with several different types of lipids, in addition to proteins and/or other components like DNA molecules and polymers. Apart from the growing complexity of the systems

---

[6] The pore will not grow indefinitely. The parabolic shape of the free energy Eq. 41 assumes a linear dependence of G on the pore area, Eq. 40, but this form is only valid for holes that are much smaller than the membrane area (but not too small to be considered as a local defect). For larger holes, the quadratic form for the elastic stretch energy Eq. 4 must be taken into account, which limits the size of the pore. See, e.g., [25].

under investigation, there have been several other trends that contributed to the rapid increase in the number of molecular simulation studies. One is obviously the increase in computing power, which makes it possible to use regular PCs for simulations that two decades ago could not have been carried even on the most advanced computer clusters. Another important development is the growing popularity of molecular dynamics simulation packages like NAMD, GROMACS, AMBER, CHARMM, and LAMMPS. These packages, whose performances have been constantly refined in recent years, allow researchers to conduct computer simulations without the need to write their own code from scratch. This has led to an explosion in the number of research groups conducting computer simulations, especially in the biochemistry community where computer simulations are increasingly regarded as "computational experiments" with the computer screen serving as a "molecular-resolution microscope" [41]. As the utilization of simulation suites does not require great familiarity with molecular dynamics algorithms, I will dedicate the first part of this section for a brief review of molecular simulation approaches, highlighting also a few technical issues that are important for the reliability and success of simulation studies.

In the second part of the section, I will discuss the different types of models existing for biological membranes, reflecting the multi-scale nature of membrane research. Biologists and biochemists, for instance, are mostly interested in processes involving relatively few lipids, where the detailed information on the conformation of the lipids (e.g., the bond angles of the hydrophobic tails, lipid tilt, local packing of the lipids, etc.) is important. Obviously, the investigations of these aspects require simulations where the lipids and other biological components are represented by all-atom descriptions. These so-called atomistic simulations are computationally expensive because each lipid consists of more than 150 atoms and, in addition, one has to simulate a large number of H_2O molecules [42]. Thus, despite the constant progress in computing power, they are limited to small membrane patches consisting of 100–1000 lipids with dimensions of about 10 nanometers. The time scales of the simulated processes in atomistic membrane simulations do not typically exceed those of a few microseconds. In order to overcome these limitations and investigate processes and systems at larger spatial and temporal scales, one has to "coarse-grain" the atomistic description [43]. This can be done, for instance, by grouping several carbon atoms of the tails and their associated hydrogens into a single "effective bead" or by replacing a water molecule by a single solvent particle. The benefit of such a coarse-grained (CG) description in terms of the computational toll is clear, but one must be aware that there is no guarantee that the CG model reproduces correctly all the relevant features of the system. Biophysical studies that do not aim to investigate specific lipids but rather explore more general properties of membranes are often based on even more CG models where lipids are generically represented as bead-spring chains of hydrophilic and hydrophobic spheres. In some models, the solvent is not explicitly included in the simulation box and, instead, effective potentials mimicking the hydrophobic attraction between the tails are used. I will briefly review these different classes of models by considering a relatively simple example problem, which is the formation of liquid-ordered domains in mixtures of saturated lipids and cholesterol.

4.1 Molecular Dynamics

Molecular Dynamics (MD) is a simulation method for tracking the movement of interacting particles by numerically integrating their equations of motion [44]. The name may be wrongly understood as if the main utility of the method is to generate computational visualizations of non-equilibrium molecular processes. Actually, MD is first and foremost a technique for sampling the configuration phase space of molecular systems at thermodynamic equilibrium. By following the dynamics of the particles, an ensemble of microscopic configurations of the equilibrium system is generated from which quantities of interest can be calculated. The most thoroughly simulated system is probably that of identical spherical particles interacting with each other via the famous two-body Lennard–Jones (LJ) 6–12 potential

$$\phi_{\mathrm{LJ}} = 4\epsilon \left[\left(\frac{\sigma}{r} \right)^{12} - \left(\frac{\sigma}{r} \right)^{6} \right]. \tag{43}$$

Here, σ is the diameter of the spherical particles, ϵ is the strength of the pair potential, and r is the distance between the centers of the spheres. The LJ potential is strongly repulsive at short distances $r < \sigma$, which represents the effect of excluded volume, i.e., the fact that the particles cannot interpenetrate each other. At long distances $r \gg \sigma$, the particles attract each other via a potential intended to model van der Waals interactions. The potential vanishes when the particles are far away from each other $r \to \infty$ and achieves a minimum value $\phi_{\min} = -\epsilon$ at $r_{\min} = 2^{1/6}\sigma \simeq 1.12\sigma$, when the particles nearly touch other. The LJ system serves as a paradigm for studying the thermodynamic transitions between the gas, liquid, and solid phases of matter. If the simulations are conducted properly, they provide quantitative information on the spatial distribution of particles, which is distinctly different at the different phases. Moreover, thermodynamic quantities such as the pressure, heat capacity, and compressibility, which are related to the mean and variance of measurable quantities, can be computed from the ensemble of configurations generated in the course of the MD simulations.

The most popular discrete-time integrator for MD simulations is the Verlet algorithm [45] that runs as follows: Denoting, respectively, by $\{\mathbf{r}_i(t)\}$ and $\{\mathbf{v}_i(t)\}$ the positions and velocities of the particles $(i = 1, \ldots, N)$ at time t, the coordinates after a small time step dt are obtained from a truncated Taylor expansion

$$\mathbf{r}_i(t + dt) = \mathbf{r}_i(t) + \mathbf{v}_i(t)dt + \frac{\mathbf{f}_i\left(\{\mathbf{r}_j(t)\}\right)}{2m}dt^2, \tag{44}$$

where m is the mass of the particle, and $\mathbf{f}_i$ is the force acting on the particle at time t that, in general, depends on the coordinates of all the particles. Once the new positions are calculated, the new forces $\mathbf{f}\left(\{\mathbf{r}_j(t+dt)\}\right)$ at time $t+dt$ can be determined. Then the velocities at time $t+dt$ are updated by

$$\mathbf{v}_i(t+dt) = \mathbf{v}_i(t) + \frac{\mathbf{f}_i\left(\{\mathbf{r}_j(t)\}\right) + \mathbf{f}_i\left(\{\mathbf{r}_j(t+dt)\}\right)}{2m}dt. \tag{45}$$

Equations 44–45 constitute the velocity Verlet algorithm.

Equation 44 is accurate to second order in dt, which means that the computed trajectory deviates from the correct one. The deviations grow exponentially with time because of the chaotic nature of the dynamics. It is, however, believed that at long times the "incorrect" computed trajectory follows a true trajectory that simply starts with different initial conditions. This "shadow trajectory" is as representative as the original one and provides an appropriate ensemble of microscopic system configurations. With the understanding that the purpose of MD is phase-space sampling rather than the calculation of actual trajectories, it is easy to understand why the simple and computationally efficient Verlet algorithm is favored over other, higher order in dt, integrators. The most important feature of the Verlet algorithm is the fact that the total energy of system is conserved over long time integration, i.e., it is bounded and does not drift at long times. This feature is directly related to the fact that mapping defined by Eqs. 44–45 is symplectic[7] and time reversible [44].

Speaking of energy conservation, the Verlet algorithm generates the so-called microcanonical ensemble, namely a sequence of configurations of an isolated system with a given energy, volume, and number of particles. The temperature of the system can be evaluated from the average kinetic energy

$$\left\langle \sum_{i=1}^{N} \frac{m\mathbf{v}_i^2}{2} \right\rangle = \frac{3Nk_BT}{2}. \tag{46}$$

This approach is, however, problematic, and the reader is advised to avoid integration methods imposing Eq. 46 via "velocity rescaling" [46]. To understand why, we must keep in mind that there are other ways to estimate the velocity of a particle. For instance, we can define the half-step velocity

$$\mathbf{v}_i(t+dt/2) = \frac{\mathbf{r}_i(t+dt) - \mathbf{r}_i(t)}{dt}, \tag{47}$$

which can be used in the leap-frog Verlet method [47]

$$\mathbf{v}_i(t+dt/2) = \mathbf{v}_i(t-dt/2) + \frac{\mathbf{f}_i\left(\{\mathbf{r}_j(t)\}\right)}{m}dt, \tag{48}$$

---

[7] The existence of a shadow trajectory also follows from this property.

$$\mathbf{r}_i(t + dt) = \mathbf{r}_i(t) + \mathbf{v}_i(t + dt/2)dt. \tag{49}$$

This version of the algorithm yields *exactly* the same spatial trajectory $\{r_i(t)\}$ as the original velocity Verlet formulation Eqs. 44–45. Thus, for the purpose of phase-space sampling, both algorithms are identical. They only differ in the auxiliary velocity variable that they use for the calculation of the position. Both Eqs. 45 and 47 are approximations of the actual velocity and may be subject to time-discretization errors. Interestingly, it turns out that the latter provides the best estimation of the temperature, when used in Eq. 46 [48].

4.2 Langevin Dynamics

In the canonical ensemble of statistical thermodynamics, the system is not thermally isolated but rather exchanges heat with its environment ("heat bath") in a manner that fixes the temperature. In this ensemble, the occurrence probability of a configuration is given by Boltzmann statistics Eq. 7. An MD algorithm that fixes the temperature of a system is called a "thermostat." There are basically two types of thermostat algorithms-deterministic (Nosé–Hoover) [49, 50] and stochastic (Langevin) (see [51], and references therein). Although these two approaches are based on different statistical–mechanical ideas, they share an important feature. They both involve the introduction of effective forces representing the coupling between the system and the external heat bath. Moreover, in both approaches, these additional forces (albeit different) depend on the velocities of the particles. This raises significant algorithmic problems because the new forces cannot be treated similarly to the physical Newtonian forces that appear in the Verlet equations 44–45 and depend on the positions of the particles only. The difficulties to use the Verlet algorithm with velocity-dependent forces may be handled with certain approximations that come at a high price: Most of the existing numerical schemes, both for deterministic and stochastic thermostats, tend to exhibit significant time-discretization errors that grow rapidly with the integration time step dt [52]. This leads to both reduction in the efficiency of the algorithms (since they must be run with a small time step) and also creates uncertainty about the accuracy of the simulation results.

A major development in the field of MD integrators is the introduction of an improved Langevin thermostat algorithm by Grønbech-Jensen and Farago (G-JF) [51]. The G-JF thermostat provides robust (i.e., insensitive to variations in the time step) and accurate configurational sampling of the phase space with larger time steps than other algorithms. Langevin thermostats are based on the idea that the coupling to the heat bath can be sampled by considering the dynamics of each particle in the system to be governed by Langevin equation [53]

$$m\ddot{\mathbf{r}} = \mathbf{f} - \alpha\mathbf{v} + \boldsymbol{\beta}(t). \tag{50}$$

Langevin Eq. 50 is a stochastic differential equation, i.e., an equation involving a non-deterministic term. Specifically, the third term on the r.h.s. of Eq. 50 is a Gaussian white noise, $\boldsymbol{\beta} = (\beta^1, \beta^2, \beta^3)$, with zero mean and delta-function auto-correlation

$$\langle \beta^m(t) \rangle = 0 \; ; \; \langle \beta^m(t)\beta^n(t') \rangle = 2k_B T\alpha\delta(t-t')\delta_{m,n}. \tag{51}$$

Together with the second term on the r.h.s., which is a friction force $-\alpha\mathbf{v}$, where $\alpha > 0$ is the friction coefficient, they account for the coupling of the system to the heat bath that sets the temperature. Notice the appearance of the friction coefficient in Eq. 51 that defines the statistical properties of the stochastic noise. This connection between the friction and noise is a manifestation of the famous fluctuation–dissipation theorem and is required for achieving the equilibrium Boltzmann distribution [54]. When the friction coefficient vanishes, we are left with only the first term on the r.h.s. of Eq. 50, which is the physical deterministic force. In this limit, Langevin equation simply becomes Newton's second law of motion.

The G-JF discrete-time integrator for Langevin equation reads

$$\mathbf{r}_i(t+dt) = \mathbf{r}_i(t) + b\left[\mathbf{v}_i(t)dt + \frac{\mathbf{f}_i(t)}{2m}dt^2 + \frac{\boldsymbol{\beta}_i(t+dt)}{2m}dt\right], \tag{52}$$

$$\mathbf{v}_i(t+dt) = a\mathbf{v}_i + \frac{1}{2m}\left[a\mathbf{f}_i(t) + \mathbf{f}_i(t+dt)\right]dt + \frac{b}{m}\boldsymbol{\beta}_i(t+dt), \tag{53}$$

where the (damping) coefficients of the algorithm are given by

$$a = \frac{1 - \frac{\alpha dt}{2m}}{1 + \frac{\alpha dt}{2m}} \tag{54}$$

and

$$b = \frac{1}{1 + \frac{\alpha dt}{2m}} \tag{55}$$

$(-1 < a \leq 1, 0 < b \leq 1)$.[8] At each time step, $3N$ *independent* random Gaussian numbers (RGNs) must be drawn for the noise terms in Eqs. 52–53:

$$\boldsymbol{\beta}_i(t+dt) = (\text{RGN}_i^1, \text{RGN}_i^2, \text{RGN}_i^3) \tag{56}$$

(one RGN for each Cartesian coordinate of each particle). The RNGs have zero mean, $\langle \text{RGN}_i^n \rangle = 0$, and variance $\langle (\text{RGN}_i^n)^2 \rangle = 2k_B T\alpha dt$.

---

[8] The Langevin dynamics of a single free particle ($\mathbf{f} = 0$) is underdamped at time scales $dt \ll m/a$ ($a, b \to 1$) and overdamped for $dt \gg m/a$ ($a \to -1, b \to 0$).

As noted above, in the absence of friction and noise ($\alpha = 0$), the Langevin equation coincides with Newton's second law of motion, and the G-JF algorithm Eqs. 52 and 53 reduces to the velocity Verlet Eqs. 44 and 45, respectively. This is true for essentially all existing Langevin integrators. But in contrast to other Langevin thermostats (with the exception of the BAOAB algorithm of Leimkuhler and Matthews [55]), the G-JF integrator provides *accurate and robust* configurational sampling for essentially any time step that is below the stability limit of the method dt_s.[9] By "accurate configurational sampling," we mean that the simulations generate a sequence of configurations that follow Boltzmann's distribution law Eq. 7. "Robust" means that the simulated distribution function does not become distorted when the integration time step is increased. This unfortunate property of other integrators causes the measured values of thermodynamic quantities of interest (e.g., the average and standard deviation of the potential energy of the system) to vary with the time step, especially when dt approaches the stability limit dt_s [52].

Finally, we return to the discussion in the previous subsection on the velocity variables and the assessment of the system's temperature via Eq. 46. For constant temperature Langevin simulations, the question is whether the average kinetic energy is *equal* to the assigned temperature T. In the case of the G-JF Eqs. 52–53, the answer is no, but this does not invalidate the method since the purpose of the simulations is correct configurational, not kinetic, sampling. Nevertheless, it has been recently shown that the G-JF method can be written in the form of a leap-frog scheme, with half-step velocity variables that do satisfy Eq. 46 to a high degree of accuracy. In fact, there is no single leap-frog version of the G-JF integrator, but infinitely many ways to define the half-step velocity in a manner that yields the correct the kinetic energy [48]. One convenient definition is

$$\mathbf{v}_i(t + dt/2) = \frac{\mathbf{r}_i(t + dt) - \mathbf{r}_i(t)}{\sqrt{b}dt}. \tag{57}$$

All the different leap-frog G-JF schemes produce exactly the same spatial trajectory as the velocity Verlet G-JF equations and, therefore, provide accurate configurational sampling of the canonical ensemble in addition to accurate kinetic temperature. They all reduce to Eq. 47 when the friction coefficient vanishes.

4.3 Multi-Scale Simulations

We arrive at the final part of this chapter that deals with the multi-scale nature of membrane biophysics and the different computational models addressing the wide

---

[9] Any numerical integration scheme becomes unstable and fails to follow the continuous-time trajectory when the integration time step becomes $dt_s \gtrsim \omega_0^{-1}$, where ω_0 is the largest vibrational frequency of the system (the resonance frequency of the strongest interaction term).

range of relevant length and time scales. As already discussed at the introduction of this chapter, multi-scale modeling and simulations is a huge topic in contemporary material science [56]. In biophysics, it is relevant not only to membrane simulations, but also to more complex systems including proteins and other biological components [57]. This subsection does not attempt to provide a comprehensive review of the topic, but simply to explain the concept of multi-scale simulations. As the whole chapter focuses on bilayer membranes, we choose a fairly simple model system of membranes consisting of DPPC lipids and cholesterol, as an example to demonstrate multi-scale studies.

In Sect. 2.5, we dealt with lipid rafts that are liquid-ordered nano-scale domains enriched in saturated lipids and cholesterol (Chol) that are present on the cell membrane. Because of the compositional complexity of the plasma membrane and due to experimental difficulties, the biophysics of such domains has been mainly studied on simple model systems consisting of a small number of components. The simplest model system featuring liquid-ordered domains is that of a binary mixture of Chol and the saturated lipid DPPC. The phase diagram of DPPC/Chol mixtures, which is shown in Fig. 10, was established theoretically [58] and experimentally [59] more than 30 years ago. A pure DPPC membrane (without Chol) undergoes a first-order melting transition at $T_m \simeq 314K$ from a gel to a liquid-disordered (L_d) phase. At high Chol concentration (above Chol mole fraction of about 0.25), the membrane is found in the liquid-ordered (L_o) phase. The interesting part of the phase diagram is at intermediate Chol concentrations, especially above the melting temperature, where coexistence between two liquid phases, L_o and L_d, is observed. This two-phase region has attracted much attention because it resembles the picture of liquid-ordered rafts in biological membranes (which are far more complex in terms of their lipid composition and the fact that they also contain certain proteins) and the surrounding sea of fluid lipids.

The biophysical mechanisms contributing to the formation of the domains are still under debate, and it is clear that their understanding requires detailed information about the structure of the liquid-ordered domains. This information can be obtained from atomistic simulations where each atom in the system is

Fig. 10 The phase diagram of mixtures of DPPC lipids and cholesterol

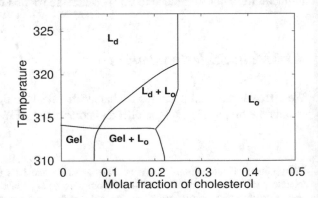

explicitly represented. The force fields in these simulations are empirical and include summation over excluded volume interactions, bonded forces (chemical bonds), bond angles, and non-bonded forces (van der Waals and electrostatic forces) [60]. The currently available computing power makes it possible to conduct atomistic simulations of lipid patches consisting of the order of 1000 amphiphilic molecules on time scales of several microseconds, which is sufficient for equilibrating and sampling these systems at the investigated length scale. Figure 11 shows snapshots from atomistic simulations of DPPC/Chol membranes performed with the GROMACS package, demonstrating their utility as a "computational microscope" with atomic resolution [61]. A word of caution should be made here that one must always question the reliability of atomistic simulations since they are not based on ab initio quantum mechanical calculations but rather on empirical force fields that may need to be tuned in certain systems or may simply not work properly. This warning is relevant to the specific example discussed herein, where many of the simulations are performed in the proximity of phase boundaries. Close to phase transitions, thermodynamic systems tend to be highly susceptible to variations in the intermolecular interactions, which also highlight the importance of setting correctly the temperature and pressure of the simulations.[10] The most interesting observation in the atomistic simulation results here is the distribution of lipids and Chol in the L_o phase, which appears to be very heterogeneous. More specifically, the atomistic simulation results show that the L_o domains consist of smaller gel-like hexagonally packed clusters of lipid chains, surrounded by liquid domains of ordered chains that are rich in Chol especially along the domain boundaries. Similar patterns of "nano-structures within nano-domains" have been speculated based on experimental neutron scattering data [63] that, in contrast to the simulations, do not provide direct visualization of the investigated membranes.

The size of the simulated system in Fig. 11 is $\lesssim 20$ nm, which is smaller than the size of most liquid-ordered domains in the system. This means that the atomistic simulations provide a glimpse into the internal structure of the L_o state, but they cannot show their dynamics and morphologies on the whole membrane scales. In order to see the larger-scale picture, we must employ CG models that are not based on computationally expensive atomic-level descriptions. Many CG models have been developed over the past twenty years for this purpose. Generally speaking, CG models can be divided into two, very distinct, classes. One class includes models attempting to mimic *specific* lipid systems. By far, the most popular model in this class is the MARTINI force field [64]. The idea in these kinds of models is to group several atoms into a "unified" particle. For instance, a water molecule of three atoms can be represented by a single "water particle." Similarly, in the hydrophobic tails

---

[10] Many simulations are conducted in the isobaric–isothermal ensemble, which differ from the canonical ensemble in that the pressure rather than the volume of the system is fixed. Algorithms for fixing the pressure in MD simulations are called "barostats," and they suffer from similar implementation problems to thermostat algorithms. A relatively robust G-JF barostat algorithm, which is based on concepts used for the derivation of the G-JF thermostat, has been presented in [62].

Fig. 11 Side and upper views of a bilayer composed of DPPC lipids and Chol, taken from atomistic simulations in the $L_d + L_o$ coexistence region of the phase space. In the side view, DPPC is shown in cyan and lime, and Chol in white. Water, ions, and lipid chain hydrogens are omitted for clarity. Red and blue boxes highlight disordered and ordered regions, respectively. In the upper view, DPPC chains and Chol are shown in green and orange, respectively. Here, one can see that the L_o domain consists of smaller gel-like hexagonally packed clusters of lipid chains. Reprinted from [61]

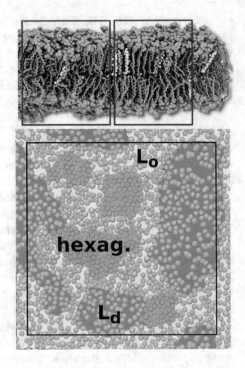

of the lipids, a group of three carbon atoms and their associated hydrogens are often represented by a single "hydrophobic unit." Fig. 12 shows a snapshot from CG MARTINI simulations of mixtures of Chol (red) with the hybrid POPC lipid (yellow). The simulations were also carried with the GROMACS package for MD simulations. The snapshot in Fig. 12 consists of about 6000 molecules and is in the L_d phase, which explains why the distribution of Chol is relatively homogeneous.

In the previous paragraph, a comment was made that the results of atomistic simulations should be taken with caution. This warning is even more relevant to CG simulations because there is no systematic way to know which atoms should be unified and what are the effective interactions between the CG particles. This explains why sometimes MARTINI simulations exhibit inconsistencies with atomistic simulations of similar system [66]. One particular problem is the chain entropy, which is heavily involved in the phase transition but is not properly represented when using a smaller number of effective beads.

The second class of CG models involves a higher degree of abstraction and is termed *generic*, in the sense that they do not focus on specific lipids but rather on the mechanisms and biophysical driving forces governing the thermodynamic behavior of membranes. In these models, the lipids are typically described as short chains (oligomers) of hydrophilic and hydrophobic beads, which are connected to each other via spring-like potentials. Many generic membrane models are implicit-solvent ones, where water is not included in the simulations, and, instead, non-bonded attractive interactions between the hydrophobic beads that mimic the

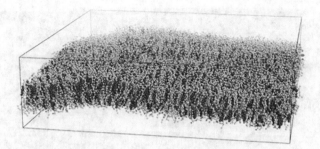

Fig. 12 A snapshot from MARTINI simulations of a bilayer consisting of POPC lipids (yellow) and Chol (red) at molar fractions 0.6 and 0.4, respectively. The total number of amphiphilic molecules in the system is ~6000. © 2015, American Chemical Society. Reprinted from [65] with permission

hydrophobic effect are introduced [67–70]. The advantage of these models is clear considering that the majority of particles in membrane simulations with explicit solvent are solvent particles. Another elegant way to reduce the computational load is to introduce "phantom" solvent, where the solvent particles interact repulsively with the hydrophobic particles, but do not interact with each other [71]. Using this approach, Meinhardt and Schmid [72] simulated a very large membrane of 20,000 amphiphilic molecules (lipids and Chol), featuring both small and large L_o domains (see Fig. 13). The distinction between small and large domains is possible because of the large scale of the generic CG simulations. In the case of the large domains, the side views of the simulated membranes show proliferation of line defects at the boundaries between the L_o and L_d phases. In these defects, the lipids are tilted with respect to the normal direction of the membrane plane, in a manner resembling the ripple phase that sometimes emerge in one component lipid membranes near the gel–liquid-disordered melting transition [73]. In the case of the small domains, the simulations indicate that they may be stabilized due to the difference in the spontaneous curvatures of the membrane monolayers in the L_o and L_d, which generates tension that breaks large domains into smaller ones.

The computationally cheapest class of CG models, which offer the highest degree of abstraction while still retaining some molecular character (i.e., excluding continuum models), are lattice-based models. In lattice simulations, the membrane is modeled as a lattice whose sites are occupied by points that represent the chains of the lipids and the Chol molecules. The "chains" can move into empty sites or exchange their places with other "chains," but this is not done with MD algorithms that only work for models where the particles move continuously in space. Lattice models are discrete, and their simulations are performed with Monte Carlo (MC) algorithms that move the points between the lattice sites in a manner ensuring proper sampling of the configuration phase space [44]. While lattice models appear to be a gross simplification of real molecular systems, they actually have an enormous role in statistical physics. The gas–liquid condensation transition, for instance, has been largely studied using the simple "lattice gas" model, where each molecule

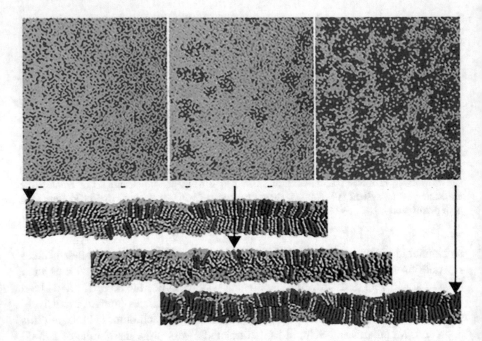

Fig. 13 Snapshots from generic CG simulations of a bilayer consisting of 20,000 amphiphilic molecules. Lipids and Chol are plotted in green/yellow and dark blue/light blue, respectively. © 2019, Royal Chemistry Society. Reprinted from [72] with permission

is represented by an occupied lattice site [74]. The transition is driven by short-range attractive interactions between particles occupying nearest-neighbor sites. The model is equivalent to the famous Ising model of magnetic materials, where on each site there is a spin pointing either "up" or "down," with a favorable interaction energy between two nearest-neighbor spins with the same orientation [74]. It is, therefore, not surprising that phase transitions and coexistence in lipid/Chol mixtures have also been studied via this approach [19]. The attractiveness of lattice simulations lies not only in their computational simplicity, but also in their minimal nature that puts the focus on the essential physical factors and mechanisms governing the thermodynamic behavior of complex systems. Figure 14a shows a snapshot from recent lattice simulations of DPPC/Chol mixtures, showing L_o domains floating in a L_d matrix. The lattice consists of almost 17,000 sites-equivalent to a linear system size of a few tens of nanometers. In contrast to MD where the time is measured in physical units (typical time steps in atomistic membrane simulations, for instance, are of the order of a few femtoseconds), in MC simulations, the time is measured in non-physical MC time units. An MC time unit is usually defined as a sequence where each particle experiences one (or a few) move attempts. MC times can be compared to physical ones by measuring quantities such as the diffusion coefficient of the lipids and comparing them to experimental data. Based on such an approach, we estimate that the lattice

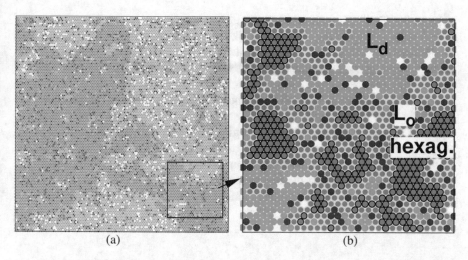

(a)   (b)

Fig. 14 (**a**) A snapshot from lattice simulations of DPPC/Chol mixtures in the $L_d + L_o$ coexistence region of the phase space. The lattice size is of nearly 17,000 sites. Ordered and disordered lipid chains, cholesterol, and empty sites (voids) are colored in green, blue, red, and white, respectively. (**b**) Zoom-in view of the region marked by a black box in (**a**). Reprinted from [75]

simulations presented in Fig. 14a are performed on time scales of hundreds of microseconds. On these macroscopic time scales, we can see the morphological dynamics of the domains, including their formation and disappearance. In the simulations, each site is occupied by either a saturated DPPC chain, a Chol molecule (red), or be empty (white) representing a small area void. Each lipid chain can be in one of two states-ordered (green) or disordered (blue). The model fully explains the formation of L_o domains via the following interactions: The disordered (d) state is entropically favored by a free energy $F_d = -\Omega_d k_B T$ compared to the ordered (o) state, but two nearest-neighbor chains in the ordered state have interaction free energy of $F_{oo} = -\epsilon_{oo}$ representing the favorable packing of chains in this state. This competition between the entropic and enthalpic contributions to the free energy is responsible for the melting (gel to L_d) transition of the DPPC membrane. Cholesterol (c) interaction free energy with a nearest-neighbor ordered chain is equal to $F_{co} = -\epsilon_{co}$, where $|\epsilon_{co}| < |\epsilon_{oo}|$. This term reflects the "duality" of the Chol, which tends to induce ordering of the chains but also disrupts their efficient packing with each other. Importantly, the simulations are conducted on a single planar lattice, which means that the observed domains are stabilized without curvature-related effects. Curvature energy can be introduced into the model by assigning, to each lattice site, an additional degree of freedom representing vertical displacement, and considering a discretized Helfrich Hamiltonian [76].

The model simulations accurately reproduce the phase diagram of the system, presented in Fig. 10. The most striking observation of the lattice simulations is the local structure of the L_o phase revealed when one zooms into the simulation's snapshots (Fig. 14b). The distribution of lipids and Chol exhibits remarkable similarity to the distribution observed in atomistic simulations (upper view in Fig. 11). Specifically, we see in Fig. 14b the gel-like hexagonal clusters that are surrounded by Chol-rich regions at the interfaces with the L_d regions. This agreement with the atomistic simulations lends credibility to the minimal biophysical picture underlined by the lattice model. The multi-scale picture provided by the lattice simulations presented here is unprecedented in terms of the wide range of length and time scales. At the same time, the picture is also limited in the sense that, contrary to atomistic and MARTINI simulations, it gives almost no information on the chain configurations or structure of the domain interfaces. The message to take here is the utility of the different modeling approaches, from the atomistic simulations to the highly CG ones, and the importance of integrating them into a full, multi-scale physical picture. This is a great challenge in many areas of material science, including membrane biophysics.

References

1. Y. Lee, D. H. Thompson, *Stimuli-responsive liposomes for drug delivery*, Wiley Interdisciplinary Reviews: Nanomedicine and Nanobiotechnology **9**, e1450 (2017).
2. B. Alberts, A. Johnson, J. Lewis, M. Raff, K. Roberts, P. Walter, *Molecular Biology of the Cell*, 4th edn. (Garland Science, New York, 2002).
3. S. J. Blundell, K. M. Blundell, *Concepts in Thermal Physics*, 2nd edn. (Oxford Press, Oxford, 2010).
4. J. Israelachvili, *Intermolecular and Surface Forces*, 2nd edn. (Academic Press, London, 1991).
5. K. A. Dill, S. Bromberg, *Molecular Driving Forces: Statistical Thermodynamics in Biology, Chemistry, Physics, and Nanoscience*, 2nd. edn. (Garland Science, Boca Raton, 2010).
6. G. Caldieri, R. Buccione, *Aiming for invadopodia: organizing polarized delivery at sites of invasion*, Trends in Cell Biology **20**, 64 (2010).
7. R. M. Lynden-Bell, S. C. Morris, J. D. Barrow, J. L. Finney, C. Harper (eds), *Water and Life: The Unique Properties of H_2O*, 1st edn. (CRC Press, Boca Raton, 2010).
8. A. Ben-Naim, *Hydrophobic Interactions* (Plenum Press, New York, 1980).
9. G. Navascues, *Liquid surfaces: theory of surface tension*, Rep. Prog. Phys. **42**, 113 (1979).
10. T. A. Witten, *Structured Fluids: Polymers, Colloids, Surfactants* (Oxford University Press, Oxford, 2010).
11. M. J. Rosen, J. T. Kunjappu, *Surfactants and Interfacial Phenomena*. 4th edn. (John Wiley & Sons, Hoboken, 2012).
12. S. Safran, *Statistical Thermodynamics of Surfaces, Interfaces, and Membranes (Frontiers in Physics)* (CRC Press, Boca Raton, 2003).
13. R. Brewster, P. A. Pincus, S. A. Safran, *Hybrid lipids as a biological surface-active component*, Biophys. J. **97**, 1087 (2009).
14. C. J. Fielding (ed), *Lipid Rafts and Caveolae: From Membrane Biophysics to Cell Biology* (Wiley-VCH, Weinheim, 2006).
15. F. G. van der Goot, T. Harder, *Raft membrane domains: From a liquid-ordered membrane phase to a site of pathogen attack*, Semin. Immunol. **13**, 89 (2001).
16. H.-J. Kaiser, D. Lingwood, I. Levental, J. L. Sampaio, L. Kalvodova, L. Rajendran, K. Simons, *Order of lipid phases in model and plasma membranes*, Proc. Natl. Acad. Sci. U.S.A **106**, 16645 (2009).
17. K. Simons, E. Ikonen, *Functional rafts in cell membranes*, Nature **387**, 569 (1997).

18. L. Pike, *Rafts defined: A report on the keystone symposium on lipid rafts and cell function*, J. Lipid Research **47**, 1597 (2006).
19. A. R. Honerkamp-Smith, S. L. Veatch, S. L. Keller, *An introduction to critical points for biophysicists; observations of compositional heterogeneity in lipid membranes*, Biochim. Biophys. Acta Biomembr. **1788**, 53 (2009).
20. S. Komura, D. Andelman, *Physical aspects of heterogeneities in multi-component lipid membranes*, Adv. Colloid Interface Sci. **208**, 34 (2014).
21. F. Schmid, *Physical mechanisms of micro- and nanodomain formation in multicomponent lipid membranes*, Biochim. Biophys. Acta Biomembr. **1859**, 509 (2017).
22. L. D. Landau, E. M. Lifshitz, *Theory of Elasticity (Volume 7 of Course in Theoretical Physics)* 3rd edn. (Butterworth-Heinemann, Oxford, 2006).
23. D. Boal, *Mechanics of the Cell*, 2nd edn. (Cambridge University Press, Cambridge, 2012).
24. E. Evans, W. Rawicz, *Entropy-driven tension and bending elasticity in condensed-fluid membranes* Phys. Rev. Lett. **64**, 2094 (1990).
25. O. Farago, P. Pincus, *The effect of thermal fluctuations on Schulman area elasticity*, Eur. Phys. J. E **11**, 399 (2003).
26. O. Farago, *Mechanical surface tension governs membrane thermal fluctuations*, Phys. Rev. E **84**, 051914 (2011).
27. P. Sens, S. A. Safran, *Pore formation and area exchange in tense membranes*, Europhys. Lett. **43**, 95 (1998).
28. P. G. de Gennes, *Scaling Concepts in Polymer Physics* (Cornell University Press, Ithaca, 1979).
29. J. F. Marko, E. D. Siggia, *Statistical mechanics of supercoiled DNA*. Phys. Rev. E. **52** 2912 (1995).
30. W. Helfrich, *Elastic properties of lipid bilayers: Theory and possible experiments*, Z. Naturforsch. **28C**, 693 (1973).
31. F. Campelo, C. Arnarez, S. J. Marrink, M. M. Kozlov, *Helfrich model of membrane bending: From Gibbs theory of liquid interfaces to membranes as thick anisotropic elastic layers*, Adv. Colloid Interface Sci. **208**, 25 (2014).
32. E. Sackmann, *Thermo-elasticity and adhesion as regulators of cell membrane architecture and function*, J. Phys.: Condens. Matter **18**, R785 (2006).
33. A. J. Ridley, *Life at the leading edge*, Cell **145**, 2012 (2011).
34. C. Monzel, K. Sengupta, *Measuring shape fluctuations in biological membranes*, J. Phys. D: Appl. Phys. **49**, 243002 (2016).
35. F. Schmid, *Fluctuations in lipid bilayers: Are they really understood?*, Biophys. Rev. Lett. **8**, 1 (2013).
36. C. Allolio, A. Haluts, D. Harries, *A local instantaneous surface method for extracting membrane elastic moduli from simulation: Comparison with other strategies*, Chem. Phys. **514**, 31 (2018).
37. O. Farago, *Mode excitation Monte Carlo simulations of mesoscopically large membranes]/*, J. Chem Phys. **128**, 184105 (2008).
38. O. Farago, P. Pincus, *Statistical mechanics of bilayer membrane with a fixed projected area*, J. Chem. Phys. **120**, 2934 (2004).
39. J. D. Litster, *Stability of lipid bilayers and red blood cell membranes*, Phys. Lett. A **53**, 193 (1975).
40. O. Farago, C. D. Santangelo, *Pore formation in fluctuating membranes*, J. Chem. Phys. **122**, 044901 (2005).
41. R. O. Dror, R. M. Dirks, J. P. Grossman, H. Xu, D. E. Shaw, *Biomolecular simulation: a computational microscope for molecular biology*, Annu. Rev. Biophys. **41**, 429 (2012).
42. V. Brádzová, D. R. Bowler, *Atomistic Computer Simulations - A Practical Guide* (Wiley-VCH, Weinheim, 2012).
43. M. Deserno, K. Kremer, H. Paulsen, C. Peter, F. Schmid, *Computational Studies of Biomembrane Systems: Theoretical Considerations, Simulation Models, and Applications*. In: T. Basché, K. Müllen, M. Schmidt (eds), From Single Molecules to Nanoscopically Structured Materials, Advances in Polymer Science, vol 260. (Springer, Cham, 2014).

44. D. Frenkel, B. Smit, *Understanding Molecular Simulations: From Algorithms to Applications*, 2nd edn (Academic Press, San Diego, 2002).
45. L. Verlet, Phys. Rev. **159**, 98 (1967).
46. H. J. C. Berendsen, J. P. M. Postma, W. F. van Gunsteren, A. DiNola, J. R. Haak, *Molecular dynamics with coupling to an external bath*, J. Chem. Phys. **81**, 3684 (1984).
47. R. W. Hockney, J. W. Eastwood, *Computer Simulation Using Particles* (McGraw-Hill, New York, 1981).
48. N. Grønbech-Jensen, O. Farago, *Defining velocities for accurate kinetic statistics in the Gronbech-Jensen Farago thermostat*. Phys. Rev. E **101**, 022123 (2020).
49. S. Nosé, *A unified formulation of the constant temperature molecular dynamics methods*, J. Chem. Phys. **81**, 511 (1984).
50. W. G. Hoover, *Canonical dynamics: Equilibrium phase-space distributions*, Phys. Rev.A **31**, 1695 (1985).
51. N. Grønbech-Jensen, O. Farago, *A simple and effective Verlet-type algorithm for simulating Langevin dynamics*,Mol. Phys. 111, 983 (2013).
52. N. Grønbech-Jensen, N. R. Hayre, O. Farago, *Application of the G-JF discrete-time thermostat for fast and accurate molecular simulations*, Comput. Phys. Commun. **185**, 524 (2014).
53. P. Langevin, *Sur la théorie du mouvement brownien (On the theory of Brownian motion)*, C. R. Acad. Sci. Paris **146**, 530 (1908).
54. R. Kubo, *The fluctuation-dissipation theorem*, Rep. Prog. Phys. **29**, 255 (1966).
55. B. Leimkuhler, C.Matthews, *Rational Construction of Stochastic Numerical Methods for Molecular Sampling*, Appl. Math. Res. Express **2013**, 34 (2013).
56. Z. X. Guo (ed), *Multiscale Materials Modeling* (Woodhead Publishing, Cambridge UK, 2007).
57. S. C. L. Kamerlin, S. Vicatos, A. Dryga, A. Warshel1, *Coarse-grained (multiscale) simulations in studies of biophysical and chemical systems*, Annu. Rev. Phys. Chem. **62**, 41 (2011).
58. J. H. Ipsen, G. Karlström, O. G. Mourtisen, H. Wennerström, M. J. Zuckermann, *Phase equilibria in the phosphatidylcholine-cholesterol system*, Biochim. Biophys. Acta Biomembr. **905**, 162 (1987).
59. M. R. Vist and J. H. Davis, *Phase equilibria of cholesterol/dipalmitoylphosphatidylcholine mixtures: Deuterium nuclear magnetic resonance and differential scanning calorimetry*, Biochemistry **29**, 451 (1990).
60. A. P. Lyubartsev, A. L. Rabinovich, *Force field development for lipid membrane simulations*, Biochim. Biophys. Acta **1858**, 2483 (2016).
61. M. Javanainen, H. Martinez-Seara, I. Vattulainen, *Nanoscale membrane domain formation driven by cholesterol*, Sci. Rep. **7**, 1143 (2017).
62. N. Grønbech-Jensen, O. Farago, *Constant pressure and temperature discrete-time Langevin molecular dynamics*, J. Chem. Phys. **141**, 194108 (2014).
63. C. L. Armstrong, D. Marquardt, H. Dies, N. Kučerka, Z. Yamani, T. A. Harroun, J. Katsaras, A.-C. Shi, M. C. Rheinstädter, *The observation of highly ordered domains in membranes with cholesterol*, PLoS ONE **8**, e66162 (2013).
64. S. J. Marrink, H. J. Risselada, S. Yefimov, D. P. Tieleman, A. H. de Vries. *The MARTINI force field: Coarse grained model for biomolecular simulations*, J. Phys. Chem. B **111**, 7812 (2007).
65. C. Díaz-Tejada, I. Ariz-Extreme, N. Awasthi, J. S. Hub, *Quantifying lateral inhomogeneity of cholesterol-containing membranes*, J. Phys. Chem. Lett. 6, 4799 (2015).
66. M. Javanainen, B. Fabian, H. Martinez-Seara, *Comment on "Capturing phase behavior of ternary lipid mixtures with a refined Martini coarse-grained force field"*, preprint arXiv:2009.07767 (2020).
67. H. Noguchi, M. Takasu, *Self-assembly of amphiphiles into vesicles: a Brownian dynamics simulation*, Phys Rev E **64**, 041913 (2001).
68. O. Farago, *"Water-free" computer model for fluid bilayer membranes*, J Chem Phys **119**, 596 (2003).
69. G. Brannigan, F. L. H. Brown, *Solvent-free simulations of fluid membrane bilayers*, J Chem Phys **120**, 1059 (2004).

70. I. R. Cooke, K. Kremer, M. Deserno, *Tunable generic model for fluid bilayer membranes* Phys Rev E **72** 011506 (2005).
71. O. Lenz, F. Schmid, *A simple computer model for liquid lipid bilayers*, J. Mol. Liq. **117**, 147 (2005).
72. S. Meinhardt, F. Schmid, *Structure of lateral heterogeneities in a coarse-grained model for multicomponent membranes*, Soft Matter **15**, 1942 (2019).
73. A. H. de Vries, S. Yefimov, A. E. Mark, S. J. Marrink, *Molecular structure of the lecithin ripple phase*, Proc. Nat. Acad. Sci. USA **102**, 5392 (2005).
74. C. N. Yang, T. D. Lee, *Statistical theory of equations of state and phase transitions II. Lattice gas and Ising model*, Phys. Rev. **87**, 410 (1952).
75. T. Sarkar, O. Farago, *Minimal lattice model of lipid membranes with liquid-ordered domains*, preprint arXiv:2103.13761 (2021).
76. N. Dharan, O. Farago, *Formation of semi-dilute adhesion domains driven by weak elasticity-mediated interactions*, Soft Matter **12**, 6649 (2016).

Self-Organization of Tissues Through Biochemical and Mechanical Signals

Georgios Misailidis, Jaroslav Ferenc, and Charisios D. Tsiairis

Abstract Self-organization enables the emergence of patterns in systems without the need of an external causal agent. Instead, the interactions and the properties of the units that constitute the system secure their organization in time and space. At the level of multicellular systems, the power of self-organization is prominent in embryogenesis and has been central in the study of in vitro developing organotypic cultures. We examine the self-organization of two multicellular systems, the cnidarian *Hydra* and the presomitic mesoderm of mouse embryo. In both cases, random aggregates of cells achieve spatial pattern around emergent organizers, characterized by high levels of Wnt signaling. Mechanical signals are involved in the process of both systems and they participate in a choreography that requires cell activity coordination. This is especially relevant for the presomitic mesoderm cells where synchronized gene expression is achieved. In both systems the ability to achieve patterning is stemming from the way cells interact and communicate, making possible the extraction of general rules that link biological with other self-organizing systems.

1 Introduction

The presence of structure and order in natural systems poses the question of the pattern's origin. Very often the cause is attributed to an agent outside the system that gets patterned. This agent does not have to be an external instructor, as blueprints,

G. Misailidis · J. Ferenc
Friedrich Miescher Institute for Biomedical Research, Basel, Switzerland

University of Basel, Basel, Switzerland

C. D. Tsiairis (✉)
Friedrich Miescher Institute for Biomedical Research, Basel, Switzerland
e-mail: charisios.tsiairis@fmi.ch

© The Author(s), under exclusive license to Springer Nature Switzerland AG 2021
J. Málek, E. Süli (eds.), *Modeling Biomaterials*, Nečas Center Series,
https://doi.org/10.1007/978-3-030-88084-2_2

pre-patterns or a pre-calculated and finely orchestrated plan can be used. While an explanation emerges, the question is then asked at a different level leading to an infinite regress of causes or in some cases leads to closed circular cause and effect chains reminiscent of a chicken-and-egg problem. Clearly, a form of an organizer of the pattern, a type of a leader, can instruct the remaining elements of the system to get patterned. However, this only leads to the question of how such an organizer is selected. If it is determined by an external factor, the ultimate cause is outside the system. An escape from these traps is offered by assigning the cause of the patterns inside the organizing system and a self-contained explanatory narrative has to trace the emergence of an organizer in the system itself. It is indeed possible to achieve pattern in a system exclusively due to the properties and interactions of its elements. Self-organization is a process in which a set of elements with well-defined characteristics and some random arrangement interact, modify their properties, and end up in a fixed configuration [1].

1.1 Self-Organization in Biological Systems

In biological systems, self-organization is observed at almost any organization scale. In fact, the philosopher Kant has described life as self-organizing and self-reproducing process [2]. Simple molecular systems generate patterns in space or in time. When microtubules and dynein molecules are mixed under proper conditions, microtubules are organized in asters, with all their minus ends collected together [3]. The circadian rhythm of cyanobacterium *Synechococcus elongatus* relies on the periodic phosphorylation of KaiC protein. It has been shown that a mixture of KaiC with KaiA and KaiB proteins, supplied with adenosine triphosphate, can recapitulate the periodic phosphorylation of KaiC [4]. In this case, the 24 h period cycle in the KaiC state is exclusively emerging as the result of the way the protein components of the system interact. Complex patterns with traveling waves appear on lipid membranes as the proteins MinD and MinE bind and unbind [5]. Normally, these waves travel along the *E. coli* bacteria and localize the division plane in the middle of the long axis. It has been shown that such spatiotemporal patterns can emerge in artificial systems through the interaction of the proteins [6]. It is noteworthy that these systems share similarities with chemical systems, where reactions and diffusion of intermediate products lead to spatiotemporal patterns as the case of Belousov-Zhabotinsky reaction shows [7].

At the level of organism populations, the concept of self-organization has been successfully applied to understand the behavior of social insects [8]. The elaborate termite nests emerge through self-organization from the local interactions between individual termites, via pheromones, and mud properties [9]. Self-organization is further a general phenomenon in social insects and ants encompassing multiple aspects of the colony's behavior, like foraging activity, coordinated defense, etc., and which emerge from the interaction of individual members [10]. In other animals a characteristic self-organization pattern is seen during swarming and flocking. There the behavior of individual is aligned with every other member of the group, so that they manage to move en masse [11, 12]. Artificial robotic systems are able to achieve swarming following the same principles as animals [13].

From the different levels of biological organization, we will focus on the one of the cells, the units of life, and the way they get organized into tissues. Self-organization of cellular systems is currently at the frontline, as the study of organoids is increasing steadily [14]. Organoids are 3D cultures that recapitulate the organization and physiology of organs. Very often the cells achieve the proper tissue organization *de novo*, under uniform culture conditions, offering a clear case of self-organization. The first such system to be described is the intestinal organoid system [15]. It is noteworthy that these organoids, recapitulating the organization unit of the gut, start from single cells. As each cell divides, the group arranges around a hollow lumen which over time breaks its symmetry and a special type of organizer cells emerge [16]. The way the symmetry is broken and the organizer spontaneously emerges is under intense study [17]. It is, however, clear that the communication and interaction between cells allows some to change their state towards organizers of the ensuing pattern.

As organoids developed to resemble more and more organs, an interesting type of self-organizing cultures emerged, attempting to recapitulate not the structural and functional organization of an organ but of an entire organism. Gastruloids are a new approach towards these directions, where undifferentiated embryonic cells are aggregated and left to develop under proper culture conditions in order to establish the main embryonic tissues [18]. In these systems the key step appears to be the spontaneous emergence of a group of cells with the properties of an organizer. The uniform clump of cells changes to an elongated shape when the cells on one pole of the sphere change their identity and signal to the other cells initiating a domino of events that resembles the embryogenesis event sequence. The organizer cells express genes that are associated with the Wnt communication cascade [19].

As is the case with any self-organizing system, the center stage belongs to the way the members of the group exchange information about their status and how they update their status upon information reception. Variable signals that can originate from random noise can trigger positive feedback loops when a specific state is enhanced further through a chain of mediating events. Alternatively, negative feedbacks can restrict a state or limit it to fewer members of the group. In this case, a chain of events leads to an inhibition of the original trigger. While these can serve as general principles, it is worth examining how they are materialized in the case of cellular systems, as the exact realizations point to opportunities and limitations.

A cell's state, its properties, reflects the exact proteins that it employs and the amounts of these proteins. The part of the genome that is expressed in a cell determines its fate. Thus, a cell state can be thought of as a position in the space of gene expression. The enthusiasm over the single cell sequencing technologies stems largely from their ability to place each cell in this space. Combined with temporal information, it is now even possible to have trajectories of cell identity changes in the gene expression space [20]. Under this lens, self-organization of cells is the spontaneous acquisition, through trajectories that may or may not be fixed, of a precise constellation of cell identities enabling the functionality of the group. It is important to point out that the events unfolding in the organization process depend on the fact that cells in a group are never exactly equivalent. Moreover, they communicate in a way that can nudge the information receivers towards specific trajectories or differentiation paths.

The state of a cell is variable. Even cells of the same type are never identical. The variability is driven by external but also internal factors. Each cell has an individual microenvironment occupying a physical location not equivalent to any other. More importantly, the internal biochemical workings in a cell are subject to noise and randomness of processes with a discrete nature. When cells divide, the distribution of molecules is random and thus cannot be expected to be exactly equal. This effect becomes more pronounced for molecules that are in low abundance. The process of gene transcription is also of a noisy nature. In fact, the promoter structure itself facilitates gene expression variability between genetically identical cells [21]. Thus, a population of cells is never uniform, and this variability can be further amplified through communication between them.

Cells exchange information with a plethora of ways employing chemical, electrical, and mechanical signals. The biochemical signals are usually proteins or small molecules that are secreted or presented to the external environment of the signaling cell. The signals' properties, and the environment determine how far is the physical distance that it can span. Diffusion is often at play and drives the signal away from the source in a signaling gradient. Sometimes the signal impacts directly the immediate neighbor, especially if it is bound or tethered to the membrane. On the receiving end, the cell should have the proper detectors, which are the receptor molecules on the cell surface. Very often, the arrival of the signal initiates a cascade of events that, after the activation of different nodes, leads to transcription of genes and the corresponding update of the cellular state. It may have been expected that signals would be extremely numerous reflecting the different outcomes on cell behavior. Surprisingly, a handful of biochemical signaling cascades are employed throughout the biological world. Some of the most widely used ones that will be a focus of further analysis are the Wnt, FGF, and Notch signaling cascades. The evolution of these communication avenues parallels the evolution of complex multicellular organization [22].

Signals can also have the form of electricity, through ions, or mechanical forces. For example, epithelial sheets of cells maintain connectivity of the cells via tight junctions. These contain pores constructed by multimers of the protein connexin that enable the transport of small molecules and ions. The charged ions can change the electric polarity of cell membranes and activate signaling cascades inside the cells leading to gene transcription changes [23]. Forces of mechanical nature are also transmitted within and between cells and are rendered into gene expression changes. Cells are probing their environment's stiffness and sense whether they are surrounded by other cells, being pushed or pulled. The intracellular cytoskeleton operates as an integrator of the mechanical forces [24]. One of the most common downstream mediators of mechanical signals is the protein

Yap which, when phosphorylated, is not able to enter the cell nucleus and drive gene expression. Yap phosphorylation is inhibited, for example, when cells are stretched or encounter a stiff substrate, and this enables Yap to drive gene expression in response to mechanical stimuli [25]. In all these cascades, there is always a point where mechanical forces are transformed into biochemical events, mostly by a conformational change caused in a protein due to mechanical changes. The piezo cell membrane channels operate in this way as they are open upon mechanical stretching of the membrane, enabling flow of ions and initiation of a biochemical chain of events leading to the cell nucleus [26].

Since the communication cascades are not insulated, it is important to realize that a significant crosstalk exists as pathways share nodes or modulate other signaling cascades through feedback interactions. A case relevant for the subsequent analysis is the protein β-catenin, important both for the Wnt signaling cascade and the ability of cells to transmit forces with their neighbors through cytoskeletal links [27]. The arrival of Wnt signals drives β-catenin to the nucleus, where it links transcription factors to transcriptional activators, thus initiating transcription of Wnt signaling target genes. At the same time, epithelial cells are connected through adherens junctions where β-catenin participates as a glue linking the junction complexes with actin filaments of the cytoskeleton [28]. Another example of a context-dependent outcome of a signaling event is the proliferation response of cells upon EGF stimulation, which is dependent on their density [29]. Thus, the effect of a signal on a cell state does not only depend on the signal identity but also on the current cellular state through a complex web of intra- and intercellular interactions.

Overall, the picture that emerges for the self-organization of cells into tissues is one of a progressive cell state update based on the information received on the state of the other cells. It is noteworthy that this mirrors theoretical exercises on the behavior of units termed cellular automata [30]. Here, fixed positions on a 1D or 2D grid can have a value, usually 0 or 1. In discrete time steps, each position's value is updated depending on the values in the neighborhood. Such simple systems are able to reach self-organization into complex patterns at the level of the field, following simple deterministic rules on the ways the values are updated [31]. One of the most studied automata is termed "game of life" and in a 2D lattice each position can take values of 0 or 1. At every time step a position maintains value 1 if there are 2 or 3 neighboring positions with value 1. Otherwise, it switches to 0. A position with 0 switches to 1 if 3 out of the 8 immediate neighboring positions are 1. Such simple rules can generate patterns from initial randomness [32]. Importantly, these approaches highlight the possibility of generating order following abstract rules that may be materialized in various systems through different mechanisms. The essence of these games captures the fundamental principles of group patterning that emerges through cell interactions and updates of their genetic identity.

For the remaining of this chapter, we will focus on two multicellular systems that manage to get organized. They can serve as case studies to demonstrate the principles and the complexity in action. Cells of the simple cnidarian *Hydra* can be assembled in a clump and get re-patterned into a new complete animal. The spatial organization of the *Hydra* cells along a single main body axis requires the cells to break their symmetry and establish an organizer that orchestrates downstream patterning events. The second system is separated from *Hydra* by 600 million years of evolution and has to do with the way mammals and other vertebrates generate the sequential pattern of bones that give them their name, the vertebrae. The embryonic tissue that produces the segments later forming the vertebrae is patterned in space and time and has the ability to self-organize when the cells are randomly mixed. Organizers that are marked by high levels of Wnt signaling production emerge in both systems.

2 The Hydra Model System

The freshwater cnidarian *Hydra* is an evolutionarily ancient organism with one of the simplest body plans among multicellular animals [33]. Its body resembles a tube with two ends that define the two poles of its single body axis (Fig. 1a). The oral (upper) pole consists of a mouth opening (hypostome) surrounded by tentacles, which the animal uses to catch prey. Once the prey has been digested inside of the body cavity, the mouth also serves as the exit route for undigested leftovers. The opposite site of the axis (aboral pole) consists of structures specialized for attaching the animal to the substrate and is termed the foot and basal disc. Two layers of epithelio-muscular tissues (ectoderm and endoderm, also known as epidermis and gastrodermis), separated by a thin layer of extracellular matrix (mesoglea), constitute the body wall [34]. Apart from the epithelia, the *Hydra* body also contains a third cell lineage known as the interstitial cells (i-cells) [35], which give rise to a variety of cells types, such as neurons, gland cells, germ cells, and the stinging cells (nematocytes) used to paralyze prey. Even though capable of sexual reproduction, these animals predominantly reproduce asexually by budding [36]. During this process, a miniature daughter *Hydra* forms perpendicular to the parental body column and, once fully developed, detaches from the parent and continues living freely. Interestingly, budding only happens in a spatially restricted region adjacent to the foot, known as the budding zone. Unlike in more complex organisms, the majority of *Hydra* cells are not terminally differentiated [37]. The whole body has a high cell turnover, similar to quickly renewing mammalian epithelia (e.g. gut or skin). Cells in the body column are constantly dividing and moving towards the extremities, where they terminally differentiate, and eventually die and are replaced by newly arriving cells. Thanks to such a big pool of stem-like cells, *Hydra* is capable of extensive regeneration. This property was first noticed and investigated by Abraham Trembley as early as in the eighteenth century, marking the dawn of experimental biology [38, 39].

Transplantation of tissue pieces from one individual to another one is easy to perform and allows altering the body architecture in a controlled manner. Such manipulations served as a powerful tool to gain insights into the principles of patterning, such as the first identification of a biological organizer, the hypostome of *Hydra* [40]. This small group of cells around the mouth opening controls patterning

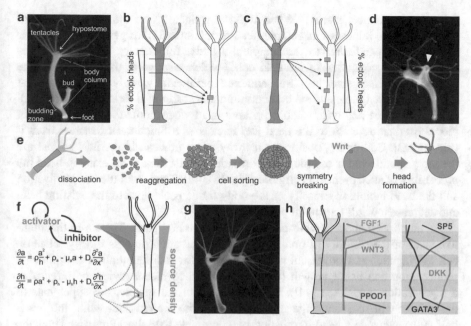

Fig. 1 The *Hydra* model system and biochemical cues for its patterning. (**a**) The morphology of *Hydra* body. Grafting experiments demonstrate head activation (**b**) and inhibition gradients (**c**) (adapted from [42, 43]). The ability to generate an ectopic head is higher in grafts originating closer to the oral pole (**b**), but the location of the graft within the host body influences the ectopic head formation success. Positions closest to the host head have the highest inhibitory potential, suggesting a head inhibition gradient (**c**). (**d**) An example of an ectopic head (arrowhead) generated by Wnt3-primed tissue transplanted from another animal (fluorescently labelled in green). (**e**) Schematic of the reaggregation experiments (adapted from [46, 47]) that demonstrated the self-organizing capabilities of the body column cells. The tissue is mechanically dissociated and the cells then reaggregated into a clump by centrifugation. After reestablishing the original tissue layers, the resulting hollow spheroid will spontaneously break symmetry and regenerate a new head. As shown in later work [48], localized expression of Wnt signaling ligands is the hallmark of molecular symmetry breaking before any noticeable morphological changes. (**f**) Summarizing the Gierer Meinhardt model with the equations adapted from [49] describe the spatiotemporal changes of activator (**a**) and inhibitor (**h**) whose production rates are ρ, degradation rates μ, and diffusion coefficients D. Both the activator and inhibitor peak in the organizer region (black dot) with the activator prevailing there. However, in the body column, head inhibition prevails. Note that if the animal is big enough, a second peak can occur (dotted lines), which would be the basis for ectopic axis generation during budding. (**g**) A Wnt3-overexpressing animal with a typical multiheaded phenotype. Even though Wnt3 is uniformly expressed at high levels throughout the entire ectoderm, the head organizer program is still only activated in discrete areas (although several of them). (**h**) Examples of gene expression profiles (max normalized) along the body axis. Some genes (left) have a spatially narrow expression domain, such as WNT3 in the head, the peroxidase PPOD1 in the foot, or the signaling factor FGF1 in tentacles. Other genes, such as SP5, DKK, and GATA3, are expressed in a graded fashion (right). Data from [79]

through biochemical signals. When it is transplanted into the body column of another animal, it induces an ectopic head formation and thus generates an ectopic body axis perpendicular to the recipient's body. Importantly, the new head is predominantly composed of the host cells, which respond to the presence of the grafted organizing center [41]. Subsequent experimental work has also shown that even transplanted pieces of the body column can sometimes form an ectopic head. Interestingly, the ability to establish a new head when grafted is the strongest in pieces originating close to the head and decays as a function of distance from it (Fig. 1) [42]. Conversely, positions far away from the head in the host animal are the most conducive for ectopic head formation, while next to an existing head the probability of such event is negligible (Fig. 1c) [43]. Thus, both the head activation and the head inhibition capacity of the body tissue peak close to the existing head and decrease towards the aboral pole (Fig. 1b–d).

As mentioned above, *Hydra* can regenerate missing parts. Thanks to the stem-like state of the majority of body cells, the regeneration process mostly relies on cell specification and movement without the need for dedifferentiation or significant proliferation and tissue growth (for a more detailed recent review of *Hydra* as a regeneration model see [44]). When bisected in the middle, the upper half of the animal will regenerate the missing foot at the injured site, while the lower half will regenerate a head, producing two complete half-sized animals. Thus, the missing organizer can also be established *de novo* under these conditions. Since the regeneration of bigger body pieces preserves the original polarity, it has been hypothesized that morphogenetic gradients spanning the entire body act as guides for proper positioning of regenerating structures [45]. However, even very small pieces (about 1/20) of the body are capable of regenerating a full animal, which suggests that the entire body pattern can be reestablished *de novo* even without the contribution of gradients. This prediction was indeed experimentally confirmed by dissociating and reaggregating the body column cells [46, 47]. Cells in such a reaggregate first sort into the two epithelial layers, while also transforming the clump into a hollow spheroid. Spontaneous symmetry breaking then follows and the spheroid establishes one or several new body axes (Fig. 1e). The number of emerging heads depends on the size of the aggregate, suggesting an intrinsic spatial scale of the patterning process [48]. Remarkably, *Hydra* cells rebuilt the animal purely through self-organization without the addition of any external signals, which is often required for similar biological models, such as organoids.

2.1   Biochemical Stimuli and Models of Patterning

Along with the experiments uncovering the foundations of *Hydra* patterning, a theoretical framework developed by Alfred Gierer and Hans Meinhardt (GM model) sought to formalize the observations into a coherent mechanism that would bridge the results from different approaches using a reaction-diffusion system [49]. It postulates two chemical morphogens (head activator and head inhibitor) produced by the organizer that diffuse into the body, generating gradients. The activator is a less diffusive molecule, providing local activation and positive feedback, while the inhibitor diffuses further away and generates long-range inhibition and negative feedback (Fig. 1f). The activator increases both the production of itself and the inhibitor, the inhibitor in turn reduces activator production. Based on other experimental results [43], the two morphogens are also assumed to have different stability (activator being more stable than the inhibitor). This simple framework can successfully explain the homeostatic morphology of the animal. Moreover, the inherent pattern length scale, dictated by the morphogen diffusion properties, would also account for the dependence of the number of emerging heads on the size of the reaggregate or the need for a freshly budded *Hydra* to grow first, before being capable of budding itself. Additionally, the GM model introduces morphogen source density, which is graded in the same way as a particular morphogen but less dynamic than its concentration. This could correspond, for example, to the presence of certain morphogen-induced cell states, which would form a positional identity gradient persisting even after removing the morphogen source. Such a gradient could then be responsible for maintaining the polarity during regeneration.

The elegance of the GM model and its ability to explain a wide range of experimental observations have served as a catalyst for efforts trying to identify the hypothesized morphogens. Even though many pioneering studies were looking for the head activator in extracts of small peptides from the *Hydra* body [50, 51], the most promising candidate turned out to be the Wnt3 protein [52]. This protein is a secreted ligand of Wnt signaling—a conserved cell communication pathway that originated along with the origins of multicellularity in animals [53]. Consistently, Wnt signaling was found to be crucial for specifying the main body axis of *Hydra* [54], as it does in most other multicellular animals [55]. Not only is the Wnt3 gene expression specific to the organizer cells, but also marks the spots of future heads in reaggregates before any morphological changes occur [48]. Wnt3 expression is also crucial for head regeneration [56] and, when ectopically induced, leads to ectopic head formation [57].

Once Wnt3 was accepted to be the head activator, the search for head inhibitor also concentrated on homologs of endogenous Wnt signaling inhibitors. The first promising candidate was the Dkk protein, a competitive inhibitor of Wnt binding to its receptor [58, 59]. While this small, secreted protein could act as a diffusible morphogen, its expression pattern in the body does not correspond with the model prediction because it peaks in the body column and not in the hypostome itself (see example expression patterns of selected factors in Fig. 1h). Another, more recent, candidate is the transcription factor Sp5, which inhibits the expression of Wnt3 yet, at the same time, is positively regulated by it [60]. When Sp5 is downregulated artificially, it phenocopies the Wnt3 activation phenotype of ectopic head formation. It is also expressed in a graded manner along the body column with a peak in the hypostome region; however, as a transcription factor, cannot diffuse between cells. Moreover, although the ectopic head phenotypes of a homogenous Sp5 suppression or Wnt3 activation clearly show their crucial role in head specification, they are not consistent with the GM model. According to the theoretical predictions, only the diffusive properties of the morphogens, but not their concentrations, can influence the pattern spacing [61]. Thus, despite decades of efforts, no molecules were found that would precisely fit the bill of head activator and inhibitor as outlined in the GM model. Conversely, any following theoretical work also now has to take into account the uncovered parts of the molecular network underlying head formation. Can these observations be reconciled with the original GM idea?

Perhaps the simplest way would be to consider the Wnt3/Sp5 feedback loop to be downstream of the actual morphogens, which remain unknown. In this case, the genetic network would function predominantly as a readout of the patterning mechanism, possibly establishing the source density gradient. Alternatively, several different modifications of the GM model try to accommodate the known molecular players as a part of the patterning mechanism. For example, by adding additional loops of slow- vs. fast-diffusing Wnt3 [62], having Wnt signaling control a competence gradient that decides which patterning centers will manifest as heads [63], or separating Wnt signaling and its main mediator β-catenin into two interconnected loops [64]. It would also be possible to consider a different realization of the long-range inhibition, such as the activator-depleted substance model, briefly introduced by Gierer and Meinhardt in their original work [49]. In this case, the axial pattern of Dkk would make it a good candidate to be such a substrate. While all of these models represent interesting conceptual developments, their assumptions and predictions also require careful evaluation, focusing on experimental setups that could distinguish between them. Yet another option would be abandoning the biochemical GM model in favor of a different mechanism. One of the proposed avenues are several mechano-chemical models, which either operate with local mechanical properties, such as curvature [65], or include tissue mechanics as a long-range factor, for example, taking tissue strain or stress into account [66]. Furthermore, recent experimental evidence indicates the presence of both of these mechanical influences in the *Hydra* patterning mechanism, hinting that taking a step beyond its purely biochemical understanding will be necessary.

2.2   *Mechanical Stimuli During Patterning*

A characteristic example highlighting the impact of mechanical properties is the finding that the orientation and topology of supracellular actin fibers can influence the organizer positioning in regenerating *Hydra* [67]. These structures are found close to the mesogleal attachment of both epithelial layers and serve as a substitute for specialized muscle cells. The actin cables in each cell within the same tissue layer have a matching orientation, enabling a directional contraction of the entire body wall. The ectodermal fibers run parallel to the body axis, while the endodermal fibers are perpendicular to it, creating circular "muscles" used for expelling the content of the body cavity. Importantly, the ectodermal fibers come together in aster-like structures at the poles of the body axis (Fig. 2a). When a small piece of the

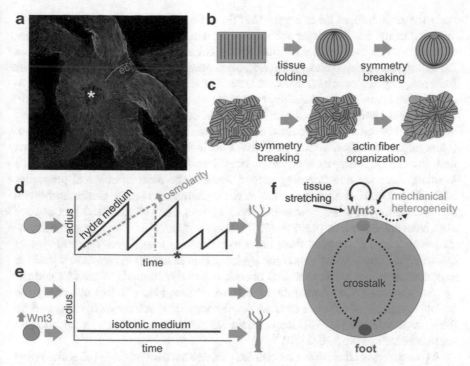

Fig. 2 Tissue mechanics in *Hydra* patterning. (**a**) Phalloidin staining reveals the organization of supracellular actin fibers in the head region of a *Hydra*. Ectodermal fibers (ec) run parallel to the axis and coalesce in the mouth (asterisk). Endodermal fibers (en) run perpendicular to them. (**b**) In rectangular pieces cut from the animal's body axis, the actin filaments remain at least partially intact and two poles with coalescing fibers are generated by tissue folding. They appear to bias the organizer positioning [71]. However, rather than directing the organizer positioning per se, the actin fiber network seems to modify the mechanical properties of the tissue, since the organized actin network is not necessary for organizer emergence. Indeed, upon reaggregation actin filaments begin to reappear in random conformations without global alignments (**c**). Symmetry is nevertheless broken, and the fibers subsequently reorganize to coalesce at the site of the new organizer (after [69]). Mechanical oscillations are evident during *Hydra* patterning as the spheroid's radius changes over time (**d**, black line). The oscillation pattern changes at the time of symmetry breaking (red asterisk). Since these oscillations are osmotically driven, in higher osmolarities (e.g. in media with added sucrose, NaCl, or sorbitol), the spheroid inflation slows down proportionately to the osmolarity increase (**d**, blue line) [75]. (**e**) Spheroids are not able to regenerate in isotonic medium where the mechanical input from oscillations is lacking, unless Wnt3 is provided externally (after [79]). (**f**) Proposed framework for how mechanical cues act during the regeneration of *Hydra* spheroids. In a freshly generated tissue spheroid, Wnt3 expression is sustained through tissue stretching as a result of mechanical oscillations. Even though all cells are being stretched and capable of differentiating to the organizer, the location of the organizer can be specified through local bias of tissue mechanical properties and the signaling crosstalk with the surrounding cells and, possibly, the developing foot organizer. Symmetry breaking happens once a small group of cells with sufficiently high Wnt3 levels is established. Wnt3 expression then becomes self-sustaining in these cells and the downstream organizer program is initiated

body tissue is cut and let to regenerate, the actin cables can be partially retained and, due to the way the tissue folds, generate similar poles in the spheroid. It has been shown that the presence of this topological arrangement is often predictive of the future organizer position [68]. However, previous experiments with spheroids from reaggregated cells have shown that such topological defects are not required to generate an organizer. On the contrary, when cells are dissociated the fibers quickly dissolve and later start reappearing in random configurations in the regenerating aggregate. However, the directional organization of fibers is not reestablished until the organizer emerges, which then drives their alignment [69]. Thus, it appears that, instead of acting as a driver of organizer formation or the memory of its position, the actin fiber topology rather changes the local mechanical properties of the tissue, making it more conducive for organizer formation. Consistently with this idea, different other mechanical perturbations, such as localized suction, can also bias the organizer position [70]. Thus, a picture of a dynamic mechano-chemical interplay emerges from these experiments, where both types of signals can serve as inputs of the patterning system. This is further evidenced by grafting experiments where mechanical and chemical signals are made to conflict and the system responds with a mixture of different outcomes, possibly reflecting the variable strength of different cues and/or the way in which they are integrated by the genetic network [71, 72]. Recent data also suggest a Wnt-driven feedback on tissue mechanical properties [73].

As mentioned earlier, tissue mechanics can also act on a more global scale during morphogenesis, such as when lumen expansion forces the surrounding tissue to stretch [74]. Regenerating *Hydra* spheroids (both made from cut tissue pieces and reaggregated cells) experience a very similar process [75]. Since *Hydra* is a fresh-water animal, cells constantly have to compensate their osmolarity by excreting surplus water entering them from the environment. In fully formed animals, this water is excreted into the body cavity and removed from there by spontaneous contractions [76]. Regenerating spheroids, however, lack a mouth. Pressure builds up with an increasing volume of water excreted into the spheroid cavity, resulting in spheroid expansion and, eventually, rupture when a threshold of tissue strength is reached. Several such cycles are typically necessary before the spheroid breaks symmetry, which leads to mechanical inflation and deflation oscillation (Fig. 2d). One of the symmetry breaking hallmarks is the stabilization of the new mouth opening [77]. This causes the oscillatory pattern to change as the mouth provides a weak point through which the accumulated water can be released. Thus, the spheroid inflates less before deflating, causing shortening of the oscillation period and decrease of amplitude. Even though the oscillations were already suggested to be important for regeneration early after they were described [77, 78], clear evidence of their role and connection to the molecular players was missing. We have recently shown that not only are the oscillations required for symmetry breaking but their role is to sustain the expression of the key organizer-defining factor Wnt3 [79]. This is mediated by tissue stretching during the inflation period, which activates Wnt3 transcription in a quantitative manner—if the spheroid inflation slows down, the Wnt3 levels drop. Conversely, supplying Wnt3 ectopically results in successful regeneration without mechanical input (Fig. 2e). Thus, a mechanism emerges where mechanical stretching can activate Wnt3 transcription in originally body tissue to enable *de novo* organizer formation. Once a critical threshold has been reached, this expression (and the organizer identity) can then sustain itself through a previously described positive feedback loop [80]. Interestingly, we have also uncovered a link of this system to the body morphogenetic gradients, since spheroids from tissue originating close to a head typically require more time (and therefore more tissue stretching) before successfully breaking symmetry.

Although the newest results provide convincing evidence about the involvement of mechanical stimuli in *Hydra* patterning, several open questions still need to be addressed. Especially, how are the mechanical cues perceived by cells, translated to biochemical signals, and integrated with the other signaling cascades. Another unclear point is how the cooperation of global and local mechanical signals achieves the spatial positioning of the organizer. We speculate, based on the available evidence, that the mechanical inhomogeneities (e.g. resulting from actin fiber organization) could bias certain tissue areas to be more amenable or sensitive to stretching, resulting in stronger Wnt activation in these spots. This could then lead to a faster activation of the self-sustaining organizer program and, conversely, its inhibition in the surrounding tissue (Fig. 2f). Ultimately, the challenge for any patterning model involving mechano-chemical crosstalk will be to provide a framework that could explain and unify all the results obtained by different experimental approaches. Moreover, the foot might not only be a passive counterpart of the head that forms on the opposite end of the axis. Even though relatively scarce, experimental evidence suggests that it could function as an independent, yet linked, organizing center [81, 82]. For example, approximately 50% of spheroids incapable of head regeneration due to the lack of mechanical stimulation can still regenerate a foot [79]. If indeed, the establishment of the two axial poles is largely independent, an additional layer of complexity emerges, because of the need for their proper spatial positioning in relation to each other.

Ultimately, the study of patterning and self-organization in *Hydra* will help us uncover not only their specific realizations in this animal but general principles applicable to other biological systems. The critical role of organizers first recognized in *Hydra* and subsequently as a general feature of animal patterning can serve as a good example [83, 84]. Similarly, the majority of signaling molecules are shared among animals and often function in similar ways [85]. In the case of Wnt, this is not only true for axis specification but also for diverse symmetry breaking events where a group of cells assumes a different identity from the rest of the tissue [53]. Moreover, the sensitivity of this pathway to a mechanical input appears to be another unifying thread that runs through most processes that rely on it [27, 86, 87]. Even though other pathways, such as the Notch [88], FGF [89], and BMP [90] signaling, are much less studied in Hydra, the existing results indicate their involvement in patterning and often a conserved mode of action. For example, the Notch signaling controls for the coordinated upregulation of genes that leads to proper boundary formation and detachment of early buds [88]. Interestingly, it also crosstalks with the head organizer and appears to play an important role in regulating tentacle patterning [91]. The best known function of FGF signaling in Hydra is controlling the cellular motility and shape changes associated with bud constriction and detachment [92]. However, since some of its ligands are also expressed in other body parts (Fig. 1), it might also have additional functions. Finally, it was recently hypothesized that the BMP signaling might have a fate-specifying role in the foot organizer, analogous to the role of Wnt signaling in the head organizer.

Even though the diversity of forms generated in animal morphogenesis is striking, it relies on a relatively restricted toolkit of underlying ancient processes, such as tissue folding, cell migration, and luminogenesis [24, 93, 94]. For example, the relationship between lumen expansion, tissue stretching, and cellular differentiation seems to be important for processes as different as the blastocyst formation [95, 96], inner ear morphogenesis [97] and the generation of lung alveoli [86]. Morphologically simple self-organizing systems amenable to experimental manipulations (such as *Hydra*, PSM cultures, and organoids) are invaluable tools to help us understand such principles. Eventually, the knowledge gained this way could be applied to more complex systems where these principal components interact in convoluted ways, with the hope of even being able to engineer them for practical applications.

3   The Presomitic Mesoderm

During early embryonic development, axial segmentation is an essential process for sculpturing the body plan of a vertebrate. This process is driven by a transient embryonic tissue called the presomitic mesoderm (PSM). The presomitic mesoderm is a tissue that is derived from the primitive streak from 6.5 days post coitum (dpc) on and is essential for the segmented anteroposterior axis of the embryo. It is the far most posterior tissue of the body and the caudal part of paraxial mesoderm, which flanks the neural tube. The functional output of the PSM is the generation of the somites. The somites are blocks of epithelial tissue that are formed from the periodic segmentation of the anterior part of the PSM [98]. While those structures are added at the anterior, the PSM elongates as more cells are being added to the posterior as a result of cell division. The somites appear sequentially in pairs and each of them consists of three major parts—the sclerotome, the myotome, and the dermatome [99]. Those components contribute to various tissues in later stages of development. The sclerotome contributes to the vertebrae, ribs, and cartilage while the myotome and dermatome add to musculature and the skin, respectively [100]. This physiological output is the result of the spatial and temporal organization of the PSM, which comprises from molecular signaling gradients and rhythmic gene expression, respectively [101]. The two are not to be considered as separate modules but rather tightly interconnected components that shape the fate of each cell.

3.1 Spatial Organization of the Presomitic Mesoderm

Spatially, the tissue is characterized by the presence of Wnt and Fibroblast Growth Factor (FGF) signaling gradients that extend from posterior towards the anterior. Retinoic acid (RA) is also distributed in an opposing graded manner [102]. The formation of the Wnt signaling gradient is the result of the expression pattern of Wnt3a, which peaks at the tail-bud, the most posterior part of the PSM. Without the expression of Wnt3a, the development of the posterior axis of the embryo is arrested [103, 104]. Binding of the Wnt3 protein to its receptor on the receiving cell's surface ignites the cascade of canonical Wnt signaling that causes the stabilization of β-catenin. This key mediator is constantly targeted by a degradation complex in the cytoplasm of the cells until Wnt signaling is activated. Upon the presence of the activating ligand, β-catenin is stabilized and translocated to the nucleus, where it promotes the regulation of expression of target genes [105]. Thus, the expression pattern of Wnt3a is shaping a gradient of nuclear localization for β-catenin from posterior to anterior [106]. The spatial point, where β-catenin levels are the lowest, indicates an important cell fate transition domain. There, cells go through mesenchymal to epithelial transition and begin expressing somite specific markers. As somites are added at the anterior and PSM elongates at the posterior, this spatial domain is gradually being displaced towards the posterior. This determination front is directly controlled by Wnt signaling through β-catenin levels [106].

The FGF signaling gradient operates in parallel and similar to Wnt gradient [107] and emerges from the increased level of expression of fgf8 at the very posterior part of the PSM. The cells that are positioned there, are loaded with fgf8 mRNA and this seems to be conserved across vertebrates. Fgf8 is gradually degraded towards the anterior part of the tissue [108]. This distribution transforms into a signaling gradient upon translation of FGF8 protein. When the FGF8 levels are increased experimentally, the segmentation is heavily undermined with few to no segments scored [107]. This, in combination with the high expression levels of fgf8 and Wnt3a at the tail-bud while the tissue elongates, points out the importance of FGF and Wnt in preserving the undifferentiated cell identity [107, 109]. Equivalently with the case of the Wnt gradient, low FGF signaling levels coincide with the determination front. It is clear that this apparent and dynamic spatial point is regulated by Wnt and FGF gradients. Furthermore, upregulation of Wnt induces the extension of FGF signaling, making evident the interplay between the two signaling pathways.

Cell state and behavior changes parallel to the molecular signaling gradients. Notably, along the anteroposterior axis of the PSM, the cells exhibit different speed of motion. Specifically, cells at the posterior of the PSM are highly motile and this property is gradually decreasing towards the anterior [110]. In addition, adhesion is also graded with the levels of adhesion proteins increasing as the cells are moving towards the anterior [111]. It is evident that the molecular gradients of Wnt and FGF generate cell identities that are characterized by different properties, such as motility and adhesion, but also gene activity that is going to be discussed in the next section. The extracellular signal-regulated kinase (ERK) and mitogen-activated protein kinase (MAPK) are the mediators of FGF signaling and their graded activity has been linked to the differential motility along the PSM [112]. In addition to the direct link with the determination front and the motility gradient, FGF signaling also connects the spatial organization of the tissue with the temporal.

3.2   Temporal Organization of the Presomitic Mesoderm

The prominent periodic deposition of somites at the anterior PSM is accompanied by periodic gene expression in the presomitic mesoderm. A milestone in the field was the identification of c-hairy1, a gene that is expressed in the PSM and its rhythmic expression matches the period of somite formation [113]. These oscillatory dynamics appear as waves of expression that initiate at the posterior PSM and sweep the tissue until they finally stop at the anterior. Each time such a wave meets the determination front at the anterior PSM, the cells cease their oscillatory gene expression. This observation was consistent with a previously proposed model to explain somitogenesis, termed the "Clock and Wavefront" model [114]. In essence, the model describes a mechanism that transforms a temporal signal into a spatial pattern by integrating the interaction of an expression wave with the determination front (Fig. 3). As such, this model links the spatial module of somitogenesis (molecular gradients) with the temporal (oscillatory gene activity). Since the identification of c-hairy1, a large network of genes has been shown to be expressed rhythmically. The vast majority of them are downstream targets of Wnt, FGF, and Delta-Notch signaling pathways [115]. This finding intuitively raised the question of how these oscillations are generated.

The quest to discover the mechanism that generates oscillations focused on Hes7, a transcription factor that forms dimers and suppresses its own expression [116]. This ability to inhibit its own expression sequence is crucial, as an autoregulatory negative feedback loop combined with delays is able to generate oscillations [117]. Lack of functional Hes7 leads to severely disrupted patterning. Removing non-coding regions of the Hes7 mRNA caused the arrest of Hes7 oscillations and major defects in somitogenesis [118]. It is a downstream target of Delta-Notch pathway as well as of FGF and Wnt signaling [119, 120]. Together, those findings position the negative feedback loop of Hes7 at the center of the mechanism that generates oscillations in the PSM. The oscillations downstream of Wnt signaling are also possibly generated via a specific negative feedback loop. Axin2 is a target gene of Wnt signaling and negative regulator of the pathway [104]. It is part of the destruction complex that targets β-catenin for degradation in the cytoplasm and its periodic expression facilitates the periodic degradation of β-catenin and the periodic activation of downstream target genes. While the mechanism that generates oscillations in the PSM has been discovered and described in detail, their properties, such as the period and amplitude, are still subject of research.

The oscillations themselves are dynamic as their characteristics change over space and time. As cells are gradually pushed towards the anterior by the newly

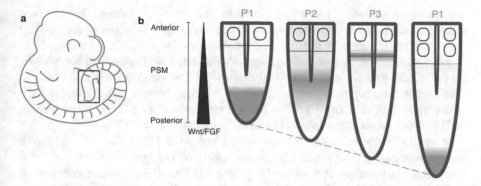

Fig. 3 The periodic formation of somites through a clock and wavefront mechanism. (**a**) The presomitic mesoderm is located at the posterior end of the vertebrate embryo and generates sequentially discrete structures at the anterior end of the tissue, the somites. (**b**) The tissue is patterned along the anterior posterior axis, and a key aspect of this pattern are Wnt and Fgf signaling gradients peaking at the posterior end. From this point waves of gene transcription sweep the entire field and once the anterior flanked is reached a new somite pair is added in the sequence. The transformation from a temporal clock into a spatially periodic pattern takes place as the oscillating phase of the periodic gene transcription freezes once the determination front (highlighted by a red line) is crossed. Snapshots of the traveling wave are presented in P1 to P3, with the blue indicating the peak of the wave. Once the wave has reached the wavefront with P3, a new wave starts with P1

added cells at the posterior, they slow down the pace of their oscillations. This generates a gradient of periods along the PSM, where cells at the posterior oscillate faster than cells at the anterior [102, 121]. Another important property of the oscillations is the amplitude, the maximum displacement from the resting point. The amplitude is gradually increasing along the PSM, following a similar trend with the period [106]. The period and the amplitude gradients correlate with Wnt and FGF that span the axis. Chemical perturbations of Wnt signaling have been reported to alter the period [119, 122, 123]. The mechanism through which this is achieved is not known. Upon genetic upregulation of Wnt signaling, the oscillations are sustained with a low amplitude, suggesting that besides FGF, high Wnt is also correlated with maintaining the undifferentiated cell identity [106]. Together, these findings propose that the cells receive inputs from Wnt and FGF as they pass through their gradients and adjust their oscillatory status accordingly.

3.3 Self-Organization of the Presomitic Mesoderm

A striking property of the cell of the PSM is the ability to self-organize. Spatiotemporal patterns emerged when the cells of the tissue are mechanically dissociated and reaggregated [121]. Soon after their aggregation, cells are able to sort out and form foci-like organizing centers. These foci appear to be distanced from each other by a rather constant length suggesting the establishment of an exclusion zone around each other. On top, immunohistochemistry reveals high β-catenin levels in those cells implying the *de novo* formation of the Wnt signaling gradient. Such organizing centers are reminiscent of the ones in *Hydra*, but also in other self-organizing systems, like gastruloids, or intestinal organoids. Tracking posterior PSM cells in these aggregates showed that they end up at those centers. What could be the mechanism that cells implement to form those centers? Is it an active or passive cell sorting? In other systems, it has been demonstrated that differential adhesion and motility between cells can have sorting outcomes [124]. Similar mechanism could be driving the sorting that is observed in PSM aggregates. Nevertheless, it is clear that the presence of a molecular gradient is important for the emergence of spatial patterns during self-organization. In the case of PSM cells, this is evident on the temporal domain as well.

Synchronous oscillations and waves of expression appear and propagate around those organizing centers [121]. Quantification of the period of the oscillations shows that it is graded. In that, cells at the center of the foci oscillate faster than those at the periphery. If the composition of the aggregates is altered by including more anterior PSM cells, the collective period is increasing. This suggests that there is a mechanism through which cells are able to average the input periods. One possible mechanism would be through molecular gradients. Wnt3a would be an appropriate candidate since its levels are distributed through cell contacts [125]. Cell-to-cell contacts are also essential for the synchronization of the oscillations upon these cultures. When Delta-Notch signaling is chemically inhibited, oscillations continue to occur but not waves [121]. Under those conditions, each cell carries on with their oscillations but the synchrony between them is lost. This could imply that the oscillatory activity that these cells exhibit is an intrinsic property.

Recent work has provided evidence that links the synchronized oscillatory expression with mechanics. Specifically, the cells utilize a mechano-sensing molecular machinery that allows them to switch from an oscillatory state to a non-oscillatory state depending on the number of neighboring cells. This Yap-mediated mechanical input can be described as a threshold regulator that controls the switch from a quiescent state to an oscillatory and vice versa [126]. These exciting findings open a new set of questions about the mechanism through which Yap and mechanical cues are tuning the oscillations of Delta-Notch pathway. Additionally, having in mind the pre-existing knowledge about the oscillations Wnt and FGF pathways, it would be very interesting to see whether mechanics affect them as well or the link is specific to Delta-Notch. In an aggregate comprising of many cells, the individual oscillations get coordinated but the mechanism is unknown. So, how do cells change their pace to stay tuned with their neighbors?

One way to approach this question is to observe the temporal patterns in the PSM. As mentioned in the previous section, the gene activity oscillations are displayed as traveling waves that originate from the posterior and sweep across the A-P axis. The generation of such a pattern suggests the existence of a coupling mechanism through which cells achieve local synchronization of their oscillations. Cells that are in close proximity with each other are on the same phase on their oscillation cycle. Delta-Notch signaling, being a cell-to-cell signaling pathway, was a strong candidate to regulate this property. Indeed, when Delta-Notch is disrupted genetically or chemically, cells lose their local synchrony and waves are replaced by uncoupled oscillations resulting in defective somite boundaries [127–129]. These findings reveal the role of Delta-Notch signaling in being the coupling avenue in the PSM.

Delta-Notch signaling is based on the interaction between a ligand presenting cell (sender) and a receptor presenting cell (receiver) [130]. Thus, the information flows from a sender to a receiver cell. Cis-inhibition between Delta and Notch in a cell guarantees the prevalence of one signaling state—sender or receiver—over the other, even when a cell expresses both the Notch receptor and the Delta ligand [131]. In the PSM, Delta-like1 (Dll1) is the ligand and it binds with high affinity to the Notch1 receptor. The expression of *Dll1* as well as *Notch1* is oscillatory in the PSM [132]. The ligand-receptor interaction causes the release of the Notch1 intracellular domain (NICD), which is translocated to the nucleus of the receiving cell and activates target genes [133]. A general mathematical framework that explains a broad spectrum of synchronization phenomena is the Kuramoto model [134]. The model has also been applied in the presomitic mesoderm and assumes that there is a constant bidirectional exchange of information between oscillators [135]. During this exchange, all oscillators utilize the input to adjust their pace until they finally reach a common phase. The oscillators that are ahead will slow down and oscillators that are lagging will speed up (Fig. 4a). Eventually the oscillators will converge to a common phase, a synchronization phase, that would be a compromise between them and from that point on they will not deviate. In the presomitic mesoderm Delta-Notch signaling is responsible for synchronization and it is important to compare the assumptions of Kuramoto with the mode of action of the pathway. During Delta-Notch interaction, the exchange of information is unidirectional, from the Delta presenting cell towards the Notch. Additionally, the PSM cells cannot exchange inputs constantly for two reasons. First, the signaling state of each cell depends on which phase on the cycle it is. This means that if the cell is Delta or Notch presenting, it will affect its signaling role. Lastly, the cells are motile and they continuously change their immediate neighbors. Those PSM specific properties suggest that there is a necessity for a model of synchronization that accounts for them.

Recent efforts from our lab's research have been focusing on constructing a model that considers the mode of action of the Delta-Notch signaling and incorporates prior and recent experimental findings. This proposed coupling mechanism is characterized by unidirectional and phase-gated communication between cells [136]. The unidirectionality stems from the Delta-Notch mode of action. In that, the information flows from the Delta to the Notch presenting oscillating cell. This exchange is phase-gated because PSM cells alternate signaling fates as they go through their oscillatory activity. There is a phase on the cycle where a cell can "talk" and a phase where it can "listen." This mechanism of information flow, termed "walkie/talkie," is compatible with two kinds of reaction responses. Upon interaction, the signal-receiving cell will either break their pace or accelerate.

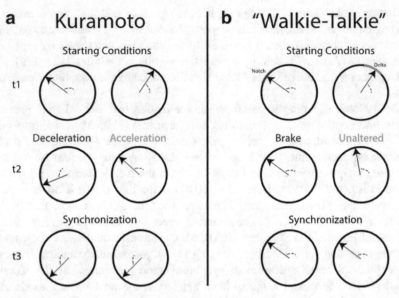

Fig. 4 Comparison of two synchronization mechanisms. (**a**) The widely employed Kuramoto mechanism assumes that two clocks that are out of phase communicate bidirectionally and constantly. The clock that is ahead will decelerate while the one lagging behind will accelerate. Eventually the two clocks will approach the same phase and will continue to cycle in synchrony. (**b**) Alternatively, the information can be unidirectional and take place only when the clocks are at specific phases. This "walkie/talkie" mechanism based on Notch signaling properties enables a cell that expresses the Notch receptor to respond to a clock that expresses the Delta ligand. The responding cell clock can then brake, enabling the other clock to reach its phase without any alterations to its pace. From that point on the clocks will continue cycling in synchrony

Monitoring two PSM cell populations that interact with each other, revealed that cells respond by transiently breaking their oscillations (Fig. 4b). Moreover, the synchronization under "walkie/talkie" is fast and achieved within 3 cycles. This is in accordance with experimental data demonstrating that PSM cells achieve synchrony within the same time frame as discussed above. Lastly, the "walkie/talkie" coupling predicts that the phase of synchronization—that is the phase on which the oscillators achieve synchrony—falls around a specific part of the cycle. In contrast, under the rules of Kuramoto any phase on the cycle could be the synchronization phase. The respective quantification for the PSM cells revealed that the synchronization phase is placed on a narrow window on the cycle. Together, those findings indicate that despite the broad applicability and explanatory elegance of the Kuramoto model, the peculiarities of the PSM cells do not fit within its capability. Instead, these cells utilize a "walkie/talkie" coupling mechanism to achieve fast and robust synchronization. Since its formulation and initial applications, the Kuramoto model for PSM synchronization has been further developing over time. In those endeavors, the focus has been given to incorporate delays [137]. The delays can be placed in different aspects of the communication pathway and represent biochemical processes. The importance of delays in the temporal organization of oscillators is intuitive and recently demonstrated. Lfng—downstream target and negative regulator of the Delta-Notch pathway—was shown to be regulating a coupling delay during synchronization in the PSM [138]. It is therefore evident that the Kuramoto model with delay needs to be tested in the context of the PSM and essentially compared with the "walkie/talkie" coupling mechanism. The challenge with this implementation would be to incorporate mathematical parameters that are biologically represented in the cell and account for feedback loops and biochemical reactions that are known to take place within the cycling activity. Moreover, the Kuramoto model needs to be adjusted beyond the addition of delays in order to accommodate the specific unidirectional and phase-gated information exchange forced by the Notch signaling dynamics.

4 Conclusions

The progress made in the understanding of how *Hydra* cells self-organize, and how presomitic mesoderm cells interact to establish spatial and temporal patterns, enable us to compare the two systems that are evolutionary positioned at the two ends of multicellular organization spectrum. *Hydra* belongs to Cnidaria, one of the oldest phyla of multicellular animals and the mouse PSM is a structure characteristic of the distant subphylum Vertebrata. Despite the evolutionary distance, both systems display self-organization properties with striking parallels. The interactions between the cells lead to the emergence of organizer centers that are characterized by similar gene expression profiles. The emergent organization axes, oral/aboral in *Hydra* and anterior/posterior in the PSM, are considered homologous [139, 140]. In fact, they describe the same main asymmetry emerging in the groups of cells. At the molecular level one pole of this axis is marked in both cases by elevated Wnt signaling. This pole is the oral end in *Hydra* and the posterior end in vertebrate PSM. Downstream of the Wnt signaling, a common set of genes, that includes the transcription factors brachyury and snail orchestrate the cell specification into organizers. Finally, beyond Wnt signaling, other cascades that play role in setting up the pattern are FGF signaling and Notch signaling. The transcription factor Hes, being central for the PSM oscillatory dynamics, is also involved in the patterning of *Hydra* [88]. The comparison between the two systems leaves also many open questions. How similar is the mechanism of spatial patterning in the PSM self-organization assays to the one in *Hydra*? It is difficult to reconcile this with a reaction-diffusion system, but maybe a role for mechanics is a common ground. The finding that mechanical stimuli are important for the ability of PSM cells to operate is important, while the differential motility is likely involved in the sorting process. On the other hand, it is unclear whether oscillatory dynamics are underlying the patterning mechanism in *Hydra*. The gene network that connects Wnt3 and Sp5 expression contains negative feedback loops, and it is potentially able to generate oscillatory gene expression. Beyond that, recent examination of the spatiotemporal dynamics in human gastruloids has shown that Wnt signaling is propagating as a wave and this aspect is critical for the downstream patterning events [141]. It has not been possible to visualize the temporal dynamics of Wnt signaling in *Hydra* and it is likely that exciting results will emerge from such an endeavor. Future research will indicate which aspects of PSM and *Hydra* patterning will further converge while others will diverge. Independent of the specific properties of the multicellular systems that have been examined here in detail, it is clear that the careful investigation of many of them can allow us to understand what rules of communication and interaction between cells are widely employed by living systems. Beyond the fact that Wnt signaling is operating in multiple cases of symmetry breaking, it will be important to understand what is special about Wnt signaling. What are the types of information that it conveys and how the information changes the behavior of the receiving cells? It is possible that there are general rules that are materialized in biological systems through the used cascades. The search of such rules for the self-organization of multicellular systems is timely as extensive and high-throughput biological experimentation has amassed quantitative data, offering tools for deeper investigations and demanding a coherent theoretical framework in which they fit.

References

1. Camazine S (2003) Self-organization in biological systems. Princeton University Press
2. Kant I (1797) Kritik der Urteilskraft
3. Verde F, Berrez JM, Antony C, Karsenti E (1991) Taxol-induced microtubule asters in mitotic extracts of Xenopus eggs: requirement for phosphorylated factors and cytoplasmic dynein. J Cell Biol 112:1177–1187
4. Nakajima M, Imai K, Ito H, Nishiwaki T, Murayama Y, Iwasaki H et al (2005) Reconstitution of circadian oscillation of cyanobacterial KaiC phosphorylation in vitro. Science 308: 414–415
5. Loose M, Fischer-Friedrich E, Herold C, Kruse K, Schwille P (2011) Min protein patterns emerge from rapid rebinding and membrane interaction of MinE. Nat Struct Mol Biol 18:577–583
6. Zieske K, Schwille P (2014) Reconstitution of self-organizing protein gradients as spatial cues in cell-free systems. eLife 3
7. Ross J, Mü ller SC, Vidal C (1988) Chemical waves. Science 240:460–465
8. Bonabeau E, Theraulaz G, Deneubourg JL, Aron S, Camazine S (1997) Self-organization in social insects. Trends Ecol Evol 12:188–193
9. Heyde A, Guo L, Jost C, Theraulaz G, Mahadevan L (2021) Self-organized biotectonics of termite nests. PNAS 118
10. Detrain C, Deneubourg J-L (2006) Self-organized structures in a superorganism: do ants "behave" like molecules? Phys of Life Reviews 3:162–187
11. Visscher PK (2003) Animal behaviour: How self-organization evolves. Nature 421:799–800
12. Sumpter DJT (2006) The principles of collective animal behaviour. Philosophical Transactions of the Royal Society B: Biological Sciences 361:5–22
13. Rubenstein M, Cornejo A, Nagpal R (2014) Robotics. Programmable self-assembly in a thousand-robot swarm. Science 345:795–799
14. Werner S, Vu HT-K, Rink JC (2017) Self-organization in development, regeneration and organoids. Current Opinion in Cell Biology 44:102–109

15. Sato T, Vries RG, Snippert HJ, van de Wetering M, Barker N, Stange DE, et al. (2009) Single Lgr5 stem cells build crypt-villus structures in vitro without a mesenchymal niche. Nature 459:262–265
16. Serra D, Mayr U, Boni A, Lukonin I, Rempfler M, Challet Meylan L, et al. (2019) Self-organization and symmetry breaking in intestinal organoid development. Nature 569:66–72
17. Keisuke I, Tanaka E M (2018) Spontaneous symmetry breaking and pattern formation of organoids. Curr Opin System Biol 11:123–128
18. Simunovic M, Brivanlou AH (2017) Embryoids, organoids and gastruloids: new approaches to understanding embryogenesis. Development 144:976–985
19. Turner DA, Girgin M, Alonso-Crisostomo L, Trivedi V, Baillie-Johnson P, Glodowski CR, et al. (2017) Anteroposterior polarity and elongation in the absence of extra-embryonic tissues and of spatially localised signalling in gastruloids: mammalian embryonic organoids. Development 144:3894–3906
20. Svensson V, Pachter L (2018) RNA velocity: molecular kinetics from single-cell RNA-Seq. Molecular Cell 72:7–9
21. Jones DL, Brewster RC, Phillips R (2014) Promoter architecture dictates cell-to-cell variability in gene expression. Science 346:1533–1536
22. Babonis LS, Martindale MQ (2017) Phylogenetic evidence for the modular evolution of metazoan signalling pathways. Philos Trans R Soc Lond, B, Biol Sci 372
23. Goodenough DA, Goliger JA, Paul DL (1996) Connexins, connexons, and intercellular communication. Annu Rev Biochem 65:475–502
24. Heisenberg C-P, Bellaïche Y (2013) Forces in tissue morphogenesis and patterning. Cell 153:948–962
25. Panciera T, Azzolin L, Cordenonsi M, Piccolo S (2017) Mechanobiology of YAP and TAZ in physiology and disease. Nat Rev Mol Cell Biol 18:758–770
26. Murthy SE, Dubin AE, Patapoutian A (2017) Piezos thrive under pressure: mechanically activated ion channels in health and disease. Nat Rev Mol Cell Biol 18:771–783
27. Fernandez-Sanchez M-E, Barbier S, Whitehead J, Bé alle G, Michel A, Latorre-Ossa H, et al (2015) Mechanical induction of the tumorigenic β-catenin pathway by tumour growth pressure. Nature 523:92–95
28. Fagotto F (2013) Looking beyond the Wnt pathway for the deep nature of β-catenin. EMBO Rep 14:422–433
29. Westermark B (1977) Local starvation for epidermal growth factor cannot explain density-dependent inhibition of normal human glial cells. PNAS 74:1619–1621
30. von Neumann J, Burks AW (1966) Theory of self-reproducing automata
31. Wolfram S (1984) Cellular automata as models of complexity. Nature 311:419–424
32. Gardner M (1970) The fantastic combinations of John Conway's new solitaire game 'life'. Scientific American 223:120–123
33. Martindale MQ, Finnerty JR, Henry JQ (2002) The Radiata and the evolutionary origins of the bilaterian body plan. Mol Phylogenet Evol 24:358–365
34. Technau U, Steele RE (2011) Evolutionary crossroads in developmental biology: Cnidaria. Development 138:1447–1458
35. David CN, Murphy S (1977) Characterization of interstitial stem cells in hydra by cloning. Developmental Biology 58:372–383
36. Sugiyama T, Fujisawa T (1977) Genetic analysis of developmental mechanisms in hydra I. Sexual reproduction of Hydra magnipapillata and isolation of mutants. Development, Growth & Differentiation 19:187–200
37. Bosch TCG, Anton-Erxleben F, Hemmrich G, Khalturin K (2010) The Hydra polyp: nothing but an active stem cell community. Development, Growth & Differentiation 52:15–25
38. Trembley A (1744) Mémoires pour servir à l'histoire d'un genre de polypes d'eau douce, à bras en forme de cornes
39. Lenhoff HM, Lenhoff SG (1988) Trembley's Polyps. Scientific American 258:108–113
40. Browne EN (1909) The production of new hydranths in Hydra by the insertion of small grafts Journal of Experimental Zoology Part a: Ecological Genetics and Physiology 7:1–23

41. Yao T (1945) Studies on the organizer problem in Pelmatohydra oligactis: I. The induction potency of the implants and the nature of the induced Hydranth. Journal of Experimental Biology 21:147–150
42. MacWilliams HK (1983) Hydra transplantation phenomena and the mechanism of Hydra head regeneration. II. Properties of the head activation. Developmental Biology 96:239–257
43. MacWilliams HK (1983) Hydra transplantation phenomena and the mechanism of hydra head regeneration. I. Properties of the head inhibition. Developmental Biology 96:217–238
44. Vogg MC, Galliot B, Tsiairis CD (2019) Model systems for regeneration: Hydra. Development 146:dev177212
45. Webster G, Wolpert L, Studies on pattern regulation in hydra. I. Regional differences in time required for hypostome determination. J Embryol Exp Morphol 16:91–104
46. Noda K (1972) Reconstitution of dissociated cells of hydra. Zool. Mag. 80:99–101
47. Gierer A, Berking S, Bode H, David CN, Flick K, Hansmann G, et al (1972) Regeneration of hydra from reaggregated cells. Nat New Biol 239:98–101
48. Technau U, Cramer von Laue C, Rentzsch F, Luft S, Hobmayer B, Bode HR, et al (2000) Parameters of self-organization in Hydra aggregates. PNAS 97:1–5
49. Gierer A, Meinhardt H (1972) A theory of biological pattern formation. Kybernetik 12:30–39
50. Schaller H, Gierer A (1973) Distribution of the head-activating substance in hydra and its localization in membranous particles in nerve cells. J Embryol Exp Morphol 29:39–52
51. Schaller HC (1973) Isolation and characterization of a low-molecular-weight substance activating head and bud formation in hydra. J Embryol Exp Morphol 29:27–38
52. Hobmayer B, Rentzsch F, Kuhn K, Happel CM, von Laue CC, Snyder P, et al (2000) WNT signalling molecules act in axis formation in the diploblastic metazoan Hydra. Nature 407:186–189
53. Loh KM, van Amerongen R, Nusse R (2016) Generating cellular diversity and spatial form: Wnt signaling and the evolution of multicellular animals. Developmental Cell 38 (2016) 643–655
54. Broun M, Gee L, Reinhardt B, Bode HR (2005) Formation of the head organizer in hydra involves the canonical Wnt pathway. Development 132:2907–2916
55. Holstein TW (2012) The evolution of the Wnt pathway. Cold Spring Harbor Perspectives in Biology 4:a007922–a007922
56. Lengfeld T, Watanabe H, Simakov O, Lindgens D, Gee L, Law L, et al (2009) Multiple Wnts are involved in Hydra organizer formation and regeneration. Developmental Biology 330:186–199
57. Gee L, Hartig J, Law L, Wittlieb J, Khalturin K, Bosch TCG, et al (2010) Beta-catenin plays a central role in setting up the head organizer in hydra. Developmental Biology 340:116–124
58. Guder C, Pinho S, Nacak TG, Schmidt HA, Hobmayer B, Niehrs C, et al (2006) An ancient Wnt-Dickkopf antagonism in Hydra. Development 133:901–911
59. Augustin R, Franke A, Khalturin K, Kiko R, Siebert S, Hemmrich G, Bosch T. C. G. (2006) Dickkopf related genes are components of the positional value gradient in Hydra. Dev Biol 296:62–70
60. Vogg MC, Beccari L, Iglesias Olle L, Rampon C, Vriz S, Perruchoud C, et al (2019) An evolutionarily-conserved Wnt3/β-catenin/Sp5 feedback loop restricts head organizer activity in Hydra. Nature Communications 10:312–15
61. Hiscock TW, Megason SG (2015) Mathematically guided approaches to distinguish models of periodic patterning. Development 142:409–419
62. Meinhardt H (2012) Modeling pattern formation in hydra: a route to understanding essential steps in development. Int J Dev Biol 56:447–462
63. Meinhardt H (2008) Models of biological pattern formation: from elementary steps to the organization of embryonic axes. Curr Top Dev Biol 81:1–63
64. Mercker M, Lengfeld T, Höger S, Tursch A, Lommel M, Holstein TW, et al. (2021) β-Catenin and canonical Wnts control two separate pattern formation systems in Hydra: Insights from mathematical modelling. bioRxiv 2021.02.05.429954

65. Mercker M, Hartmann D, Marciniak-Czochra A (2013) A mechanochemical model for embryonic pattern formation: coupling tissue mechanics and morphogen expression. PLoS ONE 8:e82617–6

66. Mercker M, Kö the A, Marciniak-Czochra A (2015) Mechanochemical symmetry breaking in Hydra aggregates. Biophysical Journal 108:2396–2407

67. Livshits A, Shani-Zerbib L, Maroudas-Sacks Y, Braun E, Keren K (2017) Structural inheritance of the actin cytoskeletal organization determines the body axis in regenerating Hydra. Cell Reports 18:1410–1421

68. Maroudas-Sacks Y, Garion L, Shani-Zerbib L, Livshits A, Braun E, Keren K (2021) Topological defects in the nematic order of actin fibres as organization centres of Hydra morphogenesis. Nat. Phys. 17:251–259

69. Seybold A, Salvenmoser W, Hobmayer B (2016) Sequential development of apical-basal and planar polarities in aggregating epitheliomuscular cells of Hydra. Developmental Biology 412:1–12

70. Sander H, Pasula A, Sander M, Giri V, Terriac E, Lautenschlaeger F, et al. (2020) Symmetry breaking and *de novo* axis formation in hydra spheroids: the microtubule cytoskeleton as a pivotal element. bioRxiv 2020.01.14.906115

71. Livshits A, Garion L, Maroudas-Sacks Y, Shani-Zerbib L, Keren K, Braun E (2021) Plasticity of body axis polarity in Hydra regeneration under constraints. bioRxiv 2021.02.04.429818

72. Wang R, Steele RE, Collins E-MS (2020) Wnt signaling determines body axis polarity in regenerating Hydra tissue fragments. Developmental Biology 467:88–94

73. Veschgini M, Petersen HO, Kaufmann S, Abuillan W, Suzuki R, Burghammer M, et al (2020) Wnt/β-catenin signaling controls spatio-temporal elasticity patterns in extracellular matrix during Hydra morphogenesis. bioRxiv 214718

74. Hannezo E, Heisenberg C-P (2019) Mechanochemical feedback loops in development and disease. Cell 178:12–25

75. Kü cken M, Soriano J, Pullarkat PA, Ott A, Nicola EM (2008) An osmoregulatory basis for shape oscillations in regenerating hydra. Biophysical Journal 95:978–985

76. Benos DJ, Kirk RG, Barba WP, Goldner MM (1977) Hyposmotic fluid formation in Hydra. Tissue and Cell 9:11–22

77. Soriano J, Rudiger S, Pullarkat P, Ott A (2009) Mechanogenetic coupling of Hydra symmetry breaking and driven Turing instability model. Biophysical Journal 96:1649–1660

78. Gamba A, Nicodemi M, Soriano J, Ott A (2012) Critical behavior and axis defining symmetry breaking in hydra embryonic development. Phys Rev Lett 108:158103

79. Ferenc J, Papasaikas P, Ferralli J, Nakamura Y, Smallwood S, Tsiairis CD (2020) Wnt3 expression as a readout of tissue stretching during Hydra regeneration. bioRxiv 2020.12.22.423911

80. Nakamura Y, Tsiairis CD, Özbek S, Holstein TW (2011) Autoregulatory and repressive inputs localize Hydra Wnt3 to the head organizer. PNAS 108:9137–9142

81. Newman SA (1974) The interaction of the organizing regions in hydra and its possible relation to the role of the cut end in regeneration. J Embryol Exp Morphol 31:541–555

82. Wenger Y, Buzgariu W, Perruchoud C, Loichot G, Galliot B (2019) Generic and context-dependent gene modulations during Hydra whole body regeneration. bioRxiv 587147

83. Sander K, Faessler PE (2001) Introducing the Spemann-Mangold organizer: experiments and insights that generated a key concept in developmental biology. Int J Dev Biol 45:1–11

84. Lenhoff HM (1991) Ethel Browne, Hans Spemann, and the discovery of the organizer phenomenon. The Biological Bulletin 181:72–80

85. Pires-daSilva A, Sommer RJ (2003) The evolution of signalling pathways in animal development. Nat Rev Genet 4:39–49

86. Li J, Wang Z, Chu Q, Jiang K, Li J, Tang N (2018) The strength of mechanical forces determines the differentiation of alveolar epithelial cells. Developmental Cell 44:297–312.e5.

87. Pukhlyakova E, Aman AJ, Elsayad K, Technau U (2018) β-Catenin-dependent mechanotransduction dates back to the common ancestor of Cnidaria and Bilateria. PNAS 115:6231–6236

88. Münder S, Käsbauer T, Prexl A, Aufschnaiter R, Zhang X, Towb P, Böttger A (2010) Notch signaling defines critical boundary during budding in Hydra. Dev Biol 344:331–345
89. Lange E, Bertrand S, Holz O, Rebscher N, Hassel M (2014) Dynamic expression of a Hydra FGF at boundaries and termini. Dev Genes Evol 224:235–244
90. Reinhardt B, Broun M, Blitz I L, Bode H R (2004) HyBMP5-8b, a BMP5-8 orthologue, acts during axial patterning and tentacle formation in hydra. Dev Biol 267:43–59
91. Münder S, Tischer S, Grundhuber M, Büchels N, Bruckmeier N, Eckert S, Seefeldt C A, Prexl A, Käsbauer T, Böttger A (383) Notch-signaling is required for head regeneration and tentacle patterning in Hydra. Dev Biol 383:146–157
92. Holz O, Apel D, Steinmetz P, Lange E, Hopfenmüller S, Ohler K, Sudhop S, Hassel M (2017) Bud detachment in hydra requires activation of fibroblast growth factor receptor and a rho-rock-myosin II signaling pathway to ensure formation of a basal constriction. Dev Dyn 246:501–516
93. Shahbazi MN (2020) Mechanisms of human embryo development: from cell fate to tissue shape and back. Development 147:dev190629
94. Teague BP, Guye P, Weiss R (2016) Synthetic morphogenesis. Cold Spring Harbor Perspectives in Biology 8:a023929
95. Chan CJ, Costanzo M, Ruiz-Herrero T, Mö nke G, Petrie RJ, Bergert M, et al. (2019) Hydraulic control of mammalian embryo size and cell fate. Nature 571:112–116
96. Ryan AQ, Chan CJ, Graner F, Hiiragi T (2019) Lumen expansion facilitates epiblast-primitive endoderm fate specification during mouse blastocyst formation. Developmental Cell 51:684–697.e4.
97. Ruiz-Herrero T, Alessandri K, Gurchenkov BV, Nassoy P, Mahadevan L (2017) Organ size control via hydraulically gated oscillations. Development 144:4422–4427
98. Pourquie O (2001) Vertebrate somitogenesis. Annu Rev Cell Dev Biol 17:311–350
99. Gossler A, Hrabě de Angelis M (1998) Somitogenesis. Curr Top Dev Biol 38:225–287
100. Barresi MJF, Gilbert SF (2019) Developmental biology. Sinauer Associates, Incorporated
101. Aulehla A, Pourquie O (2008) Oscillating signaling pathways during embryonic development. Current Opinion in Cell Biology 20:632–637
102. Oates AC, Morelli LG, Ares S (2012) Patterning embryos with oscillations: structure, function and dynamics of the vertebrate segmentation clock. Development 139:625–639
103. Greco TL, Takada S, Newhouse MM, McMahon JA, McMahon AP, Camper SA (1996) Analysis of the vestigial tail mutation demonstrates that Wnt-3a gene dosage regulates mouse axial development. Genes & Development 10:313–324
104. Aulehla A, Wehrle C, Brand-Saberi B, Kemler R, Gossler A, Kanzler B, et al. (2003) Wnt3a plays a major role in the segmentation clock controlling somitogenesis. Developmental Cell 4:395–406
105. Niehrs C (2012) The complex world of WNT receptor signalling. Nat Rev Mol Cell Biol 13:767–779.
106. Aulehla A, Wiegraebe W, Baubet V, Wahl MB, Deng C, Taketo M, et al. (2008) A beta-catenin gradient links the clock and wavefront systems in mouse embryo segmentation. Nat Cell Biol 10:186–193.
107. Dubrulle J, McGrew MJ, Pourquie O (2001) FGF signaling controls somite boundary position and regulates segmentation clock control of spatiotemporal Hox gene activation. Cell 106:219–232
108. Dubrulle J, Pourquie O (2004) fgf8 mRNA decay establishes a gradient that couples axial elongation to patterning in the vertebrate embryo. Nature 427:1–4
109. Takada S, Stark KL, Shea MJ, Vassileva G, McMahon JS, McMahon AP (1994) Wnt-3a regulates somite and tailbud formation in the mouse embryo. Genes & Development 8:174–189
110. Bé nazé raf B, Francois P, Baker RE, Denans N, Little CD, Pourquie O (2010) A random cell motility gradient downstream of FGF controls elongation of an amniote embryo. Nature 466:248–252

111. Duband JL (1980) Adhesion molecules during somitogenesis in the avian embryo., J Cell Biol 104:1–14
112. Delfini MC, Dubrulle J, Malapert P, Chal J, Pourquie O (2005) Control of the segmentation process by graded MAPK ERK activation in the chick embryo. PNAS 102:1–6
113. Palmeirim I, Henrique D, Ish-Horowicz D, Pourquie O (1997) Avian hairy gene expression identifies a molecular clock linked to vertebrate segmentation and somitogenesis. Cell 91:639–648
114. Cooke J, Zeeman EC (1976) A clock and wavefront model for control of the number of repeated structures during animal morphogenesis. Journal of Theoretical Biology 58:455–476
115. Dequeant M-L, Glynn E, Gaudenz K, Wahl M, Chen J, Mushegian A, et al (2006) A complex oscillating network of signaling genes underlies the mouse segmentation clock. Science 314:1595–1598
116. Bessho Y, Hirata H, Masamizu Y, Kageyama R (2003) Periodic repression by the bHLH factor Hes7 is an essential mechanism for the somite segmentation clock. Genes & Development 17:1451–1456
117. Lewis J (2003) Autoinhibition with transcriptional delay: a simple mechanism for the zebrafish somitogenesis oscillator. Current Biology 13:1398–1408
118. Niwa Y, Masamizu Y, Liu T, Nakayama R, Deng C-X, Kageyama R (2007) The initiation and propagation of Hes7 oscillation are cooperatively regulated by Fgf and notch signaling in the somite segmentation clock. Developmental Cell 13:298–304
119. González A, Manosalva I, Liu T, Kageyama R (2013) Control of Hes7 expression by Tbx6, the Wnt Pathway and the Chemical Gsk3 Inhibitor LiCl in the Mouse Segmentation Clock. PLoS ONE 8:e53323–8.
120. Anderson MJ, Magidson V, Kageyama R, Lewandoski M (2020) Fgf4 maintains Hes7 levels critical for normal somite segmentation clock function. eLife. 9
121. Tsiairis CD, Aulehla A (2016) Self-organization of embryonic genetic oscillators into spatiotemporal wave patterns. Cell 164:656–667
122. Gibb S, Zagorska A, Melton K, Tenin G, Vacca I, Trainor P, et al. (2009) Interfering with Wnt signalling alters the periodicity of the segmentation clock. Developmental Biology 330:21–31
123. Bajard L, Morelli LG, Ares S, Pecreaux J, Julicher F, Oates AC (2014) Wnt-regulated dynamics of positional information in zebrafish somitogenesis. Development 141:1381–1391
124. Beatrici CP, Brunnet LG (2011) Cell sorting based on motility differences. Phys Rev E Stat Nonlin Soft Matter Phys 84:031927
125. Alexandre C, Baena-Lopez A, Vincent J-P (2013) Patterning and growth control by membrane-tethered Wingless. Nature 505:180–185
126. Hubaud A, Regev I, Mahadevan L, Pourquie O (2017) Excitable dynamics and yap-dependent mechanical cues drive the segmentation clock. Cell 171:668–682.e11.
127. Jiang YJ, Aerne BL, Smithers L, Haddon C, Ish-Horowicz D, Lewis J (2000) Notch signalling and the synchronization of the somite segmentation clock. Nature 408:475–479
128. Okubo Y, Sugawara T, Abe-Koduka N, Kanno J, Kimura A, Saga Y (2012) Lfng regulates the synchronized oscillation of the mouse segmentation clock via trans- repression of Notch signalling. Nature Communications 3:1141–9
129. Riedel-Kruse IH, Muller C, Oates AC (2007) Synchrony dynamics during initiation, failure, and rescue of the segmentation clock. Science 317:1911–1915
130. Artavanis-Tsakonas S, Rand MD, Lake RJ (1999) Notch signaling: cell fate control and signal integration in development. Science 284:770–776
131. Sprinzak D, Lakhanpal A, LeBon L, Santat LA, Fontes ME, Anderson GA, et al (2010) Cis-interactions between Notch and Delta generate mutually exclusive signalling states. Nature 465:86–90
132. Shimojo H, Isomura A, Ohtsuka T, Kori H, Miyachi H, Kageyama R (2016) Oscillatory control of Delta-like1 in cell interactions regulates dynamic gene expression and tissue morphogenesis. Genes & Development 30:102–116
133. Bray SJ (2006) Notch signalling: a simple pathway becomes complex. Nat Rev Mol Cell Biol 7:678–689

134. Kuramoto Y (2012) Chemical oscillations, waves, and turbulence. Springer Science & Business Media, Berlin, Heidelberg.
135. Strogatz SH (2000) From Kuramoto to Crawford: exploring the onset of synchronization in populations of coupled oscillators. Physica D: Nonlinear Phenomena 143:1–20
136. Roth G, Misailidis G, Tsiairis CD (2020) Cellular synchronisation through unidirectional and phase-gated signalling. bioRxiv 3:1–24
137. Wu H, Kang L, Liu Z, Dhamala M (2018) Exact explosive synchronization transitions in Kuramoto oscillators with time-delayed coupling. Sci Rep 8:15521–8
138. Yoshioka-Kobayashi K, Matsumiya M, Niino Y, Isomura A, Kori H, Miyawaki A, et al (2020) Coupling delay controls synchronized oscillation in the segmentation clock. Nature 580:119–123
139. Petersen C P, Reddien P W (2009) Wnt signaling and the polarity of the primary body axis. Cell 139:1056–1068
140. Rentzsch F, Guder C, Vocke D, Hobmayer B, Holstein T W (2007) An ancient chordin-like gene in organizer formation of Hydra. PNAS 104:3249–3254
141. Chhabra S, Liu L, Goh R, Kong X, Warmflash A (2019) Dissecting the dynamics of signaling events in the BMP, Wnt, and Nodal cascade during self-organized fate patterning in human gastruloids. PLOS Biol 17:e3000498

Foundations of Viscoelasticity and Application to Soft Tissue Mechanics

Michele Righi and Valentina Balbi

Abstract Soft tissues are complex media; they display a wide range of mechanical properties such as anisotropy and non-linear stress-strain behaviour. They undergo large deformations and they exhibit a time-dependent mechanical behaviour, i.e. they are viscoelastic. In this chapter we review the foundations of the linear viscoelastic theory and the theory of Quasi-Linear Viscoelasticity (QLV) in view of developing new methods to estimate the viscoelastic properties of soft tissues through model fitting. To this aim, we consider the simple torsion of a viscoelastic Mooney-Rivlin material in two different testing scenarios: step-strain and ramp tests. These tests are commonly performed to characterise the time-dependent properties of soft tissues and allow to investigate their stress relaxation behaviour. Moreover, commercial torsional rheometers measure both the torque and the normal force, giving access to two sets of data. We show that for a step test, the linear and the QLV models predict the same relaxation curves for the torque. However, when the strain history is in the form of a ramp function, the non-linear terms appearing in the QLV model affect the relaxation curve of the torque depending on the final strain level and on the rising time of the ramp. Furthermore, our results show that the relaxation curve of the normal force predicted by the QLV theory depends on the level of strain both for a step and a ramp tests. To quantify the effect of the non-linear terms, we evaluate the maximum and the equilibrium (as $t \to \infty$) values of the relaxation curves. Our results provide useful guidelines to accurately fit QLV models in view of estimating the viscoelastic properties of soft tissues.

M. Righi
Department of Mathematics and Statistics, University of Limerick, Limerick, Ireland
e-mail: michele.righi@ul.ie

V. Balbi (✉)
School of Mathematics, Statistics and Applied Mathematics, National University of Ireland Galway, Galway, Ireland
e-mail: vbalbi@nuigalway.ie

© The Author(s), under exclusive license to Springer Nature Switzerland AG 2021   71
J. Málek, E. Süli (eds.), *Modeling Biomaterials*, Nečas Center Series,
https://doi.org/10.1007/978-3-030-88084-2_3

1   Introduction

Soft tissues, such as the brain, the skin, tendons and ligaments, are viscoelastic materials, their mechanical behaviour is therefore time-dependent. Two typical experiments that show the time-dependent nature of soft tissues consist in stress relaxation and creep tests. In a stress relaxation test the tissue is suddenly stretched and then held in position for a certain time while the resulting stress is measured. Conversely, in a creep test, the load is applied to the tissue and the resulting deformation is measured. For many soft tissues the stress relaxation curve has a decaying exponential form. Stress relaxation has been observed in the brain [2, 3], in ligaments and tendons [5, 6] and in the skin [15]. At the microscale, the physical mechanisms behind stress relaxation differ from tissue to tissue. In tendons, for example, it has been observed that crimping and un-crimping of the hierarchical structures that build up the tissues, i.e. the individual collagen fibrils, are responsible for the stress relaxation of the tissue [9]. In the skin, the interaction between collagen and elastic fibres plays a crucial role in determining the time-dependent behaviour of the tissue. When the tissue is deformed, the cross-links maintain the structure and allow the elastic fibres to stretch and relax [10].

However, in practice there is no machine that can instantaneously deform a tissue. A more realistic test is indeed a ramp test, where the tissue is deformed in a finite time and then held in that position. The duration of the ramp phase is called rising time t^*. When the rising time of the ramp is nearly zero, the ramp test can be well approximated by a step-strain test. However, if t^* is not small (compared to the characteristic time constants of the material) modelling the ramp test as a step test can introduce errors in the estimation of the viscoelastic parameters.

From the modelling viewpoint, the simplest constitutive theory that can be used to describe the time-dependent behaviour of soft tissues is the linear viscoelastic theory, where the stress is related to the strain by a time-dependent function which in turn depends on the tissue's viscoelastic parameters. Linear models are based on three main assumptions:

1. the tissue remembers the past deformation history through a *fading memory*, so that contributions to recent strain increments are more important than past contributions. A typical form of the time-dependent parameters that satisfies this assumption is a decaying exponential form;
2. according to the Boltzmann superposition principle, the total stress at the current time t is given by the sum of all past stress contribution;
3. the deformation applied to the tissue is small.

In early times, linear models have been employed to predict the viscoelastic behaviour of soft tissues. However, soon scientists have realised that these models do not provide accurate predictions, mainly because in reality soft tissues undergo large deformations. To overcome this limitation, Fung proposed what is now called the Quasi-Linear Viscoelastic (QLV) theory, which is the simplest extension of the linear theory to large deformations [8]. QLV models can capture stress relaxation and creep, the strain-rate dependent response and account for large deformations. Moreover, the governing equations of the viscoelastic problem can be solved analytically for the most common modes of deformations used in experiments (e.g. tension, compression, equi-biaxial tension, simple shear and torsion). Therefore, the constitutive parameters can be directly estimated through fitting of the experimental data, by implementing a minimisation algorithm. Linear and QLV models have a common limitation: being based on the linear superposition principle, they cannot account for the coupling between different time-scales, which is a limitation, especially for tissues with hierarchical structures. Although more complex non-linear models that account for this coupling have been proposed, they are numerically costly when it comes to model fitting and material parameters estimation [16, 17]. Another class of non-linear models goes under the umbrella of internal variable or rate-type models which have recently gained popularity among the biomechanical community [12–14]. These models are based on thermodynamics foundations. According to the multiplicative decomposition, the gradient of the deformation is split into an elastic and a viscous part. The resulting stress is then split into the sum of an elastic and a viscous term. The elastic stress is generally written with respect to an elastic strain energy function. The viscous stress is written with respect to a number of internal variables, whose evolution laws are dictated by the second law of thermodynamics and motivated by the linear theory [18]. This approach has the advantage of allowing an easy implementation of the constitutive model into finite element codes. However, when it comes to model fitting, the resulting equations are in implicit forms and need to be solved numerically, even for simple deformation modes.

Finally, differential-type models formulate the time-dependent constitutive equation in terms of the derivatives of the right stretch tensor evaluated at the current time [11]. Despite being computationally easy to implement, these models do not allow for an explicit form with respect to the relaxation functions, therefore they are less straightforward to fit with experimental data. We conclude this brief introductory review by noting that viscoelasticity is not the only time-dependent property of soft tissues. Rate-type effects, such as stiffening and softening as a results of cyclic loading and unloading and ageing, are other common effects displayed by biological tissues [29, 30].

In this chapter, we focus on viscoelasticity with the aim of providing useful guidelines on model fitting and estimation of the viscoelastic parameters for linear and QLV models. We consider two main experimental scenarios, the step-and-hold test and the ramp-and-hold test for the torsion of a cylindrical tissue. These tests are common experimental protocols used to investigate the viscoelastic properties of soft tissues, in particular stress relaxation. In Sect. 2 we review the standard linear viscoelastic theory and its rheological interpretation. In Sect. 3 we review the QLV theory following the formulation proposed in [19]. In Sect. 4, we consider the simple torsion of a cylindrical sample. This deformation can be performed with commercially available rheometers which measure both the torque and the normal force required to twist a cylindrical sample, giving access to two independent sets of data. Torsion has been successfully used to characterise the elastic properties

of the brain in large deformations [20] suggesting that the tissue behaves as a
Mooney-Rivlin material. The same behaviour was previously observed in simple
shear experiments [21]. In view of applications to brain mechanics, we therefore
solve the equilibrium equations for a viscoelastic material whose elastic stress obeys
a Mooney-Rivlin law and we calculate the expressions for the torque and the normal
force. In Sect. 5 we compare the predictions of the QLV model in the scenario of a
step-strain test and of a ramp test. We conclude the chapter by discussing our results
and by summarising the main findings.

2   Linear Viscoelastic Models

Linear viscoelastic constitutive models are formulated by introducing the time
dependency in the material parameters, a sort of fading memory which remembers
the strain history of the material up to the current configuration. Accordingly,
Hooke's law $\boldsymbol{\sigma} = \mathbb{K} : \boldsymbol{\varepsilon}$ rewrites as follows:

$$\sigma(t) = \int_{-\infty}^{t} \mathbb{K}(t - \tau) : \frac{d\boldsymbol{\varepsilon}(\tau)}{d\tau} \, d\tau \,, \tag{1}$$

where $\boldsymbol{\sigma}$ is the stress tensor, $\boldsymbol{\varepsilon}$ is the infinitesimal strain tensor, and $\mathbb{K}(t)$ is called
the *tensorial relaxation function* and is a fourth-order tensor whose entries are the
time-dependent material parameters. The symbol : denotes the double contraction
between a fourth-order tensor $\mathbb{Y}$ and a second-order tensor $\mathbf{Z}$ such that $(\mathbb{Y} : \mathbf{Z})_{ab} = Y_{abcd} Z_{cd}$.

Equation (1) is based on the Boltzmann superposition principle [22–24]. Accordingly, the total stress at the current time t can be written as sum of past stress
contributions up to the time t. In a one-dimensional setting, for the generic
component σ we can then write $\sigma = \sum_i \Delta\sigma_i$, as sketched in Fig. 1. Each $\Delta\sigma_i$
is the stress response to the step increment $\Delta\varepsilon_i = \frac{d\varepsilon}{dt}\Delta t_i$ and is governed by the
relaxation function $k_{\text{step}}(t - t_i)$. Therefore, each stress increment can be written as
$\Delta\sigma_i = k_{\text{step}}(t - t_i)\Delta\varepsilon_i$. By assuming that the strain history is continuous over time,
the sum can be converted into an integral over time. The stress component σ at time
t is then given by the following convolution integral:

$$\sigma(t) = \int_{-\infty}^{t} k_{\text{step}}(t - \tau) \frac{d\varepsilon(\tau)}{d\tau} \, d\tau \,, \tag{2}$$

for a given deformation history $\varepsilon(t)$.

Equation (1) is the tensorial version of Eq. (2). The components of the tensor
$\mathbb{K}(t)$ are the different relaxation functions of the tissue, i.e. the time-dependent
mechanical parameters. We will show in Sect. 2.1 that the tensor $\mathbb{K}(t)$ can be split
into its components according to a set of fourth-order bases. Thus, the resulting

Fig. 1 Boltzmann superposition principle: the strain history ε is approximated as sum of steps $\Delta\varepsilon_i$ and the resulting total stress response σ is the sum of the stress responses $\Delta\sigma_i$ to each step increment

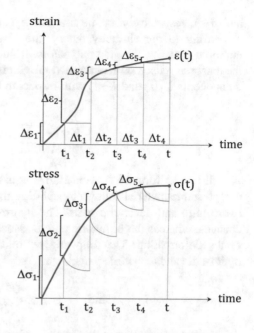

constitutive equation can be written with respect to different sets of relaxation functions according to the choice of bases.

From an experimental viewpoint, the choice of the bases might be dictated by which mechanical properties we want to estimate. For instance, if we are interested in estimating the time-dependent shear modulus we will perform a simple shear test or a torsion test, whereas if we want to estimate the time-dependent Young's modulus we will perform a tensile test. Moreover, in order to be able to estimate the components of $\mathbb{K}(t)$ we first have to specify their functional form with respect to time. In Sect. 2.2 we will review a common form used in the biomechanics community, i.e. the Prony series form. Furthermore, we will discuss two common experimental protocols that are performed to estimate the components of $\mathbb{K}(t)$, namely the step-strain and the ramp tests.

2.1 Bases Decomposition for the Tensor $\mathbb{K}(t)$

In this section, we focus on the tensorial nature of the relaxation function $\mathbb{K}(t)$ and we show that the constitutive equation (1) can be written with respect to different sets of components of $\mathbb{K}(t)$, i.e. the time-dependent mechanical properties of the tissue, according to different choices of fourth-order tensorial bases. To simplify the analysis, in this chapter we restrict out attention to homogeneous isotropic tissues. The mechanical behaviour of such tissues is fully described by two independent mechanical parameters, e.g. the bulk and the shear modulus, $\kappa(t)$

and $\mu(t)$, respectively, or the first Lamé parameter $\lambda(t)$ and the shear modulus. According to the elasticity theory, the elasticity tensor $\mathbb{C}$ for a homogeneous isotropic material depends only on two independent elastic constants. Similarly, the tensorial relaxation function $\mathbb{K}(t)$ for an isotropic material has two independent components $K_1(t)$ and $K_2(t)$ with respect to two bases $\mathbb{I}_1$ and $\mathbb{I}_2$, respectively, such that:

$$\mathbb{K}(t) = \sum_{n=1,2} K_n(t)\mathbb{I}_n. \tag{3}$$

A well-known form of the constitutive equation for a homogeneous isotropic and compressible material follows by splitting the infinitesimal strain tensor ε into its hydrostatic and deviatoric parts. The hydrostatic part is associated with volume changes, whereas the deviatoric part is associated with the volume-preserving part of the deformation. The following set of bases splits the strain tensor into its hydrostatic and deviatoric parts:

$$\mathbb{I}_{1abcd} = \frac{1}{3}\delta_{ab}\delta_{cd} \quad \text{and} \quad \mathbb{I}_{2abcd} = \frac{1}{2}(\delta_{ac}\delta_{bd} + \delta_{ad}\delta_{bc}) - \frac{1}{3}\delta_{ab}\delta_{cd}. \tag{4}$$

Accordingly, Eq. (1) takes the following form:

$$
\begin{aligned}
\sigma(t) &= \int_{-\infty}^{t} \sum_{n=1,2} K_n(t-\tau)\mathbb{I}_n : \frac{d\varepsilon(\tau)}{d\tau}\, d\tau \\
&= \int_{-\infty}^{t} K_1(t-\tau)\frac{d\mathbb{I}_1 : \varepsilon(\tau)}{d\tau}\, d\tau + \int_{-\infty}^{t} K_2(t-\tau)\frac{d\mathbb{I}_2 : \varepsilon(\tau)}{d\tau}\, d\tau \\
&= \int_{-\infty}^{t} K_1(t-\tau)\frac{d}{d\tau}\left(\frac{1}{3}\operatorname{tr}(\varepsilon(\tau))\,\mathbf{I}\right) d\tau + \int_{-\infty}^{t} K_2(t-\tau)\frac{d}{d\tau}\left(\varepsilon(\tau)-\frac{1}{3}\operatorname{tr}(\varepsilon(\tau))\,\mathbf{I}\right) d\tau \\
&= \int_{-\infty}^{t} \kappa(t-\tau)\frac{d}{d\tau}(\operatorname{tr}(\varepsilon(\tau))\,\mathbf{I})\, d\tau + 2\int_{-\infty}^{t} \mu(t-\tau)\frac{d}{d\tau}(\operatorname{dev}(\varepsilon(\tau)))\, d\tau,
\end{aligned}
\tag{5}
$$

where δ_{ab} is the Kronecker delta ($\delta_{ab} = 1$ if $a = b$ and $\delta_{ab} = 0$ if $a \neq b$), $\mathbf{I}$ is the second-order identity tensor and $\operatorname{dev}\varepsilon = \varepsilon - \frac{1}{3}\operatorname{tr}(\varepsilon)\,\mathbf{I}$ is the deviatoric part of the second-order tensor ε. The bases $\mathbb{I}_1$ and $\mathbb{I}_2$ defined in Eq. (5) act on a second-order tensor by splitting the tensor into its spherical and deviatoric parts, respectively. The associated material parameters $\kappa(t)$ and $\mu(t)$ are the time-dependent bulk and shear modulus, respectively.

For incompressible materials, i.e. materials that deform by keeping their volume constant, the bulk modulus is much greater than the shear modulus ($\kappa(t) \gg \mu(t)$ for $\forall t$). Moreover, the following assumptions are true: $\operatorname{tr}\varepsilon(t) \to 0$, $\kappa(t) \to \infty$, $\forall t$. In these limits, Eq. (5) reduces to:

$$\sigma = -p(t)\mathbf{I} + 2\int_{-\infty}^{t} \mu(t-\tau)\frac{d}{d\tau}(\operatorname{dev}\varepsilon(\tau))\, d\tau, \tag{6}$$

where we have introduced the Lagrange multiplier $p(t)$:

$$- p(t) = \lim_{\text{tr}\,\varepsilon(t) \to 0} \lim_{\kappa(t) \to \infty} \int_{-\infty}^{t} \kappa(t - \tau) \frac{\mathrm{d}}{\mathrm{d}\tau} (\text{tr}\,\varepsilon(\tau))\, \mathrm{d}\tau . \tag{7}$$

The scalar $p(t)$ can be interpreted as a hydrostatic pressure and can be calculated by solving the governing equations of motions for a continuum body, upon imposing the boundary conditions. Note that the stress component $-p(t)\mathbf{I}$ represents a workless reaction with respect to the kinematic constraint of the deformation field. No dissipation is involved in the isochoric deformation of the body. Hence, for materials that can be treated as incompressible, only the deviatoric part of the stress exhibits a viscoelastic nature.

Similarly, by choosing the following bases:

$$\mathbb{J}_{1abcd} = \delta_{ab}\delta_{cd} \quad \text{and} \quad \mathbb{J}_{2abcd} = \frac{1}{2}\left(\delta_{ac}\delta_{bd} + \delta_{ad}\delta_{bc}\right), \tag{8}$$

$\mathbb{K}(t) = \sum_{n=1,2} A_n(t)\mathbb{J}_n$ and Eq. (1) writes as follows:

$$\boldsymbol{\sigma}(t) = \int_{-\infty}^{t} \sum_{n=1,2} A_n(t - \tau)\mathbb{J}_n : \frac{\mathrm{d}\boldsymbol{\varepsilon}(\tau)}{\mathrm{d}\tau}\, \mathrm{d}\tau = \cdots =$$

$$= \int_{-\infty}^{t} A_1(t - \tau)\frac{\mathrm{d}}{\mathrm{d}\tau}(\text{tr}\,\boldsymbol{\varepsilon}(\tau)\mathbf{I})\, \mathrm{d}\tau + \int_{-\infty}^{t} A_2(t - \tau)\frac{\mathrm{d}}{\mathrm{d}\tau}(\boldsymbol{\varepsilon}(\tau))\, \mathrm{d}\tau \tag{9}$$

$$= \int_{-\infty}^{t} \lambda(t - \tau)\frac{\mathrm{d}}{\mathrm{d}\tau}(\text{tr}\,\boldsymbol{\varepsilon}(\tau)\mathbf{I})\, \mathrm{d}\tau + 2\int_{-\infty}^{t} \mu(t - \tau)\frac{\mathrm{d}}{\mathrm{d}\tau}(\boldsymbol{\varepsilon}(\tau))\, \mathrm{d}\tau .$$

Now, we have $A_1(t) = \lambda(t)$, which is the time-dependent first Lamé parameter and $A_2(t) = 2\mu(t)$. Moreover, note that $\mathbb{J}_1 = 3\mathbb{I}_1$ and $\mathbb{J}_2 = \mathbb{I}_2 - \mathbb{I}_1$ and the following link is true $\lambda(t) = \kappa(t) - \frac{2}{3}\mu(t)$, $\forall t$. Clearly, Eqs. (5) and (9) are equivalent forms of the constitutive equation (1) and predict the same stress response to a general strain input $\boldsymbol{\varepsilon}(t)$. The choice of the bases and therefore the final form of the constitutive model is usually dictated by what type of material properties we want to determine and the type of experimental devices available for testing (e.g. tensile machines, rheometers, bi-axial devices, etc.). In Sect. 3 we will use the bases decomposition of Eq. (4) to write the constitutive equation for the QLV model. In the next section, we focus on the mathematical form of the components of the tensor $\mathbb{K}(t)$.

2.2 Rheological Models for the Relaxation Function

In order to use the constitutive equation (1) for model fitting and parameter estimations, a mathematical form for the components of the tensorial function $\mathbb{K}(t)$

has to be chosen. On the one hand, the form of the relaxation functions is restricted by the following physical principles: positive strain energy and satisfaction of the second law of thermodynamics. Imposing the energy density to be non-negative during the tissue relaxation results in requiring the relaxation function to be positive $\forall t$ [7]. Furthermore, the second law of thermodynamics (dissipation inequality) requires that the relaxation function decreases monotonically with time [7]. On the other hand, any function that satisfies the physical constraints and replicates the shape of the observed stress relaxation curve can be used. From the experimental viewpoint, a classical experiment that can be done to determine the relaxation curve is to apply a displacement to the tissue, then hold the tissue in position (i.e. maintain a constant level of strain) for a certain time and measure the resulting stress curve. For most soft tissues, the measured stress relaxation curve has a decaying exponential behaviour [4–6].

The simplest form for the relaxation function that captures the exponential decaying behaviour and satisfies the physical constraints is the so-called Prony series. The Prony series has its origin in one-dimensional rheological models [7, 8]. Such models are represented by an arrangement of linear springs and linear dash-pots. The layout of such arrangements of elements provides the qualitative behaviour of the system, e.g. solid-like or fluid-like behaviour, while the values of the constants characterise the quantitative behaviour. In Fig. 2 we sketch the generalised Maxwell scheme, which is used to model the viscoelastic response of solid materials. The isolated spring k_∞ represents the residual (long-term) elasticity of the tissue.

One-dimensional rheological models can be described by a linear ordinary differential equation in the variables σ (the stress) and ε (the strain). The stress response σ of the system to a step-strain input ε provides the form of the relaxation function k_{step} in (2). A convenient way to derive the response of rheological models, especially when a large number of elements is involved, is to employ the Laplace transform. The approach involves the following steps:

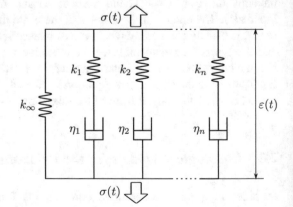

Fig. 2 Rheological models: springs and dash-pots arrangement of the generalised Maxwell model. The parameters k_i are spring constants (Pa) and η_i are dash-pot constants (Pas), or viscosities. σ is the stress response of the system to the applied strain ε

1. Write the constitutive equations for all the elements in the system: $\sigma_i = k_i \varepsilon_i$ for springs and $\sigma_i = \eta_i \dot{\varepsilon}_i$ for dash-pots. Then write the equilibrium equations $\sigma = \sum_i \sigma_i$ for elements in parallel that experience the same strain and $\varepsilon = \sum_i \varepsilon_i$ for elements in series that experience the same stress.
2. By applying the Laplace transform to each equations at point 1, convert the system of mixed differential and algebraic equations into a system of only algebraic equations.
3. Apply a variable elimination procedure to the system derived at point 2. The system has $3n + 2$ equations in $3n + 3$ unknowns, i.e. ε_i and σ_i. This reduces the system to the single equation: $\bar{\sigma}(s) = F(s)\bar{\varepsilon}(s)$, where $\bar{\sigma}$ and $\bar{\varepsilon}$ are the Laplace transforms of $\sigma(t)$ and $\varepsilon(t)$, respectively. $F(s)$ is the transfer function in the complex domain represented by the complex variable s.
4. By applying the inverse Laplace transform to the equation $\bar{\sigma}(s) = F(s)\bar{\varepsilon}(s)$, obtain the constitutive equation in the time domain.

From point 3, the transfer function $F(s)$ of the generalised Maxwell model in Fig. 2 is given by:

$$F(s) = \frac{\bar{\sigma}(s)}{\bar{\varepsilon}(s)} = k_\infty + \sum_{i=1}^{n} \frac{s\eta_i}{1 + s\tau_i}, \tag{10}$$

where the constants $\tau_i = \frac{\eta_i}{k_i}$ are called relaxation times of the model.

To calculate the response of a generalised Maxwell system to a step input, we write the strain $\varepsilon(t) = \varepsilon_0 H(t)$, where $H(t)$ is the Heaviside function: $H(t) = 1$, $\forall t \geq 0$ and $H(t) = 0$, $\forall t < 0$ and ε_0 is the amplitude of the step. Then we calculate the Laplace transform $\bar{\varepsilon}$ and substitute the result into Eq. (10) obtaining:

$$\bar{\sigma}(s) = \left(\frac{k_\infty}{s} + \sum_{i=1}^{n} \frac{\eta_i}{1 + s\tau_i} \right) \varepsilon_0. \tag{11}$$

According to point 4, by transforming back into the time domain, we obtain the stress response to a step-strain input with amplitude ε_0:

$$\sigma(t) = \left(k_\infty + \sum_{i=1}^{n} k_i e^{-\frac{t}{\tau_i}} \right) \varepsilon_0. \tag{12}$$

From Eq. (12) we can then calculate the stress response to a step-strain input, i.e. the relaxation function $k_{\text{step}}(t)$, by dividing $\sigma(t)$ by the amplitude of the step, as follows:

$$k_{\text{step}}(t) = \frac{\sigma(t)}{\varepsilon_0} = k_\infty + \sum_{i=1}^{n} k_i e^{-\frac{t}{\tau_i}}. \tag{13}$$

Fig. 3 Stress response of the generalised Maxwell model to a step-strain input

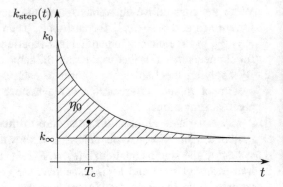

The function (13) is called Prony series and is a sum of exponential terms, each corresponding to a branch of the generalised Maxwell in Fig. 2. Note that the relaxation function $k_{\text{step}}(t)$ does not depend on the strain ε.

The constants k_∞, k_i and η_i can be determined by fitting Eq. (13) to the stress relaxation curve experimentally measured from a step-strain test. The sample is suddenly deformed up to the strain ε_0 and held in position for a certain amount of time. In the limit $t \to \infty$ Eq. (12) recovers the elastic equilibrium stress ($\sigma_\infty = k_\infty \varepsilon_0$) and the relaxation function in Eq. (13) reduces to the long-term elastic modulus k_∞:

$$k_\infty = \lim_{t \to \infty} k_{\text{step}}(t). \tag{14}$$

Experimentally, this limit is equivalent to a very slow ramp test, i.e. a quasi-static test, where the final value of strain ε_0 is attained as $t \to \infty$.

On the other hand, in the limit $t \to 0$, Eq. (13) reduces to the instantaneous elastic modulus k_0:

$$k_0 = k_{\text{step}}(0) = k_\infty + \sum_{i=1}^{n} k_i. \tag{15}$$

The value in Eq. (15) corresponds to the maximum of the relaxation function, see Fig. 3. Experimentally, this limit corresponds to the application of an instantaneous strain, which is practically impossible to perform. The constants k_0 and k_∞ are the elastic parameters of the constitutive model and describe the instantaneous and long-term elastic behaviours of the tissue, respectively.

The viscous behaviour of the tissue is associated with the parameters η_i and τ_i. To get some insights on the viscous parameters, we define the function:

$$\tilde{k}_{\text{step}}(t) = k_{\text{step}}(t) - k_\infty = \sum_{i=1}^{n} k_i e^{-\frac{t}{\tau_i}} \tag{16}$$

and we integrate over the whole time spectrum:

$$\eta_0 = \int_0^\infty \tilde{k}_{step}(t)\, dt = \sum_{i=1}^n \eta_i. \tag{17}$$

The value η_0 has an important geometrical interpretation since it represents the area between the curve $k_{step}(t)$ and the asymptotic line k_∞, see Fig. 3. The bigger the area, the more viscous the material.

Moreover, we can compute the mean relaxation time T_c of the tissue as follows:

$$T_c = \frac{\int_0^\infty t\tilde{k}_{step}(t)\, dt}{\int_0^\infty \tilde{k}_{step}(t)\, dt} = \frac{\sum_{i=1}^n \eta_i \tau_i}{\sum_{i=1}^n \eta_i}. \tag{18}$$

Geometrically, T_c represents the centroid of the shaded area below the relaxation function $k_{step}(t)$ and the asymptotic value k_∞ and can be interpreted as the average relaxation time.

Ramp Tests

Now, we recall that Eqs. (13–18) are valid for a strain input in the form of a step function (i.e. $\varepsilon(t) = H(t)\varepsilon_0$), where the strain value ε_0 is attained instantaneously. However, such experiment is not feasible in laboratory since that would require a testing machine able to reach an infinite rate of deformation. Real tests are much closer to a ramp test, where the strain ε_0 is reached after a finite rising time $t^* > 0$, as shown in Fig. 4.

In view of providing an analytical expression for the relaxation curve to be fitted with the experimental data, in this section we derive the stress response of a generalised Maxwell system to a ramp input. The strain input for a ramp test takes the following form:

$$\varepsilon(t) = \frac{\varepsilon_0}{t^*} t - \frac{\varepsilon_0}{t^*}(t - t^*) H(t - t^*), \tag{19}$$

Fig. 4 Strain history for a ramp-and-hold experiment. The constant strain value ε_0 is reached at the end of the loading phase ($t = t^*$), where the strain increases at a constant rate

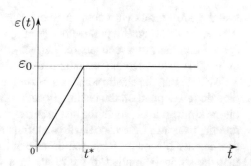

where $H(t)$ is the Heaviside function. We can calculate the stress response to the input in Eq. (19) by substituting Eqs. (13) and (19) into Eq. (2). We then obtain:

$$\sigma(t) = k_{\mathrm{ramp}}(t)\varepsilon_0, \qquad \text{for} \quad t > t^*, \tag{20}$$

where:

$$k_{\mathrm{ramp}}(t) = \left(k_\infty + \frac{1}{t^*}\sum_{i=1}^{n}\tau_i k_i e^{-\frac{t}{\tau_i}}\left(e^{\frac{t^*}{\tau_i}} - 1\right)\right), \qquad \text{for} \quad t > t^*. \tag{21}$$

Note that by taking the limit for $t \to \infty$ in Eq. (21), we recover the long-term elastic modulus k_∞. Similarly, the equilibrium stress as $t \to \infty$ is given by $\sigma_\infty = k_\infty\varepsilon_0$ as for the step test.

On the other hand, the stress response at $t = t^*$ is now affected by the previous deformation history at $t < t^*$. To quantify the effect of the rising time of the ramp on the instantaneous elastic and viscous response, we calculate the function $\tilde{k}_{\mathrm{ramp}}(t)$:

$$\tilde{k}_{\mathrm{ramp}}(t) = k_{\mathrm{ramp}}(t) - k_\infty = \sum_{i=1}^{n}\frac{k_i}{v_i}e^{-\frac{t}{\tau_i}}\left(e^{v_i} - 1\right), \tag{22}$$

where $v_i = t^*/\tau_i$ are the ratio between the rise time and the characteristic time constants of the tissue τ_i. Note that by taking the limit $t^* \to 0$ (i.e. a step input), Eq. (22) recovers Eq. (16):

$$\lim_{t^*\to 0}\tilde{k}_{\mathrm{ramp}}(t) = \tilde{k}_{\mathrm{step}}(t) = \sum_{i=1}^{n}k_i e^{-\frac{t}{\tau_i}}. \tag{23}$$

Moreover, we define the modified instantaneous elastic modulus $k_{0\mathrm{ramp}}$ as the function (22) evaluated at the end of the ramp phase, i.e. at $t = t^*$:

$$k_{0\mathrm{ramp}} = \tilde{k}_{\mathrm{ramp}}(t^*) = \sum_{i=1}^{n}k_i\zeta_i = \mathbf{k}\cdot\boldsymbol{\zeta}, \quad \text{with} \quad \zeta_i = v_i^{-1}\left(1 - e^{-v_i}\right). \tag{24}$$

$\mathbf{k}$ and $\boldsymbol{\zeta}$ are vectors with components k_i and ζ_i, $i = \{1, \ldots, n\}$, respectively. Since $0 \le v_i < \infty$, then the parameters ζ_i range between $0 < \zeta_i \le 1$. When $v_i = 0$, $\zeta_i = 1$ and the ramp recovers the perfect step-strain input and the elastic modulus in Eq. (24) reduces to the instantaneous modulus in Eq. (15).

We note that the elastic constants k_i are intrinsic properties of the tissue, therefore they do not depend on the testing procedure nor on the form of the strain history or on the strain-rate at which the test is performed. On the other hand it is well-known that the response of a viscoelastic material strongly depends on the strain-rate. The coefficients ζ_i account for the strain-rate of the deformation process, i.e. for the fact that the strain is applied to the tissue in a finite time t^*. The vector $\boldsymbol{\zeta}$ allows to isolate the effect of the strain-rate from the constant elastic moduli k that describe the material. Therefore, fitting stress relaxation data by using the relaxation function of a step instead of the ramp will result in underestimating the elastic moduli. The smaller the strain-rate (i.e. the greater t^*), the smaller k_{0ramp}. In the limit of a quasi-static deformation, only the infinite modulus k_∞ is recoverable.

Similarly to Eq. (17) we can now compute the total viscosity of the material using the modified function $\tilde{k}_{ramp}(t)$ in Eq. (22):

$$\eta_{0ramp} = \int_{t^*}^{\infty} \tilde{k}_{ramp}(t) \, dt = \sum_{i=1}^{n} \eta_i \zeta_i = \boldsymbol{\eta} \cdot \boldsymbol{\zeta}. \tag{25}$$

Equation (25) highlights that the same set of coefficients ζ_i that link the elastic moduli also link the viscous constants. Since for cases of practical interest $0 < \zeta_i < 1$, it follows that $\eta_{0ramp} < \eta_0$. Therefore, neglecting the influence of the deformation rate will result in underestimating the total viscosity of the material. This is in agreement with the result $\tilde{k}_{0ramp} < \tilde{k}_0$, with $\tilde{k}_0 = k_0 - k_\infty$.

In conclusion, by fitting a stress relaxation curve obtained from a ramp test with Eq. (13) we can obtain a correct estimation of the infinite modulus k_∞, which is strain-rate independent. However, the peak of the relaxation curve, which is related to the instantaneous response of the tissue and therefore to its instantaneous elastic modulus k_0, depends on the strain-rate and it is given by Eq. (24). In particular, the lower the strain-rate, the lower the peak. The area enclosed between the relaxation function and its horizontal asymptote is related to the total viscosity of the material. Since k_∞ is not affected by the rate of deformation, k_{0ramp} decreases with the area represented by η_{0ramp}.

These preliminary synthetic information (k_∞, k_{0ramp}, η_{0ramp}) derived from observation of the experimental relaxation function can be used as a starting point of the fitting procedure to determine the constitutive parameters of the rheological model (k_∞, k_i, τ_i).

3 QLV Model

The linear model in Eq. (1) predicts accurate results only in the small deformation regime, i.e. when $\varepsilon \approx 0$. However, it fails to accurately predict the stress response when a tissue is subjected to a large deformation. To account for large deformations, Fung originally proposed the theory of Quasi-Linear Viscoelasticity (QLV) [8], which is the extension of the linear theory we reviewed in the previous section to the large deformation regime. In this section we review the QLV theory, following [19] to derive the constitutive equation for isotropic compressible and incompressible soft tissues.

The QLV theory is based on the same assumptions of the linear theory, i.e. the Boltzmann superposition principle and the assumption of fading memory. Moreover,

Fung postulated that the total stress is separable into the product of a function of time, i.e. the relaxation function, and a function of the deformation, i.e. the elastic stress. The former accounts for the time-decaying relaxation of the stress and the latter accounts for the non-linear elastic response of the tissue. In the QLV formulation, the relation between the elastic stress and the strain is non-linear. To write the constitutive equation for a QLV model, we start by rewriting the linear model in Eq. (1) in the following equivalent form:

$$\sigma(t) = \int_0^t \mathbb{G}(t - \tau) : \frac{d\sigma^e(\tau)}{d\tau} \, d\tau \,, \tag{26}$$

where the tensor $\mathbb{G}(t)$ is now a fourth-order tensor whose components are non-dimensional and such that $G_n(0) = 1$ for $n = \{1, 2\}$. We call $\mathbb{G}(t)$ the reduced relaxation tensor. Note that in Eq. (26) we have assumed that the deformation history starts at $t = 0$. The stress term $\sigma^e = \mathbb{K}(0)\varepsilon$ is the linear elastic stress. Fung proposed to replace the linear stress σ^e by the corresponding instantaneous elastic stress in large deformation and rewrite the constitutive equation (26) as follows:

$$\mathbf{\Pi}(t) = \int_0^t \mathbb{G}(t - \tau) : \frac{d\mathbf{\Pi}^e(\tau)}{d\tau} \, d\tau \,. \tag{27}$$

The tensor $\mathbf{\Pi}(t)$ is the second Piola-Kirchhoff stress tensor and $\mathbf{\Pi}^e(t)$ is the elastic second Piola-Kirchhoff stress tensor defined as follows:

$$\mathbf{\Pi}^e = J\mathbf{F}^{-1}\mathbf{T}^e\mathbf{F}^{-T}. \tag{28}$$

$\mathbf{T}^e$ is the elastic Cauchy stress, $\mathbf{F} = \mathbf{I} - \nabla\mathbf{u} = \partial\mathbf{x}/\partial\mathbf{X}$ is the deformation gradient associated with the large deformation $\mathbf{x} = \chi(\mathbf{X})$, and $J = \det\mathbf{F}$. $\mathbf{x}$ and $\mathbf{X}$ are the position vectors in the undeformed and deformed configurations, respectively. We use the notation $\mathbf{T}$ to avoid confusion with the linear stress tensor σ. In the small deformation regime the undeformed and deformed configurations coincide since $\nabla\mathbf{u} \approx 0$ and $\mathbf{F} \approx \mathbf{I}$ and therefore the second Piola-Kirchhoff and the Cauchy stress tensors also reduce to the same stress tensor.

Now, we can use the bases in Eq. (9) to split the tensor $\mathbb{G}$ of Eq. (27). We call the associated components $\mathcal{H}(t)$ and $\mathcal{D}(t)$, respectively, and we rewrite Eq. (27) in the following form:

$$\mathbf{\Pi}(t) = \int_0^t \mathcal{H}(t - \tau) \frac{d\mathbf{\Pi}_H^e(\tau)}{d\tau} \, d\tau + \int_0^t \mathcal{D}(t - \tau) \frac{d\mathbf{\Pi}_D^e(\tau)}{d\tau} \, d\tau \,. \tag{29}$$

By following [19], we define:

$$\mathbf{\Pi}_H^e = J\mathbf{F}^{-1} \left(\frac{1}{3}\mathrm{tr}(\mathbf{T}^e)\mathbf{I} \right) \mathbf{F}^{-T} \quad \text{and} \quad \mathbf{\Pi}_D^e = J\mathbf{F}^{-1} \left(\mathrm{dev}(\mathbf{T}^e) \right) \mathbf{F}^{-T}, \tag{30}$$

so that $\mathbf{\Pi}^e = \mathbf{\Pi}^e_H + \mathbf{\Pi}^e_D$. Note that the relaxation functions $\mathcal{H}(t)$ and $\mathcal{D}(t)$ are associated with the Piola transformations of the hydrostatic and deviatoric parts of the Cauchy stress, $\mathbf{\Pi}^e_H$ and $\mathbf{\Pi}^e_D$, respectively. Moreover, by comparing Eqs. (26) and (27) we see that they are both written with respect to the same tensorial relaxation function $\mathbb{G}(t)$. Therefore, the components $\mathcal{H}(t)$ and $\mathcal{D}(t)$ can be determined by performing step-strain tests in the linear regime. Upon a closer inspection of Eq. (5) we can also note that $\mathcal{H}(t) = \kappa(t)/\kappa_0$ and $\mathcal{D}(t) = \mu(t)/\mu_0$ are the non-dimensional version of the relaxation functions $\kappa(t)$ and $\mu(t)$, where $\mu_0 = \mu(0)$ and $\kappa_0 = \kappa(0)$ are the instantaneous elastic bulk and shear modulus, respectively.

Finally, the Cauchy stress tensor follows from applying the transformation $\mathbf{T} = J^{-1}\mathbf{F}\mathbf{\Pi}\mathbf{F}^T$ to Eq. (29) and is given by:

$$\mathbf{T}(t) = J^{-1}(t)\mathbf{F}(t)\left(\int_0^t \mathcal{H}(t-\tau)\frac{d\mathbf{\Pi}^e_H(\tau)}{d\tau}\,d\tau + \int_0^t \mathcal{D}(t-\tau)\frac{d\mathbf{\Pi}^e_D(\tau)}{d\tau}\,d\tau\right)\mathbf{F}^T(t). \tag{31}$$

Equation (29) is the QLV form of the constitutive equation for an isotropic compressible viscoelastic material. In the incompressible limit $J \to 1$ and $\kappa(t) \to \kappa_0 \to \infty$, $\forall t$, therefore Eq. (31) reduces to the following form:

$$\mathbf{T}(t) = \mathbf{F}(t)\left(\int_0^t \mathcal{D}(t-\tau)\frac{d\mathbf{\Pi}^e_D(\tau)}{d\tau}\,d\tau\right)\mathbf{F}^T(t) - p(t)\mathbf{I}, \tag{32}$$

where the Lagrange multiplier $p(t)$ is given by:

$$p(t) = \lim_{k(t) \to \infty}\lim_{J \to 1}\left(J^{-1}(t)\mathbf{F}(t)\left(\int_0^t \frac{\kappa(t-\tau)}{\kappa_0}\frac{d\mathbf{\Pi}^e_H(\tau)}{d\tau}\,d\tau\right)\mathbf{F}^T(t)\right). \tag{33}$$

Equation (32) is the QLV form of the constitutive equation for an isotropic incompressible material.

In the next section we consider the simple torsion of a solid cylinder. We derive the analytical expressions of the torque and the normal force required to twist the cylinder, both in the linear and the large deformations regime. We then derive the analytical expression for the relaxation curves of the torque and the normal force in two experimental scenarios: the step-strain test and the ramp test.

4 Simple Torsion

In this section we consider the problem of simple torsion of a solid cylinder. We start by defining the coordinates of the cylinder in the reference configuration $\mathcal{B}_0$ and in the deformed configuration $\mathcal{B}(t)$ as $\{R, \Theta, Z\}$ and $\{r(t), \theta(t), z(t)\}$, respectively. We assume that the deformation starts at time $t = 0$ and take the reference configuration as the initial configuration $\mathcal{B}_0 = \mathcal{B}(0)$. The displacement vectors $\mathbf{X}$ and

$\mathbf{x}(t)$ in $\mathcal{B}_0$ and $\mathcal{B}(t)$, respectively, are defined with respect to the bases $\{\mathbf{E}_R, \mathbf{E}_\Theta, \mathbf{E}_Z\}$ and $\{\mathbf{e}_r, \mathbf{e}_\theta, \mathbf{e}_z\}$, so that $\mathbf{X} = R\mathbf{E}_R + \Theta\mathbf{E}_\Theta + Z\mathbf{E}_Z$ and $\mathbf{x}(t) = r(t)\mathbf{e}_r + \theta(t)\mathbf{e}_\theta + z(t)\mathbf{e}_z$. The deformation can then be written as follows:

$$r(t) = R, \qquad \theta(t) = \Theta + \phi(t)Z, \qquad z(t) = Z, \tag{34}$$

where $\phi(t) = \alpha(t)/l$ is the amount of twist experienced by the cylinder at time t, defined as the angle of rotation $\alpha(t)$ per unit length. l is the length of the cylinder which remains constant at all times. The strain $\gamma(r, t)$, a non-dimensional measure of the deformation is:

$$\gamma(r, t) = \frac{r\alpha(t)}{l} = r\phi(t). \tag{35}$$

The deformation gradient $\mathbf{F}(r, t) = \dfrac{\partial \mathbf{x}(t)}{\partial \mathbf{X}}$ is given by:

$$\mathbf{F}(r, t) = \begin{pmatrix} 1 & 0 & 0 \\ 0 & 1 & r\,\phi(t) \\ 0 & 0 & 1 \end{pmatrix} \tag{36}$$

and the left Cauchy-Green tensor $\mathbf{B}(r, t) = \mathbf{F}(r, t)\mathbf{F}(r, t)^{\mathrm{T}}$ and its inverse are given by:

$$\mathbf{B}(r, t) = \begin{pmatrix} 1 & 0 & 0 \\ 0 & 1 + r^2\phi^2(t) & r\phi(t) \\ 0 & r\phi(t) & 1 \end{pmatrix} \quad \text{and} \quad \mathbf{B}(r, t)^{-1} = \begin{pmatrix} 1 & 0 & 0 \\ 0 & 1 & -r\phi(t) \\ 0 & -r\phi(t) & 1 + r^2\phi^2(t) \end{pmatrix}. \tag{37}$$

Note that the deformation gradient depends on the spatial variable r, i.e. the deformation is non-homogeneous and the stress distribution will depend on the radial position as well.

The principal stretches and the principal directions associated with the torsion deformation are the eigenvalues and the eigenvectors of the tensor $\mathbf{B}$, respectively. Upon diagonalising $\mathbf{B}$, we find that the principal stretches are given by:

$$\lambda_1 = 1, \quad \lambda_{2,3}(r, t) = \sqrt{1 + \frac{\gamma(r, t)}{2}\left(\gamma(r, t) \pm \sqrt{\gamma^2(r, t) + 4}\right)}. \tag{38}$$

λ_2 and λ_3 are the greatest and the smallest stretch, respectively, and the associated eigenvectors are the directions where λ_2 and λ_3 occur. Note that both λ_2 and λ_3 depend on the spatial variable r. Moreover, λ_2 is maximum at the outer surface $r = r_o$. It is useful to define the strain $\gamma_o(t)$ as the strain at the outer surface of the cylinder at time t:

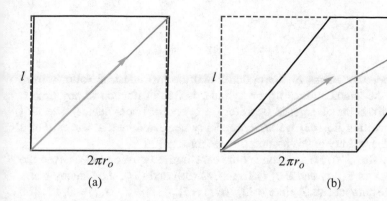

Fig. 5 Sketch of the lateral surface of a cylinder with length l and radius r_o twisted with $\gamma_o = 0.02$ (**a**) and $\gamma_o = 0.8$ (**b**). The green arrow shows the principal direction associated with the maximum stretch λ_2 in Eq. (38). The dashed and the solid lines represent the undeformed and the deformed cylinder, respectively

$$\gamma_o(t) = \gamma(r_o, t) = r_o \phi(t) \tag{39}$$

so that:

$$\max_r \lambda_2(r, t) = \lambda_2(r_o, t) = \sqrt{1 + \frac{\gamma_o(t)}{2}\left(\gamma_o(t) + \sqrt{\gamma_o^2(t) + 4}\right)}. \tag{40}$$

In Fig. 5 we show the principal direction (green arrow) associated with the maximum stretch (i.e. $\lambda_2(r_o, t)$) on the external surface of the cylinder, when the cylinder experiences a strain $\gamma_o(t) = 0.02$ (Fig. 5a) and $\gamma_o(t) = 0.8$ (Fig. 5b). When $\gamma_o(t) \ll 1$, i.e. in the small deformation regime, the principal direction is aligned with the diagonal of the rectangle and the maximum stretch $\lambda_2(r_o, t)$ can be approximated by the following expansion:

$$\lambda_2(r_o, t) = 1 + \frac{\gamma_o(t)}{2} + O(\gamma_o^2(t)), \tag{41}$$

which recovers the relation between stretch λ used for large deformations and the infinitesimal strain ε used in small deformations $\lambda = \varepsilon + 1$, with $\varepsilon = \gamma/2$.

However, Fig. 5b shows that in the large deformation regime, the principal direction associated with the maximum stretch $\lambda_2(r_o, t)$ is not aligned with the diagonal and $\lambda_2(r_o, t)$ is given by Eq. (40).

We can now write the governing equations for the simple torsion of a solid cylinder. Assuming that the inertia is negligible, the equilibrium equations at any time $t > 0$ are given by:

$$\begin{cases} \mathrm{div}\mathbf{T}(r, t) = \mathbf{0} \\ T_{rr}(r, t) = 0 \qquad \text{at} \qquad r = r_o, \end{cases} \tag{42}$$

where r_o is the outer radius of the cylinder and the last equation follows from imposing that the lateral surface of the cylinder is free of traction at any time t. The operator div is the divergence operator (in cylindrical coordinates), see [25] for details. In writing Eq. (42) we have implicitly neglected inertial forces. For a discussion on the validity of this assumption we refer to [26, 27].

The Cauchy stress $\mathbf{T}(r, t)$ is given by the constitutive equation (32). Given the form of the tensors $\mathbf{F}(r, t)$ and $\mathbf{B}(r, t)$ from Eqs. (36) and (37), the Cauchy stress tensor will have components $T_{r\theta}(r, t) = T_{\theta r}(r, t) = T_{rz}(r, t) = T_{zr}(r, t) = 0$, $\forall t$. The remaining non-zero components are:

$$T_{rr}(r, t) = \int_0^t \mathcal{D}(t - \tau) \frac{\partial}{\partial \tau} \Pi_{\mathrm{D}11}^{\mathrm{e}}(r, \tau) \, d\tau - p(r, \theta, z, t),$$

$$T_{\theta\theta}(r, t) = \int_0^t \mathcal{D}(t - \tau) \frac{\partial}{\partial \tau} \Pi_{\mathrm{D}22}^{\mathrm{e}}(r, \tau) \, d\tau + 2r\phi(t) \int_0^t \mathcal{D}(t - \tau) \frac{\partial}{\partial \tau} \Pi_{\mathrm{D}23}^{\mathrm{e}}(r, \tau) \, d\tau$$

$$+ r^2 \phi(t)^2 \int_0^t \mathcal{D}(t - \tau) \frac{\partial}{\partial \tau} \Pi_{\mathrm{D}33}^{\mathrm{e}}(r, \tau) \, d\tau - p(r, \theta, z, t),$$

$$T_{zz}(r, t) = \int_0^t \mathcal{D}(t - \tau) \frac{\partial}{\partial \tau} \Pi_{\mathrm{D}33}^{\mathrm{e}}(r, \tau) \, d\tau - p(r, \theta, z, t),$$

$$T_{\theta z}(r, t) = \int_0^t \mathcal{D}(t - \tau) \frac{\partial}{\partial \tau} \Pi_{\mathrm{D}23}^{\mathrm{e}}(r, \tau) \, d\tau + r\phi(t) \int_0^t \mathcal{D}(t - \tau) \frac{\mathrm{d}}{\partial \tau} \Pi_{\mathrm{D}33}^{\mathrm{e}}(r, \tau) \, d\tau. \tag{43}$$

Therefore, the governing equations reduce to:

$$\begin{cases} \dfrac{\partial T_{rr}(r, t)}{\partial r} + \dfrac{T_{rr}(r, t) - T_{\theta\theta}(r, t)}{r} = 0 \\[2mm] \dfrac{\partial T_{\theta\theta}(r, t)}{\partial \theta} = 0 \\[2mm] \dfrac{\partial T_{zz}(r, t)}{\partial z} = 0. \end{cases} \tag{44}$$

From the last two equations in Eqs. (44) we can conclude that the Lagrange multiplier p only depends on the spatial variable r and at any time t the governing problem reduces to a single Ordinary Differential Equation (ODE) in the argument r:

$$\begin{cases} \dfrac{\mathrm{d}T_{rr}}{\mathrm{d}r} + \dfrac{T_{rr} - T_{\theta\theta}}{r} = 0 \\[2mm] \qquad\qquad T_{rr} = 0 \qquad \text{at} \qquad r = r_o. \end{cases} \tag{45}$$

Now, we restrict our attention to soft tissues whose elastic behaviour can be considered hyperelastic. For such tissues a strain energy function W can be defined. Here, we choose W in the form of the Mooney-Rivlin model:

$$W(I_1, I_2) = (\frac{\mu_0}{2} - c_2)(I_1 - 3) + c_2(I_2 - 3), \tag{46}$$

where $I_1 = \operatorname{tr} \mathbf{B}$ and $I_2 = 1/2 \left((\operatorname{tr} \mathbf{B})^2 - \operatorname{tr} \mathbf{B}^2\right)$ are the first and second invariants of the tensor $\mathbf{B}$, respectively, μ_0 is the infinitesimal shear modulus and c_2 is the second Mooney-Rivlin parameter. The Neo-Hookean model is recovered by setting $c_2 = 0$. The choice for W is motivated by many experimental observations on soft tissues. In particular, it has been observed that the brain behaves as a Mooney-Rivlin material in torsion [20] and in simple shear [21]. Moreover, the Mooney-Rivlin model has the key feature of predicting a linear elastic response in torsion, i.e. the torque required to twist the cylinder depends linearly on the strain. The elastic Cauchy stress $\mathbf{T}^e$ for an incompressible material is given by the following relation:

$$\mathbf{T}^e = 2W_1\mathbf{B} - 2W_2\mathbf{B}^{-1} - p^e\mathbf{I}, \tag{47}$$

where $W_i = \partial W/\partial I_i$, $i = \{1, 2\}$ and p^e is the elastic Lagrange multiplier [25]. By combining Eqs. (46), (47) and (30) we can calculate the components of the tensor $\mathbf{\Pi}_D^e$:

$$
\begin{aligned}
\Pi_{D11}^e(r, t) &= 1/3(4c_2 - \mu_0)r^2\phi^2(t), \\
\Pi_{D22}^e(r, t) &= -2/3(c_2 + 2\mu_0)r^2\phi^2(t) - 1/3(2c_2 + \mu_0)r^4\phi^4(t), \\
\Pi_{D33}^e(r, t) &= -1/3(2c_2 + \mu_0)r^2\phi^2(t), \\
\Pi_{D23}^e(r, t) &= \mu_0 r\phi(t) + 1/3(2c_2 + \mu_0)r^3\phi^3(t).
\end{aligned}
\tag{48}
$$

By combining Eqs. (43) and (48) and substituting into Eq. (45) we obtain an ODE for the variable p:

$$
\begin{aligned}
\frac{dp}{dr} = {}&\frac{r}{3}(14c_2 + \mu_0)\int_0^t \mathcal{D}(t - \tau)\frac{d}{d\tau}\phi^2(\tau)d\tau + \frac{r^3}{3}(2c_2 + \mu_0)\int_0^t \mathcal{D}(t - \tau)\frac{d}{d\tau}\phi^4(\tau)\,d\tau \\
&- \phi(t)\left(2r\mu\int_0^t \mathcal{D}(t - \tau)\frac{d}{d\tau}\phi(\tau)\,d\tau + \frac{2}{3}r^3(2c_2 + \mu)\int_0^t \mathcal{D}(t - \tau)\frac{d}{d\tau}\phi^3(\tau)\,d\tau\right) \\
&+ \phi^2(t)\frac{r^3}{3}(2c_2 + \mu_0)\int_0^t \mathcal{D}(t - \tau)\frac{d}{d\tau}\phi^2(\tau)\,d\tau
\end{aligned}
\tag{49}
$$

with the initial condition:

$$p = \frac{r_o^2}{3}(4c_2 - \mu_0)\int_0^t \mathcal{D}(t - \tau)\frac{d}{d\tau}\phi^2(\tau)\,d\tau \qquad \text{at } r = r_o \tag{50}$$

whose solution is:

$$
\begin{aligned}
p = {} & \frac{1}{6}\left((14c_2 + \mu_0)r^2 - 3(2c_2 + \mu_0)r_o^2\right)\int_0^t \mathcal{D}(t-\tau)\frac{\mathrm{d}}{\mathrm{d}\tau}\phi^2(\tau)\,\mathrm{d}\tau \\
& + \frac{1}{12}(2c_2 + \mu_0)(r^4 - r_o^4)\int_0^t \mathcal{D}(t-\tau)\frac{\mathrm{d}}{\mathrm{d}\tau}\phi^4(\tau)\,\mathrm{d}\tau \\
& - \left(\mu_0(r^2 - r_o^2)\int_0^t \mathcal{D}(t-\tau)\frac{\mathrm{d}}{\mathrm{d}\tau}\phi(\tau)\,\mathrm{d}\tau \right. \\
& + \frac{1}{6}(2c_2 + \mu_0)(r^4 - r_o^4)\int_0^t \mathcal{D}(t-\tau)\frac{\mathrm{d}}{\mathrm{d}\tau}\phi^3(\tau)\,\mathrm{d}\tau \Bigg)\phi(t) \\
& + \left(\frac{1}{12}(2c_2 + \mu_0)(r^4 - r_o^4)\int_0^t \mathcal{D}(t-\tau)\frac{\mathrm{d}}{\mathrm{d}\tau}\phi^2(\tau)\,\mathrm{d}\tau \right)\phi^2(t).
\end{aligned}
\tag{51}
$$

Finally, the components of the stress $\mathbf{T}(r, t)$ can be obtained by substituting Eq. (51) into Eqs. (43) to fully determine the final stress distribution in the cylinder.

The torque $T(t)$ required to twist the cylinder can be computed as:

$$
T(t) = \int_0^{2\pi}\int_0^{r_o} T_{\theta z}(r, t)r^2\,\mathrm{d}r\,\mathrm{d}\theta ,
\tag{52}
$$

and the normal force N necessary to keep to cylinder length constant reads:

$$
N(t) = \int_0^{2\pi}\int_0^{r_o} T_{zz}(r, t)r\,\mathrm{d}r\,\mathrm{d}\theta .
\tag{53}
$$

The components $T_{\theta z}(r, t)$ and $T_{zz}(r, t)$ are given by Eqs. (43) upon substituting Eqs. (48) and (51). The final expressions for the torque and the normal force read:

$$
\begin{aligned}
T(t) = {} & \frac{\pi}{2}\mu_0 r_o^4 \int_0^t \mathcal{D}(t-\tau)\frac{\mathrm{d}}{\mathrm{d}\tau}\phi(\tau)\,\mathrm{d}\tau \\
& + \frac{\pi}{9}(2c_2 + \mu_0)r_o^6\left(\int_0^t \mathcal{D}(t-\tau)\frac{\mathrm{d}}{\mathrm{d}\tau}\phi^3(\tau)\,\mathrm{d}\tau - \phi(t)\int_0^t \mathcal{D}(t-\tau)\frac{\mathrm{d}}{\mathrm{d}\tau}\phi^2(\tau)\,\mathrm{d}\tau\right)
\end{aligned}
\tag{54}
$$

and

$$
\begin{aligned}
N(t) = {} & -\frac{\pi}{2}\mu_0 r_o^4\phi(t)\int_0^t \mathcal{D}(t-\tau)\frac{\mathrm{d}}{\mathrm{d}\tau}\phi(\tau)\,\mathrm{d}\tau - \frac{\pi}{4}(2c_2 - \mu_0)r_o^4\int_0^t \mathcal{D}(t-\tau)\frac{\mathrm{d}}{\mathrm{d}\tau}\phi^2(\tau)\,\mathrm{d}\tau \\
& + \frac{\pi}{18}(2c_2 + \mu_0)r_o^6\left(\int_0^t \mathcal{D}(t-\tau)\frac{\mathrm{d}}{\mathrm{d}\tau}\phi^2(\tau)\,\mathrm{d}\tau\right)\phi^2(t) \\
& - \frac{\pi}{9}(2c_2 + \mu_0)r_o^6\left(\int_0^t \mathcal{D}(t-\tau)\frac{\mathrm{d}}{\mathrm{d}\tau}\phi^3(\tau)\,\mathrm{d}\tau\right)\phi(t) \\
& + \frac{\pi}{18}(2c_2 + \mu_0)r_o^6\int_0^t \mathcal{D}(t-\tau)\frac{\mathrm{d}}{\mathrm{d}\tau}\phi^4(\tau)\,\mathrm{d}\tau ,
\end{aligned}
\tag{55}
$$

respectively.

Note that Eqs. (54) and (55) are written with respect to the twist $\phi(t)$ which is a dimensional measure of the deformation. In view of comparing the predictions of the QLV theory with those of the linear theory, it is useful to rewrite (54) and (55) in terms of the strain $\gamma_o(t)$, defined in Eq. (39), which is the strain at the outer surface of the cylinder and is a non-dimensional measure of the deformation. Then, Eqs. (54) and (55) rewrite as follows:

$$
T(t) = \frac{\pi}{2} r_o^3 \int_0^t \mu(t-\tau) \frac{\mathrm{d}}{\mathrm{d}\tau} \gamma_o(\tau) \, \mathrm{d}\tau
$$
$$
+ \frac{\pi}{9} \left(\frac{2c_2}{\mu_0} + 1 \right) r_o^3 \left(\int_0^t \mu(t-\tau) \frac{\mathrm{d}}{\mathrm{d}\tau} \gamma_o^3(\tau) \, \mathrm{d}\tau - \gamma_o(t) \int_0^t \mu(t-\tau) \frac{\mathrm{d}}{\mathrm{d}\tau} \gamma_o^2(\tau) \, \mathrm{d}\tau \right)
$$

$$(56)$$

and

$$
N(t) = -\frac{\pi}{2} r_o^2 \gamma_o(t) \int_0^t \mu(t-\tau) \frac{\mathrm{d}}{\mathrm{d}\tau} \gamma_o(\tau) \, \mathrm{d}\tau - \frac{\pi}{4} \left(\frac{2c_2}{\mu_0} - 1 \right) r_o^2 \int_0^t \mu(t-\tau) \frac{\mathrm{d}}{\mathrm{d}\tau} \gamma_o^2(\tau) \, \mathrm{d}\tau
$$
$$
+ \frac{\pi}{18} \left(\frac{2c_2}{\mu_0} + 1 \right) r_o^2 \left(\int_0^t \mu(t-\tau) \frac{\mathrm{d}}{\mathrm{d}\tau} \gamma_o^2(\tau) \, \mathrm{d}\tau \right) \gamma_o^2(t)
$$
$$
- \frac{\pi}{9} \left(\frac{2c_2}{\mu_0} + 1 \right) r_o^2 \left(\int_0^t \mu(t-\tau) \frac{\mathrm{d}}{\mathrm{d}\tau} \gamma_o^3(\tau) \, \mathrm{d}\tau \right) \gamma_o(t)
$$
$$
+ \frac{\pi}{18} \left(\frac{2c_2}{\mu_0} + 1 \right) r_o^2 \int_0^t \mu(t-\tau) \frac{\mathrm{d}}{\mathrm{d}\tau} \gamma_o^4(\tau) \, \mathrm{d}\tau ,
$$

$$(57)$$

where we have used the connection $\mathcal{D}(t) = \mu(t)/\mu_0$.

In the next section we will use Eqs. (56) and (57) to calculate the predictions of the QLV model in the experimental scenarios of a step-strain test and a ramp test.

5 Results

In view of predicting the relaxation behaviour of a tissue in simple torsion, we consider two scenarios which are important from the experimental viewpoint: the step-strain test and the ramp test. By using Eqs. (56) and (57), we then derive the analytical expressions of the relaxation curves for the torque and the normal force.

5.1 Small Deformations

We start by considering the torsion of a cylindrical tissue in the small deformation regime and we calculate the torque and the normal force predicted by the linear viscoelastic theory presented in Sect. 2. We assume that the viscoelastic response of the tissue can be modelled as that of a generalised Maxwell model, i.e. a system of spring and dash-pots arranged as shown in Fig. 2. For small deformations, the only non-zero component of the infinitesimal strain tensor $\boldsymbol{\varepsilon}$ is $\varepsilon_{\theta z} = \gamma(r, t)/2$, thus it follows that $\mathrm{dev}\boldsymbol{\varepsilon} = \boldsymbol{\varepsilon}$ and the constitutive equation (6) reduces to the following equation:

$$\sigma_{\theta z}(r, t) = \int_{-\infty}^{t} \mu(t - \tau) \frac{\mathrm{d}}{\mathrm{d}\tau} \gamma(r, \tau) \, \mathrm{d}\tau . \tag{58}$$

The governing equations (42) are automatically satisfied. Equation (58) is a one-dimensional equation in the same form of Eq. (2), where $\mu(t)$ is now the relaxation function. According to the formula $T_{\mathrm{lin}} = 2\pi \int_0^{r_o} \sigma_{\theta z} r^2 \mathrm{d}r$, the torque is then given by:

$$T_{\mathrm{lin}}(t) = \frac{\pi}{2} r_o^3 \int_{-\infty}^{t} \mu(t - \tau) \frac{\mathrm{d}}{\mathrm{d}\tau} \gamma_o(\tau) \, \mathrm{d}\tau , \tag{59}$$

where $\gamma_o(t)$ is the shear evaluated at the outer radius r_o, see Eq. (39). Note that Eq. (59) is the linearised version of Eq. (56) for $\gamma_o \ll 1$.

Following Sect. 2.2, we can calculate the analytical expressions of the relaxation curve for the torque in response to a step and a ramp input, respectively.

For the step-strain test, we consider the following form for the strain $\gamma_o(t)$:

$$\gamma_o(t) = \frac{r_o}{l} \alpha_0 H(t) = \gamma_{o,0} H(t), \tag{60}$$

where $\gamma_{o,0} = \frac{r_o}{l} \alpha_0$ is the amplitude of the step and $H(t)$ is the Heaviside function as defined in Sect. 2.2. By substituting Eq. (60) into Eq. (59) and upon integrating, we obtain:

$$T^{\mathrm{lin}}(t) = \frac{1}{2} \pi r_o^3 \mu_{\mathrm{step}}(t) \gamma_{o,0}, \tag{61}$$

where the relaxation function $\mu_{\mathrm{step}}(t)$ is the following Prony series:

$$\mu_{\mathrm{step}}(t) = \frac{2 T^{\mathrm{lin}}(t)}{\pi r_o^3 \gamma_{o,0}} = \mu_\infty + \sum_{i=1}^{n} \mu_i e^{-\frac{t}{\tau_i}} . \tag{62}$$

The parameters μ_i in Eq. (62) are the shear moduli of the n branches of the generalised Maxwell model, see Fig. 2, and $\tau_i = \frac{\eta_i}{\mu_i}$ are the associated relaxation times. The instantaneous shear modulus is obtained by evaluating the maximum of the relaxation function $\mu_{\mathrm{step}}(t)$ in $t = 0$, as follows:

$$\mu_0 = \mu_{\text{ramp}}(0) = \mu_\infty + \sum_{i=1}^{n} \mu_i. \tag{63}$$

We now consider a strain history in the form of a ramp as follows:

$$\gamma_o(t) = \frac{\gamma_{o,0}}{t^*} t - \frac{\gamma_{o,0}}{t^*} (t - t^*) H(t - t^*), \tag{64}$$

where t^* is the rising time of the ramp. Substituting Eq. (64) into Eq. (59) and computing the integral provides an analogous expression for the torque as Eq. (61). The relaxation curve of the torque is now given by:

$$\mu_{\text{ramp}}(t) = \frac{2T^{\text{lin}}(t)}{\pi r_o^3 \gamma_{o,0}} = \mu_\infty + \frac{1}{t^*} \sum_{i=1}^{n} \tau_i \mu_i e^{-\frac{t}{\tau_i}} \left(e^{\frac{t^*}{\tau_i}} - 1 \right), \qquad \text{for} \quad t > t^*. \tag{65}$$

Equation (65) describes the relaxation curve of the torque in response to the ramp function in Eq. (64). The right-hand side term in Eq. (65) is the response of the generalised Maxwell system to a ramp input. Equations (62) and (65) can then be used to estimate the viscoelastic parameters μ_i, μ_∞, and τ_i by fitting the data from a step test and a ramp test, respectively. The left-hand side of Eq. (65) can be computed from the experimental data, i.e. the measured torque T_{lin} and the imposed strain $\gamma_{o,0}$ and the radius of the sample r_o. The right-hand side is the analytical expression to be fitted in order to estimate the viscoelastic parameters.

Note that the expression on the right-hand side of Eq. (65) does not depend on the amount of shear $\gamma_{o,0}$, therefore Eq. (65) will predict accurate results for those tissues that display the same relaxation response when subjected to different levels of strain. Moreover, we define the following non-dimensional function:

$$\tilde{\mu}_{0\text{ramp}} = \frac{\mu_{\text{ramp}}(t^*) - \mu_\infty}{\mu_0} = \frac{1}{t^*} \sum_{i=1}^{n} \frac{\tau_i \mu_i}{\mu_0} \left(1 - e^{-\frac{t^*}{\tau_i}} \right) = \sum_{i=1}^{n} \frac{v_i^{-1} \mu_i}{\mu_0} \left(1 - e^{-v_i} \right), \tag{66}$$

which provides an estimate of the maximum value of the relaxation curve in Eq. (65).

In the next section, we derive the corresponding expressions for the relaxation curves of the torque and the normal force for the QLV model.

5.2 Large Deformations

In this section we derive the relaxation curves for the torque and the normal force for a step test and a ramp test in the large deformation regime. We first address the ramp test scenario and then by taking the limit case $t^* \rightarrow 0$ we consider the step test scenario.

5.2.1 Torque

To compute the analytical expression of the relaxation curve for the torque, we substitute the strain history (64) and the relaxation function (62) into Eq. (56). Upon integrating we obtain:

$$T(t, \gamma_{o,0}) = \frac{1}{2} \pi r_o^3 \, \mu_{\text{ramp}}^{\text{QLV}}(t, \gamma_{o,0}) \, \gamma_{o,0}, \qquad \text{for} \quad t > t^*, \tag{67}$$

where:

$$\mu_{\text{ramp}}^{\text{QLV}}(t, \gamma_{o,0}) = \mu_{\text{ramp}}(t) + \frac{2(1 + 2c_2/\mu_0)}{9t^{*3}} \sum_{i=1}^{n} \tau_i \mu_i e^{-\frac{t}{\tau_i}} \Big(-2\tau_i(t^* + 3\tau_i)$$

$$+ e^{\frac{t^*}{\tau_i}}(t^{*2} - 4t^*\tau_i + 6\tau_i^2)\Big)\gamma_{o,0}^2, \tag{68}$$

which is valid for $t > t^*$. Equation (68) shows that the relaxation curve for the torque predicted by the QLV model is the sum of the relaxation function $\mu_{\text{ramp}}(t)$ in Eq. (65) and a term proportional to the square of the imposed strain $\gamma_{o,0}$. Therefore, in the limit $\gamma_{o,0} \ll 1$, i.e. small deformation regime, Eq. (68) recovers the result in (65), which is the relaxation function predicted by the linear theory for a ramp test.

A less straightforward comment is that the quadratic term in Eq. (68) arises as a consequence of the time-dependent nature of the strain history. In other words, if the tissue is deformed infinitely fast (as it is the case for a perfect step-strain), Eq. (68) recovers the function $\mu_{\text{step}}(t)$ in (62), without making the assumption of small deformations! By taking the limit for $t^* \to 0$ in Eq. (68) we indeed obtain:

$$\lim_{t^* \to 0} \mu_{\text{ramp}}^{\text{QLV}}(t, \gamma_{o,0}) = \mu_\infty + \sum_{i=1}^{n} \mu_i e^{-\frac{t}{\tau_i}} = \mu_{\text{step}}(t), \tag{69}$$

which is the relaxation curve predicted by the linear theory for a step test. Moreover, opposite to the linear viscoelastic theory, the relaxation curve of the torque predicted by the QLV model, i.e. $\mu_{\text{ramp}}^{\text{QLV}}(t, \gamma_{o,0})$, depends on both the time and the level of strain $\gamma_{o,0}$. This is in agreement with the original assumption made by Fung when he first formulated the QLV theory [8].

With the aim of plotting the relaxation curve $\mu_{\text{ramp}}^{\text{QLV}}$ in Eq. (68), we now consider a simplified version of the generalised Maxwell model with only one branch. This layout is also called Standard Linear Solid model, and it is the simplest arrangement of elements which is able to describe the behaviour of a viscoelastic solid. Note that the QLV theory, despite taking into account large deformations, obeys the superposition principle in time. Therefore, adding more branches to the generalised Maxwell model will increase the accuracy of the model without producing mixed higher order terms as, for example, in the multiple integral formulation [28]. We can

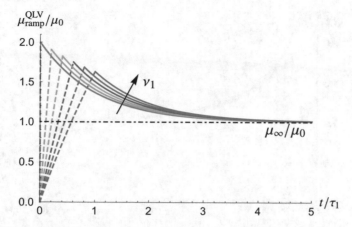

Fig. 6 Effect of the rising time t^* on the relaxation curve $\mu_{\text{ramp}}^{\text{QLV}}(t, \gamma_{o,0})/\mu_0$ in Eq. (68). The parameter $\nu_1 = t^*/\tau_1$ spans over $\{0, 0.2, 0.4, 0.6, 0.8, 1\}$. The following parameters have been fixed: $n = 1$, $\tau_1 = 1$, $\mu_1 = \mu_\infty = 1$, $c_2/\mu_0 = 2/3$ and the strain level $\gamma_{o,0} = 0.02$

then set $n = 1$ in Eq. (68) without loss of generality. Furthermore, we set $c_2 = c_1/2$ according to the observed values of c_1 and c_2 for brain tissues [20].

In Fig. 6 we plot the relaxation curve $\mu_{\text{ramp}}^{\text{QLV}}(t, \gamma_{o,0})$ in Eq. (68) for six different ramp histories. To quantify the effect of the rising time on the profile of the relaxation curve we vary the parameter $\nu_1 = t^*/\tau_1$. We note that Eq. (68) is valid for $t > t^*$. Experimentally, the ramp phase ($t < t^*$) is the noisy part of the data and cannot be used to perform the model fitting. Therefore, we plot the ramp phase of the curves in Fig. 6 with dashed lines. The dashed lines can be obtained by integrating Eq. (56) with the strain history (64) in the time interval $0 \leq t < t^*$.

From Fig. 6 we can conclude that the faster the ramp phase, the higher the peak of the relaxation curve. However, the limiting value of the curves as t approaches ∞ is μ_∞/μ_0 and is not affected by the rising time t^*.

To quantify the effect of the strain level reached at the end of the ramp phase, in Fig. 7 we plot Eq. (68) for different values of $\gamma_{o,0}$ at fixed $\nu_1 = 0.5$.

We now look at two limiting values of Eq. (68), namely the long-term equilibrium and the instantaneous values of $\mu_{\text{ramp}}^{\text{QLV}}(t, \gamma_{o,0})$. First, by taking the limit for $t \to \infty$ we obtain:

$$\lim_{t \to \infty} \mu_{\text{ramp}}^{\text{QLV}}(t, \gamma_{o,0}) = \mu_\infty. \tag{70}$$

From Eq. (70) we observe that the equilibrium value of the relaxation curve of the torque predicted by the QLV model is not affected by the quadratic terms in $\gamma_{o,0}$. In other words, after an infinite time, i.e. at the equilibrium, the response of the tissue is dominated by the long-term shear modulus μ_∞. Therefore, time and its non-linear effect through the quadratic terms in Eq. (68) have no influence on the long-term modulus. This effect is also observable from the horizontal asymptotes of Figs. 6 and 7.

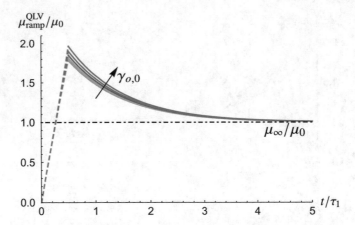

Fig. 7 Effect of the final strain level $\gamma_{o,0}$ on the relaxation curve $\mu_{\text{ramp}}^{\text{QLV}}(t, \gamma_{o,0})/\mu_0$ in Eq. (68). The strain $\gamma_{o,0}$ spans over $\{0.02, \pi/5, \pi/4, \pi/3, \pi/2, \pi\}$. The following parameters have been fixed: $n = 1, \tau_1 = 1, \mu_1 = \mu_\infty = 1, c_2/\mu_0 = 2/3$ and the parameter $\nu_1 = t^*/\tau_1 = 0.5$

Then, we look at the instantaneous value of the function $\mu_{\text{ramp}}^{\text{QLV}}(t, \gamma_{o,0})$, defined as $\mu_{\text{ramp}}^{\text{QLV}}(t^*, \gamma_{o,0})$. The point $\mu_{\text{ramp}}^{\text{QLV}}(t^*, \gamma_{o,0})$ is the maximum of the relaxation curve. In the linear model of Eq. (62), the maximum of the relaxation curve (obtained for a step-strain test) is equal to the instantaneous shear modulus μ_0 in Eq. (63). For a ramp test, the instantaneous response is related to the instantaneous modulus through the formula in Eq. (24), where the elastic constants k_i are replaced by μ_i for $i = \{1, \ldots, n\}$, respectively.

We can now investigate how the non-linear terms in Eq. (68) affect the instantaneous response of the torque. To quantify this effect, we compute the following function:

$$
\tilde{\mu}_{\text{0ramp}}^{\text{QLV}} = \frac{\mu_{\text{ramp}}^{\text{QLV}}(t^*, \gamma_{o,0}) - \mu_\infty}{\mu_0}
$$

$$
= \nu_1^{-1}\left(1 - e^{-\nu_1}\right) + \frac{2}{9}(1 + 2c_2/\mu_0)\nu_1^{-3}e^{-\nu_1}\left(e^{\nu_1}\left(\nu_1^2 - 4\nu_1 + 6\right) - 2(\nu_1 + 3)\right)\gamma_{o,0}^2
\tag{71}
$$

where $\nu_1 = t^*/\tau_1$.

In Fig. 8, we plot Eq. (71) with respect to $\nu_1 = t^*/\tau_1$ and for different values of $\gamma_{o,0}$. The parameter ν_1 spans from 0 to 1, where the value 0 corresponds to perfect step input ($t^* = 0$) and the value 1 corresponds to a ramp input when the rising time of the ramp is equal to the characteristic relaxation time of the tissue ($t^* = \tau_1$).

The curves in Fig. 8 show that the higher $\gamma_{o,0}$, the higher is the peak of the relaxation curve. For $\gamma_{o,0} = 0.02$ (solid blue line), Eq. (71) recovers the relaxation curve in Eq. (66). In the small deformation regime $\gamma_{o,0} \ll 1$, the non-linear terms in Eq. (71) are very small and as expected the predictions of the QLV model recover those of the linear model.

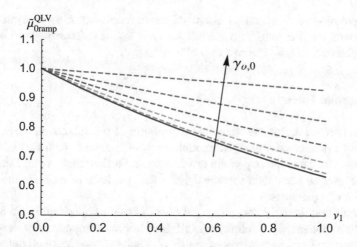

Fig. 8 Effect of the rising time t^* and of the strain level $\gamma_{o,0}$ on the maximum of the relaxation curve of the torque. The function in Eq. (71) is plotted with respect to $\nu_1 = t^*/\tau_1$ for different values of strain $\gamma_{o,0} = \{0.02, \pi/5, \pi/4, \pi/3, \pi/2, \pi\}$. The solid blue line is obtained from the linear model, see Eq. (24) and does not depend on the level of strain

When $t^* \to 0$ (i.e. $\nu_1 \to 0$), the ramp is infinitely fast and is very close to be a step. In this limit Eq. (68) reduces to Eq. (62) and the linear and quasi-linear viscoelastic theories predict the same results for any value of $\gamma_{o,0}$. This limit corresponds to the point $(0, 1)$ in Fig. 8.

The effect of the rising time t^* (and therefore of the strain-rate if $\gamma_{o,0}$ is fixed) on the relaxation curve is accounted by the parameter ν_1: the slower the test, the lower the peak of the experimental relaxation curve in Eq. (68) and the bigger is the difference between the predictions of the linear and the QLV model.

When $t^* \to \infty$, i.e. for very slow (quasi-static) tests, $\tilde{\mu}_{0ramp}^{QLV} \to 0$, which means that only the quasi-static elastic properties of the material, i.e. μ_∞, can be estimated. This would correspond to a quasi-static experiment.

Moreover, from Fig. 8 we can quantify the influence of the quadratic term in Eq. (71). For small values of ν_1, the quadratic contribution is minimal even for large values of strain. Its effect increases with both ν_1 and $\gamma_{o,0}$, as expected.

Note that $\tilde{\mu}_{0ramp}^{QLV}$ decreases as ν_1 increases but increases with $\gamma_{o,0}$. These two opposite effects can lead to an apparent compensation, i.e. constant value of $\tilde{\mu}_{0ramp}^{QLV}$, when slow experiments are performed in the large deformation range.

Finally, we note that the second Mooney-Rivlin parameter c_2 enters Eq. (68) through the quadratic term $\gamma_{o,0}^2$ of the function $\mu_{ramp}^{QLV}(t, \gamma_{o,0})$. As we highlighted above, the quadratic term vanishes in both the limits $t^* \to 0$ (step-strain test) and $t \to \infty$ (elastic equilibrium). Therefore, it is difficult to determine the parameter c_2 from the torque data. The identification of c_2 requires information on the normal force.

In conclusion, the relaxation curve of the torque allows us to estimate the instantaneous and the long-term moduli μ_0 and μ_∞, respectively, the moduli μ_i and the relaxation times τ_i from a step or a ramp test.

5.2.2 Normal Force

In this section we derive the analytical expression of the relaxation curve for the normal force predicted by the QLV model. As in the previous section, we consider both the ramp test and the step-strain test scenarios. Furthermore, we will show how to use the normal force data measured from the two tests to estimate the second Mooney-Rivlin parameter c_2.

We recall that the only non-zero component of the infinitesimal strain tensor is the shear component $\varepsilon_{\theta z}$. It follows that all three normal components of the Cauchy stress tensor are zero, particularly $\sigma_{zz} = 0$, $\forall t$. Therefore, in the small deformation regime the linear theory predicts a zero normal force response for any strain history input.

On the other hand, in the large deformation regime the QLV theory predicts a non-zero normal force for an incompressible Mooney-Rivlin viscoelastic material under torsion, see Eq. (57). By substituting the deformation history (64) into Eq. (57), we calculate the relaxation curve for the normal force in response to a ramp input. Upon integrating, we obtain:

$$N(t, \gamma_{0,0}) = -\frac{\pi}{4} r_o^2 \left((1 + 2c_2/\mu_0)\, \mu_\infty + 2 \sum_{i=1}^{n} \mu_i v_i^{-2} e^{-\frac{t}{\tau_i}} \right.$$

$$\left. \left(2c_2/\mu_0 \left(e^{v_i}\, (v_i - 1) + 1 \right) + e^{v_i} - v_i - 1 \right) \right) \gamma_{0,0}^2$$

$$(72)$$

$$- \frac{\pi}{9} (1 + 2c_2/\mu_0)\, r_o^2 \sum_{i=1}^{n} \mu_i v_i^{-4} e^{-\frac{t}{\tau_i}}$$

$$\left(e^{v_i} \left(v_i^2 - 6v_i + 12 \right) - v_i^2 - 6v_i - 12 \right) \gamma_{0,0}^4,$$

which is valid for $t > t^*$ and where $v_i = t^*/\tau_i$. From Eq. (72) we observe that the normal force N is given the sum of two terms proportional to $\gamma_{0,0}^2$ and $\gamma_{0,0}^4$, respectively. Therefore, in the small deformation limit ($\gamma_{0,0} \to 0$) Eq. (72) recovers the predictions of the linear theory, i.e. $N = 0$, $\forall t$. According to the definition in (13), the relaxation function associated with the normal force is $\sigma_{zz}(t, \gamma_{0,0})/\varepsilon_0$, where $\sigma_{zz}(t, \gamma_{0,0}) = -N(t, \gamma_{0,0})/(\pi r_o^2)$ and $\varepsilon_0 = \gamma_{0,0}/2$, namely:

$$\mu_N(t, \gamma_{0,0}) = -\frac{2N(t, \gamma_{0,0})}{\pi r_o^2 \gamma_{0,0}}.$$

$$(73)$$

In the limit $t^* \to 0$ (step-strain input), Eq. (73) reduces to:

$$\lim_{t^* \to 0} \mu_N(t, \gamma_{o,0}) = \frac{\gamma_{o,0}}{2} \left(1 + 2\frac{c_2}{\mu_0}\right)\mu(t), \quad \text{with} \quad \mu(t) = \mu_\infty + \sum_{i=1}^{n} \mu_i e^{-\frac{t}{\tau_i}}.$$
(74)

The factor $\gamma_{o,0}/2$ represents the dependence on the deformation, while the factor $(1 + 2c_2/\mu_0)\mu(t)$ represents the dependence on time, which is indeed strain-independent. Thus, it follows that different relaxation curves obtained for different values of the strain $\gamma_{o,0}$ do not overlap. The functional dependence of μ_N on the strain $\gamma_{o,0}$ is dictated by the form of the elastic constitutive model. In this particular case, the choice of a Mooney-Rivlin model yields a linear dependence on $\gamma_{o,0}$ in Eq. (74). However, the relaxation curves display the same exponential decay in time, which is dictated by the choice of the rheological model. For a generalised Maxwell model, the decay is exponential according to the Prony series in Eq. (74). If the experimental curves do not display the same decay in time, then the QLV model will not fit the data accurately and more advanced/non-linear models should be considered [1].

Since $N(t)$ depends on the deformation through quadratic terms in $\gamma_{o,0}$ and the aim here is to illustrate how the relaxation curve of the normal force is affected by the level of deformation and the ramp phase, we choose to introduce the following function:

$$f_N(t, \gamma_{o,0}) = -\frac{2N(t)}{\pi r_o^2 \gamma_{o,0}^2}$$

$$= \frac{1+2c_2/\mu_0}{2}\mu_\infty + \sum_{i=1}^{n} \mu_i v_i^{-2} e^{-\frac{t}{\tau_i}} \left(2c_2/\mu_0 \left(e^{v_i}(v_i-1)+1\right) + e^{v_i} - v_i - 1\right)$$

$$+ \frac{1+2c_2/\mu_0}{9} \sum_{i=1}^{n} \mu_i v_i^{-4} e^{-\frac{t}{\tau_i}} \left(e^{v_i}\left(v_i^2 - 6v_i + 12\right) - v_i^2 - 6v_i - 12\right)\gamma_{o,0}^2,$$
(75)

which is valid for $t > t^*$. By taking the limit $t \to \infty$, Eq. (75) reduces to:

$$f_{N_\infty} = \lim_{t \to \infty} f_N(t, \gamma_{o,0}) = (1/2 + c_2/\mu_0)\,\mu_\infty.$$
(76)

The limiting value in Eq. (76) explicitly depends on the second Mooney-Rivlin coefficient c_2. Therefore, experimentally we can determine c_2 by performing a ramp (or a step) test and by measuring the asymptotic value of the normal force curve as $t \to \infty$. We point out that the parameters c_2 appears explicitly in the limiting value (as $t \to \infty$) of the normal force only (it does not appear in the asymptotic value on the torque). This is a consequence of the fact that the vertical force is associated with the change in area of the section of the cylinder. Furthermore, this is consistent

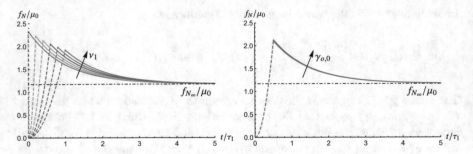

Fig. 9 Effect of the rising time (**a**) and of the strain level (**b**) on the relaxation curve of the normal force in Eq. (75). The following parameters are fixed: $n = 1, \tau_1 = 1, \mu_1 = \mu_\infty = 1, c_2/\mu_0 = 2/3$. In (**a**) we fix $\gamma_{o,0} = 0.02$ and let ν_1 spanning over $\{0, 0.2, 0.4, 0.6, 0.8, 1\}$. $\nu_1 = 0$ represents a perfect step test and $\nu_1 = 1$ represents a ramp test with rising time $t^* = \tau_1$. In (**b**) we fix $\nu_1 = 1/2$ and let $\gamma_{o,0} = \{0.02, \pi/5, \pi/4, \pi/3, \pi/2, \pi\}$

with the fact that c_2 is the parameter associated with the second invariant I_2 of the strain tensor **B**, which indeed accounts for the changes in the area of the material due to the deformation.

Finally, to show the influence of the rising time t^* and of the level of strain at the end of the ramp $\gamma_{o,0}$ on the function f_N, we plot f_N/μ_0 for different values of ν_1 and $\gamma_{o,0}$ in Fig. 9a,b, respectively.

Figure 9b shows that the final level of strain reached at the end of the ramp has a negligible effect on the relaxation curves of the normal force. However, for a fixed strain level, the slower is the ramp, the lower is the peak of the relaxation curve, similarly as we observed for the relaxation curves of the torque (Fig. 6).

In conclusion to fully characterise the viscoelastic behaviour of a soft tissue in torsion (that obeys a Mooney-Rivlin hyperelastic law), according to the QLV theory, we only need to perform a single step-strain test. This test can be performed by using a rheometer that gives access to two sets of data, the torque and the normal force. Moreover, as we showed in Figs. 8 and 9b, the level of strain does not affect the relaxation curves of the torque and the normal force for a step test. Therefore, if the rheometer can achieve high strain-rates and if we are only interested in estimating μ_0, μ_∞, μ_i and τ_i (for instance, if the material is Neo-Hookean, i.e. $c_2 = 0$), the test can be performed in the small deformation regime and the parameters can be estimated by fitting the torque data with Eq. (62). Otherwise, the test should be modelled as a ramp test in the large deformation regime (see Sect. 5.2). In this case, from the relaxation curve of the torque in Eq. (68) we can estimate the long-term shear modulus μ_∞ from the value of the data as $t \to \infty$, according to Eq. (70). We can obtain all the other moduli μ_i and the relaxation times τ_i by fitting the relaxation curve in Eq. (68). Finally, we can estimate the parameter c_2 from the value of the normal force data as $t \to \infty$, according to Eq. (76).

6 Conclusions

In this chapter we reviewed the foundations of linear viscoelasticity and the theory of Quasi-Linear Viscoelasticity (QLV). With the aim of providing a fitting procedure for the QLV model and estimating the viscoelastic properties of a soft tissue, we considered the torsion of a soft solid cylinder and we wrote the governing equations of the viscoelastic problem for a tissue that elastically behaves as a Mooney-Rivlin material, such as the brain. We derived the analytical predictions of the relaxation curves for the torque and the normal force necessary to twist a cylindrical sample. We considered two experimental scenarios: the step test, where the tissue is instantaneously deformed and held in position, and the ramp test, where the tissue is deformed in a finite time and then held in position. These tests are commonly performed to characterise the time-dependent properties of soft tissues and allow to investigate their stress relaxation behaviour. We investigated the effect of the strain level and rising time of the ramp on the relaxation curves of the torque and the normal force. Our results show that in a step test, the linear and the QLV models predict the same relaxation curves for the torque. However, when the strain input is in the form of a ramp function, the non-linear terms appearing in the QLV model affect the relaxation curve of the torque depending on the strain level attained at the end of the loading phase (see Fig. 8). In particular, the higher is the strain level, the higher is the maximum of the relaxation curve, whilst the equilibrium value remains constant and unchanged.

The linear model predicts a zero normal force $\forall t$, whilst the QLV model predicts a non-zero normal force that depends on $\gamma_{o,0}^2$ and $\gamma_{o,0}^4$. Our results show that the relaxation curve of the normal force depends on the level of strain both for a step test (see Eq. (74)) and a ramp test (see Eq. (73)). Although the contributions of the non-linear terms are negligible for a Mooney-Rivlin material (see Fig. 9b), their effect might be relevant for materials that obey a different elastic law (e.g. Fung, Ogden, Gent, etc.).

Finally, our results provide useful guidelines to accurately fit QLV models in view of estimating the viscoelastic properties of soft tissues. We showed how to use the data from a ramp test in torsion (i.e. the relaxation curves of the torque and the normal force) to estimate the constitutive parameters of a Mooney-Rivlin viscoelastic tissue.

References

1. Christensen R, (2012) Theory of viscoelasticity: an introduction. Elsevier.
2. Budday S, Sommer G, Haybaeck J, Steinmann P, Holzapfel G A, Kuhl E. (2017). Rheological characterization of human brain tissue. Acta biomaterialia, 60, 315–329.
3. Rashid B, Destrade M, Gilchrist M D (2014). Mechanical characterization of brain tissue in tension at dynamic strain rates. Journal of the mechanical behavior of biomedical materials, 33, 43–54.

4. Chatelin S, Constantinesco A, Willinger R (2010) Fifty years of brain tissue mechanical testing: from in vitro to in vivo investigations. Biorheology, 47:255–276.
5. Pack E, Dubik J, Snyder W, Simon A, Clark S, De Vita R (2020) Biaxial Stress Relaxation of Vaginal Tissue in Pubertal Gilts. Journal of Biomechanical Engineering, 142(3).
6. Shearer T, Parnell W J, Lynch B, Screen H R, Abrahams D I (2020) A Recruitment Model of Tendon Viscoelasticity That Incorporates Fibril Creep and Explains Strain-Dependent Relaxation. Journal of Biomechanical Engineering, 142(7).
7. Lakes R, (2009) Viscoelastic Materials. Cambridge University Press.
8. Fung Y C (2013) Biomechanics: mechanical properties of living tissues. Springer Science and Business Media.
9. Rigby B J, Hirai N, Spikes J D, Eyring H (1959) The mechanical properties of rat tail tendon. The Journal of general physiology, 43(2), 265–283.
10. Everett J S, Sommers M S. (2013). Skin viscoelasticity: physiologic mechanisms, measurement issues, and application to nursing science. Biological research for nursing, 15(3), 338–346.
11. Pucci E, Saccomandi G (2012) On the nonlinear theory of viscoelasticity of differential type, Mathematics and mechanics of solids, 17(6), 624–630.
12. Reese, S (2003) A micromechanically motivated material model for the thermo-viscoelastic material behaviour of rubber-like polymers. International Journal of Plasticity, 19(7), 909–940.
13. Reese S, Govindjee S (1998) A theory of finite viscoelasticity and numerical aspects. International journal of solids and structures, 35(26–27), 3455–3482.
14. Lion A (1997) A physically based method to represent the thermo-mechanical behaviour of elastomers. Acta Mechanica, 123(1), 1–25.
15. Yazdi S J M, Cho K S, Kang N (2018). Characterization of the viscoelastic model of in vivo human posterior thigh skin using ramp-relaxation indentation test. Korea-Australia Rheology Journal, 30(4), 293–307.
16. Green A E, Rivlin R S (1957). The mechanics of non-linear materials with memory. Archive for Rational Mechanics and Analysis, 1(1), 1–21.
17. Findley W N, Davis F A. (2013). Creep and relaxation of nonlinear viscoelastic materials. Courier Corporation.
18. Valanis K C. (2014). Irreversible thermodynamics of continuous media: internal variable theory (Vol. 77). Springer.
19. De Pascalis R, Abrahams I D, Parnell W J (2014) On nonlinear viscoelastic deformations: a reappraisal of Fung's quasi-linear viscoelastic model. Proceedings of the Royal Society A: Mathematical, Physical and Engineering Sciences, 470(2166), 20140058.
20. Balbi V, Trotta A, Destrade M, Annaidh A N (2019) Poynting effect of brain matter in torsion. Soft matter, 15(25), 5147–5153.
21. Rashid B, Destrade M, Gilchrist M D (2013) Mechanical characterization of brain tissue in simple shear at dynamic strain rates. Journal of the mechanical behavior of biomedical materials, 28, 71–85.
22. Boltzmann L (1874) Zur Theorie der elastischen Nachwirkungen. Sitzungsberichte der Kaiserlichen Akademie der Wissenschaften. Mathematisch-Naturwissenschaftliche Classe, 70 (II) 275.
23. Markovitz H (1977) Boltzmann and the beginnings of linear viscoelasticity. Transactions of the Society of Rheology, 21(3), 381–398.
24. Falcioni M, Vulpiani A (2015) Ludwig Boltzmann: a tribute on his 170th birthday. Lettera Matematica, 2(4), 171–183.
25. Ogden R W (1997) Non-linear elastic deformations. Courier Corporation.
26. Balbi V, Shearer T, Parnell W J (2018) A modified formulation of quasi-linear viscoelasticity for transversely isotropic materials under finite deformation. Proceedings of the Royal Society A: Mathematical, Physical and Engineering Sciences, 474(2217), 20180231.
27. Gilchrist M D, Rashid B, Murphy J G, Saccomandi G (2013) Quasi-static deformations of biological soft tissue. Mathematics and Mechanics of Solids, 18(6), 622–633.
28. Lockett F J (1972) Nonlinear viscoelastic solids. Academic Press.

29. Afsar-Kazerooni N, Srinivasa A R, Criscione J C (2020) Experimental investigation of the inelastic response of pig and rat skin under uniaxial cyclic mechanical loading. Experimental Mechanics, 1–17.
30. Kazerooni N A, Srinivasa A R, Criscione J C (2020) A multinetwork inelastic model for the hysteretic response during cyclic loading of pig and rat skin. International Journal of Non-Linear Mechanics, 126, 103555.

Modelling of Biomaterials as an Application of the Theory of Mixtures

Václav Klika

Abstract An overview of the mixture theory is provided while building upon similarities with the classical single continuum theory. The mixture theory can be formulated on different levels of description, in terms of different state variables. The second law of thermodynamics is used as a fundamental constraint for obtaining the constitutive relations, the closures. For this purpose, one can either use a definition of entropy (Gibbs' relation) or a definition of temperature, which is used to identify the entropy production. We discuss the significance and role of coupling in a model formulation and illustrate it using examples stemming from biology.

The theory is applied to the formulation of a biphasic model of cartilage. The superiority of mixture theory over the single continuum framework is evident, but there is a trade-off in terms of more parameters that need to be estimated and the number of boundary conditions. In the latter, the difficulties are inherent to the theory and remain an open problem. They are not derivable and require further modelling, although there are situations where boundary conditions can be assessed. Upscaling methods might provide answers in certain situations as well as a new idea within GENERIC framework.

1 Introduction

This chapter is based on a series of lectures given at EMS School on Modeling of Biomaterials, held in February 2020 in Kácov, Czech Republic, where multiple approaches to mathematical modelling of biomaterials were presented. In particular, this chapter provides an overview of a continuum mechanics approach while focusing on the so-called mixture theory (theory of interacting continua), its advantages, and shortcomings.

V. Klika (✉)
Department of Mathematics, Faculty of Nuclear Sciences and Physical Engineering, Czech Technical University in Prague, Prague, Czech Republic
e-mail: vaclav.klika@cvut.cz; vaclav.klika@fjfi.cvut.cz

© The Author(s), under exclusive license to Springer Nature Switzerland AG 2021   105
J. Málek, E. Süli (eds.), *Modeling Biomaterials*, Nečas Center Series,
https://doi.org/10.1007/978-3-030-88084-2_4

This text intends to provide an overview of what such an approach and framework have to offer. Hence, we start with the classical single continuum concept of formulating thermodynamically consistent models while highlighting certain aspects of the framework allowing us a better understanding of the next step, the mixture theory. We pause on the significance of conserved quantities in the standard models of continuum thermodynamics and how one can obtain constitutive relations for simple problems (like Newtonian fluids). On the other hand, we mention the difficulties that one faces when more details (structural or material) are necessary to be added.

Consequently, we present an alternative to the formulation of complex single continuum models: the theory of mixtures. It allows one to generalise some of the concepts of the single continua straightforwardly, allowing one for a straightforward description of complex phenomena within a rather simple framework. For example, a rather complex viscoelastic response of a material can be obtained as an outcome of two simple incompressible interacting continua—a solid with linear stress–strain response and a Newtonian fluid. We discuss the essential novel feature that mixture theory provides: the coupling phenomena, i.e., how one process can drive another process against its "natural" direction (as, for example, in Maxwell–Stefan model).

We then proceed to apply the mixture theory to a two-constituent (also referred to as a biphasic) model of cartilage and explore, in certain simplified settings, its behaviour and richness of phenomena. Finally, the formulation of boundary conditions for the mixture is identified as challenging and remains an open problem for several decades now.

Theory of mixtures is a potent theory with wide applications not only in physical but also biological disciplines. We hope that this text helps to share this view.

2　General Framework: Single Continuum

We start with a single continuum nonequilibrium thermodynamic framework not only as a reminder but also as an illustration of the main concepts that perhaps might be formulated from a different perspective than usual in classical texts [10, 20].

We denote with X the Lagrangian coordinates (coordinates in the reference body configuration), x the Eulerian coordinates (coordinates in the actual body configuration), while we assume that these two coordinates are linked with a smooth enough map as is standard in the classical treatment of continuum mechanics. The requirement for a smooth mapping $x^i = \xi^i(t, X)$ between the configurations can be removed (which might be relevant in biological applications) in a slightly different treatment of nonequilibrium thermodynamics using the distortion matrix, where the focus lies in the consistent transitions among levels of description (see "GENERIC" thermodynamics below Sect. 3.2). Deformation gradient is then $F^{iI} = \frac{\partial \xi^i}{\partial X^I}$, usually written as $F^{iI} = \frac{\partial x^i}{\partial X^I}$, while by velocity we mean $v^i(t, x) = \frac{\partial \xi^i(t,X)}{\partial t}\big|_{X=\xi^{-1}(t,x)}$.

Fig. 1 An arbitrary volume
with a discontinuity surface
$\Gamma(t)$ of a quantity ψ

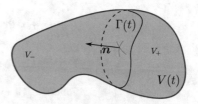

Note that sometimes a third configuration, called a natural configuration, is added to account for a prestress (via accounting for an associated stress-free state) or growth (to distinguish whether the deformation originates from growth or stress). Typically, the concept of multiplicative decomposition is used to add growth to the deformation, but it should be noted that it is not clear whether such an approach is correct. To illustrate this, we remark that this decomposition into two subsequent events means that the newly formed material cannot have different material properties than the original material at that point. Thus, multiplicative decomposition cannot be used in the case when the newly formed tissue is deposited with a different stress than the already existing material, at a different velocity or when ageing of the tissue is of relevance. Again, the distortion matrix might provide an alternative, but this has not been shown yet.

Let us now consider a quantity ψ and an arbitrary volume of the body with a potential discontinuity surface $\Gamma(t)$ in the considered volume as shown in Fig. 1. A general balance law (balancing possible sources and fluxes in a given volume) for the *volumetric property* ψ of a system reads

$$\frac{\mathrm{d}}{\mathrm{d}t} \int_{V(t)} \psi \, \mathrm{d}V = - \int_{\partial V(t) \smallsetminus \Gamma(t)} \boldsymbol{\varphi} \cdot \boldsymbol{n} \, \mathrm{d}A + \int_{V(t)} \sigma \, \mathrm{d}V + \int_{\Gamma(t)} \widehat{\boldsymbol{\psi}}^{s} \, \mathrm{d}A,$$

where $\boldsymbol{n}$ stands for outer normal; we denoted with $\boldsymbol{\varphi}$ the flux of the quantity ψ, while the volumetric source is σ and the source at the surface of discontinuity is $\widehat{\boldsymbol{\psi}}^{s}$. We remark that the left-hand side could also in principle contain a surface density term to reflect the surface phenomena.

Using Gauss and Reynolds transport theorem, noting that the velocity advects fields

$$\frac{\mathrm{d}}{\mathrm{d}t} \int_{V(t)} \psi \, \mathrm{d}V = \int_{V(t) \smallsetminus \Gamma(t)} \left(\frac{D\psi}{Dt} + \psi \operatorname{div} \boldsymbol{v} \right) \mathrm{d}V + \int_{\Gamma(t)} \boldsymbol{n} \cdot [\![(\boldsymbol{v} - \boldsymbol{v}_\Gamma) \otimes \psi]\!] \, \mathrm{d}A,$$

where we introduced the so-called material derivative

$$\frac{D\bullet}{Dt} = \frac{\partial \bullet}{\partial t} + (\boldsymbol{v} \cdot \nabla) \bullet,$$

which tracks changes in a given material point and $\boldsymbol{n} \cdot [\![\bullet]\!]$ denotes a jump of the quantity in the brackets in the direction of the normal vector.

A local form of the balance law can be obtained as

$$\frac{D}{Dt}\psi + \psi\,\mathrm{div}\boldsymbol{v} = \sigma - \mathrm{div}\boldsymbol{\varphi} \quad \text{in } V(t) \smallsetminus \Gamma(t)$$

$$\text{or } \frac{\partial}{\partial t}\psi + \mathrm{div}(\psi \otimes \boldsymbol{v}) = \sigma - \mathrm{div}\boldsymbol{\varphi} \quad \text{in } V(t) \smallsetminus \Gamma(t)$$

$$\text{and } \boldsymbol{n} \cdot [\![(\boldsymbol{v} - \boldsymbol{v}_\Gamma) \otimes \psi - \boldsymbol{\varphi}]\!] = \widehat{\psi^s} \quad \text{on } \Gamma(t).$$

We shall simplify our exposition by disregarding discontinuities, while we shall assume conservative external forces, $\boldsymbol{f} = -\nabla\phi$. The latter allows us to include the effects of charge (and introduce the electrochemical potential) below, which is frequently crucial in biological tissues.

If we now identify what are the sources σ and fluxes φ of a given quantity ψ, we immediately obtain a governing equation for its evolution. In particular, the local balance equation of mass is

$$\frac{\partial\rho}{\partial t} = -\mathrm{div}(\rho\boldsymbol{v}),$$

as the mass cannot be produced, only transported.

Similarly, what are the sources σ and fluxes φ of linear momentum? Volumetric sources of linear momentum are the volumetric forces, while there is a flux of forces inside the body, which we typically refer to as pressure tensor P (which is equal to the minus Cauchy stress τ; the name pressure tensor refers to the fact that pure hydrostatic pressure corresponds to $\mathsf{P} = p\mathsf{I}$). Hence, the governing equation for linear momentum is

$$\frac{\partial\rho\boldsymbol{v}}{\partial t} = -\mathrm{div}(\rho\boldsymbol{v} \otimes \boldsymbol{v}) - \mathrm{div}\mathsf{P}^T + \rho\boldsymbol{f}.$$

One can proceed in the same manner to obtain the balance of energy, which is the sum of kinetic, potential, and internal energy (we denote with ε the specific internal energy):

$$\frac{\partial}{\partial t}\left(\frac{1}{2}\rho v^2 + \rho\phi + \rho\varepsilon\right) = -\mathrm{div}\left(\rho\boldsymbol{v}\left(\frac{1}{2}v^2 + \varepsilon + \phi\right)\right) - \mathrm{div}\left(\boldsymbol{j}_q + \mathsf{P}^T \cdot \boldsymbol{v}\right) + \rho\frac{\partial\phi}{\partial t},$$

where $\boldsymbol{j}_q$ is the heat flux.

Have we included all conserved quantities? We have not employed the angular momentum, but one can show that it translates into the requirement of the symmetry of the pressure tensor P.

A closer look at the balance equations reveals that: (i) we have more unknowns than equations, (ii) still some fundamental physical principles are missing—the "thermodynamic" principles, i.e., the first and second laws of thermodynamics. We use the second point to address the first, i.e., we shall use the second law of thermodynamics to close the system of equations with thermodynamically consistent constitutive laws. We remark that the first law of thermodynamics has been already used in the balance of energy, i.e., the balance of energy can be reformulated into the more usual form of the first law [12].

When invoking the form of entropy balance (s is the specific entropy)

$$\frac{\partial \rho s}{\partial t} = -\text{div}(\rho \boldsymbol{v}s + \boldsymbol{j}_s) + \sigma_s,$$

the second law of thermodynamics (reflecting a time arrow in the system's evolution) requires the entropy production to be non-negative, $\sigma_s \geq 0$. In particular, reversible processes do not raise entropy, $\sigma_s = 0$, and indeed entropy can be shown to be closely related to the concept of irreversibility [51]. Further, the determination of time arrow in a given system allows an assessment of *stability*: an equilibrium corresponds to $dS = 0$, where the entropy reaches its maximum, i.e., $d^2 S \leq 0$. In addition, Prigogine showed that under certain assumptions (mainly that the system is closed and the phenomenological coefficients are constants), a minimum entropy production principle applies, resulting in the observation that even nonequilibrium steady states are always stable in a linear regime. Recently, a more general approach has been proposed, extending the analysis of stability even to open systems and giving the "energy method" used to estimate its proper thermodynamic meaning [7].

The evolution of entropy can be expressed in two forms. One is the balance equation introduced above, while another form can be obtained by calculating the time derivative from Gibbs' definition of entropy. We will follow the latter avenue, and hence, we have that $T\,ds = d\varepsilon + p\,d(1/\rho)$, where p is the equilibrium pressure. This expression is valid in equilibrium and follows from the assumption of having entropy as a well-defined function of the state variables $s = s(\rho, \varepsilon)$. If we further adopt the concept of local equilibrium assumption, meaning that the system reaches equilibrium within small enough elements, we can use this expression and extend it into the nonequilibrium setting while considering it only locally valid. Hence, the (local-in-space) time evolution of entropy is assumed to take the following form for a small element of the material moving with the barycentric velocity $\boldsymbol{v}$ [12, Ch3.2] (and see [40, Ch2.2] for further discussion of this point):

$$T\frac{Ds}{Dt} = \frac{D\varepsilon}{Dt} + p\frac{D\frac{1}{\rho}}{Dt}.$$

By comparing these two forms, the entropy flux and production can be identified

$$\sigma_s = \boldsymbol{j}_q \cdot \nabla\frac{1}{T} - \frac{1}{T}\mathsf{P}_{\text{dis}} : \nabla\boldsymbol{v} \geq 0,$$

where $\mathsf{P} = p\mathsf{I} + \mathsf{P}_{\text{dis}}$ (and we see that $p\mathsf{I}$ corresponds to the equilibrium pressure) and the symbol : denotes the contraction in both indices.

Alternatively to Gibbs' definition of entropy, one may encounter a framework (e.g., in the rational thermodynamics approach) where a definition of temperature as a function of entropy and the state variables is assumed instead.

Entropy production should be objective in the sense that it is material frame indifferent; otherwise, the entropy production condition would not be a sensible constraint [6, 12, 42, 45]. As a special case, an invariance to Galilean transformation has to be valid. Note that there is an interesting relation between conservation laws and symmetries in the Hamiltonian evolution equations: not only that a conservation law generates a symmetry, but one can assess directly its particular form. Note that such observations together with Noether's theorem identify an equivalence between symmetry and conservation laws. For example, a conservation of energy leads to the invariance to time translation; conservation of momentum leads to the invariance of the fields to translations in space. If we know that in addition "Galilean booster" is conserved, we automatically have invariance to Galilean transformation, see [54] for more details. Hence, certain invariances are automatically satisfied within this framework of conservation laws, hence adding yet another appeal to this procedure.

Note that both the Fourier law and Newtonian fluid constitutive law can be obtained from the entropy production relation when assuming linear flux–force relations. Consider, for example, an isothermal system. Then the entropy production reads $\sigma_s = -\frac{1}{T}\mathsf{P}_{\mathrm{dis}} : \nabla v \geq 0$, from where we get the Newtonian fluid description $\mathsf{P}_{\mathrm{dis}} \propto -\nabla v$. Similarly, if we focus on phenomena without flow, $v = 0$, we get the Fourier law of heat conduction $j_q \propto \nabla \frac{1}{T}$.

One can obtain more complex models via the same procedure but where typically one considers more complex definitions of free energy or, equivalently, entropy (e.g., being dependent on the gradient of the deformation gradient or its time derivative).

Finally, the evolution equations determining the system's behaviour can be obtained by substituting the constitutive relations back into the balance equations. This is how, for example, the Fourier heat law transforms into the classical parabolic partial differential equation for the evolution of temperature.

To summarise, this standard derivation of the set of governing equations based on conservation laws has an appeal. First principles (conservation laws and the knowledge of sources and fluxes) are included. Is there some limitation in such an approach? There is, as all these fields (mass, linear and angular momentum, energy) are very particular—we can assess their conservation properties, we can identify their sources and fluxes. Hence, if one is interested in a more complex model, for example, with internal or structural variables, one can formulate balance equations for these new quantities, but one cannot rely on an insight regarding the identification of their sources and fluxes (and hence introducing further unknowns into the evolution equations). Thus, in such situations, the added value of this framework is limited and the internal or structural variables should not be introduced unless they have a very clear physical meaning (this comment is, however, rather general and is not limited to this framework; see for example [40] for discussion of some of the variants of nonequilibrium thermodynamics).

3 General Framework: Multiple Continua, Theory of Mixtures

We shall formulate the theory of mixtures in an analogous way as in the preceding section concerning a single continuum. Please see the classical texts, e.g., [6, 12, 13, 20], or the detailed lecture notes [41]. We remark that there are several variants of the theory and certain open issues regarding the definitions of the whole mixture quantities, both of which the interested reader can find in [25].

The concept stays indeed the same: the balance equations of relevant quantities (state variables), Gibbs' definition of entropy, and the balance of total entropy identifying the fundamental entropy production inequality that, in turn, yields constitutive relations.

We introduce the same concept of configurations as we did before but with the distinction that we do so for n constituents labelled by α and with the assumption that at every point of the mixture, each constituent is present. We may see the build-up of the framework in layers as follows: first, we introduce kinematics (define all necessary bodies and functions connecting them including the deformation gradients); second, specify the state variables and formulate balance laws for them; third, introduce entropy and identify the second law constraint via non-negative entropy production; finally, constitutive theory satisfying the non-negative entropy production condition.

Partial quantities, quantities related to each constituent α, will be indexed with α as in the partial deformation gradient $\mathbf{F}_\alpha = \frac{\partial \xi_\alpha^i}{\partial X_\alpha^I} = \frac{\partial x_\alpha^i}{\partial X_\alpha^I}$ or the partial velocity $v_\alpha^i(t, \mathbf{x}) = \frac{\partial \xi_\alpha^i(t, X_\alpha)}{\partial t}\Big|_{X_\alpha = \xi_\alpha^{-1}(t, \mathbf{x})}$.

A general balance law for a property of a system where we assume that there is no surface of discontinuity reads

$$\frac{d}{dt} \int_{V(t)} \psi_\alpha dV = - \int_{\partial V(t)} \boldsymbol{\varphi}_\alpha \cdot \mathbf{n}_{out} dA + \int_{V(t)} \sigma_\alpha dV + \int_{V(t)} \widehat{\psi}_\alpha dV,$$

where $\widehat{\psi}_\alpha$ is a volume source from different constituents $\beta \neq \alpha$ and σ_α is the remaining volume source. The partial material derivative is defined as

$$\frac{D_\alpha \bullet}{Dt} = \frac{\partial \bullet}{\partial t} + (\boldsymbol{v}_\alpha \cdot \nabla) \bullet.$$

By employing Gauss and Reynolds transport theorem, a local form of the balance law can be obtained

$$\frac{D_\alpha}{Dt} \psi_\alpha + \psi_\alpha \mathrm{div} \boldsymbol{v}_\alpha = \widehat{\psi}_\alpha + \sigma_\alpha - \mathrm{div} \boldsymbol{\varphi}_\alpha \quad \text{in } V(t)$$

or

$$\frac{\partial}{\partial t}\psi_\alpha + \mathrm{div}(\psi_\alpha \otimes v_\alpha) = \widehat{\psi}_\alpha + \sigma_\alpha - \mathrm{div}\varphi_\alpha \quad \text{in } V(t).$$

Let us consider the following state variables: partial densities ρ_α, barycentric velocity v, total energy e, and total entropy s. Note that this choice is not unique and in particular whether one considers all partial velocities v_α or only the barycentric velocity v (or other form of averaged velocity) is essential. Here we make the usual choice [12, 25, 41, 53] as just described, being the easiest extension of a single continuum description, while we demonstrate in the biphasic model below how to proceed if we prefer to track the partial velocities. Note that it is not always beneficial to include all details (consider as many state variables as possible) as one faces the issue of parameter estimation. Hence, this is exactly the step where a tight link between theory and experiment is necessary: are the chosen state variables experimentally accessible? For example, if one cannot trace the individual partial velocities, it can be recommended to include just the averaged velocity among the state variables (although it might be beneficial to develop a more detailed theory, see below).

Balance of mass depends on the choice of the state variables in the following way: are v_α among the state variables (experimentally available)?

If the answer to this question is positive, then the balance of mass reads

$$\frac{\partial \rho_\alpha}{\partial t} = -\mathrm{div}(\rho_\alpha v_\alpha) + \widehat{\rho}_\alpha.$$

On the other hand, if the answer to the above question is negative, then one considers a single (here it is the barycentric) velocity only, and hence,

$$\frac{\partial \rho_\alpha}{\partial t} = -\mathrm{div}(\rho_\alpha v + j_{\mathrm{D}\alpha}) + \widehat{\rho}_\alpha,$$

where $v = \sum_\alpha \rho_\alpha v_\alpha / \rho$ and $j_{\mathrm{D}\alpha}$ denote the diffusion fluxes. Note that these two forms are, of course, the same: one derives the second form via $j_{\mathrm{D}\alpha} = \rho_\alpha(v_\alpha - v)$ but then abandons this relation as v_α are inaccessible. Instead, one considers the diffusion fluxes as unknowns while noting the constraint $\sum_\alpha j_{\mathrm{D}\alpha} = 0$.

The purpose of this distinction, of the choice of the velocity type state variables, becomes apparent below when we discuss the boundary conditions. This remark does not mean that if one cannot measure the partial velocities, one cannot formulate a more detailed model that would include such quantities. On the contrary, such models might reveal new phenomena and provide interpretations of the observed system's behaviour by estimating processes inside the mixture (on a different level of description). The caveat is, however, that specifying appropriate boundary conditions, say for the partial velocities or the related variables, will typically be not clear.

The term $\widehat{\rho}_\alpha$ represents the bulk production of ρ_α due to the interaction with other constituents (for example, phase transition, chemical reactions) and hence adopts the form $\widehat{\rho}_\alpha = \sum_{i=1}^m v_{i\alpha} M_i r_\alpha$, where $v_{i\alpha}$ denote the stoichiometric coefficients, M_i is the molar mass, and r_i is the reaction rate of i-th reaction given in $[molm^{-3}s^{-1}]$ [6, 12]. The structure of the source term together with its relation to the conservation of atoms (stoichiometry) translates into

$$\sum_\alpha \widehat{\rho}_\alpha = 0,$$

as is very nicely described in [6]. Note that the total mass $\rho = \sum_\alpha \rho_\alpha$ satisfies

$$\frac{\partial \rho}{\partial t} = -\mathrm{div}(\rho v).$$

As mentioned above, we shall proceed with one velocity only, the barycentric velocity v, but in the application below, the biphasic model, we shall concern ourselves with keeping v_α; hence, we list the balance of momentum in both its forms as above with the balance of masses.

Balances of partial linear momenta (relevant when partial velocities are among the state variables) are

$$\frac{\partial \rho_\alpha v_\alpha}{\partial t} = -\mathrm{div}(\rho_\alpha v_\alpha \otimes v_\alpha) - \mathrm{div}P_\alpha{}^T + \rho_\alpha f_\alpha + \widehat{p}_\alpha, \qquad (1)$$

where the sum of all sources of momentum has to vanish $\sum_\alpha \widehat{p}_\alpha = 0$, P_α represent partial pressure tensors, and f_α the volumetric forces acting on the α constituent.

If, however, there is only a single velocity (in our case the barycentric velocity) considered among the state variables, the momentum balance reads

$$\frac{\partial \rho v}{\partial t} = -\mathrm{div}(\rho v \otimes v) - \mathrm{div}P^T + \rho f,$$

where P and f are the total pressure tensor and total volumetric force. Note that the balance equation for mixture quantities can take different forms based on the weighting (barycentric velocity is obtained via mass fraction weighting). The choice of barycentric velocity as the mixture velocity is reasonable (it is the velocity where the sources and fluxes in its balance equations are clearly interpretable via the balance of forces), but it is not clear whether it is the correct representation of what is experimentally measured as a mixture velocity. The argument that the barycentric velocity is a correct choice because then the balance equation for linear momentum involving barycentric velocity is the same as for a single continuum does not seem to be sufficient as the single continuum theory should be a special case of mixture theory and not vice versa [25].

The balance of total mixture energy is linked to the balance of partial energies (as a sum)

$$\frac{\partial}{\partial t} \sum_{\alpha} \left(\frac{1}{2} \rho_\alpha \boldsymbol{v}_\alpha^2 + \rho_\alpha \varepsilon_\alpha + \rho_\alpha \phi_\alpha \right) = -\mathrm{div} \left(\sum_{\alpha} \rho_\alpha \boldsymbol{v}_\alpha \left(\frac{1}{2} \boldsymbol{v}_\alpha^2 + \varepsilon_\alpha + \phi_\alpha \right) \right) -$$

$$- \mathrm{div} \left(\boldsymbol{j}_q + \sum_{\alpha} \mathsf{P}_\alpha{}^T \cdot \boldsymbol{v}_\alpha \right) + \sum_{\alpha} \rho_\alpha \frac{\partial \phi_\alpha}{\partial t},$$

where we assumed conservative forces $\boldsymbol{f}_a = -\nabla \phi_\alpha$, but the structural details of the energy flux and source are of no use as they involve partial quantities that we do not consider among the state variables and hence are inaccessible.

Hence, instead, we shall introduce all mixture quantities as barycentric, and then the total mixture energy reads

$$\frac{\partial}{\partial t} \left(\frac{1}{2} \rho \boldsymbol{v}^2 + \rho \varepsilon + \rho \phi \right) = -\mathrm{div} \left(\rho \left(\frac{1}{2} \boldsymbol{v}^2 + \varepsilon + \phi \right) \boldsymbol{v} + \mathsf{P} \cdot \boldsymbol{v} + \sum_{\alpha} \phi_\alpha \boldsymbol{j}_{\mathrm{D}\alpha} + \boldsymbol{j}_q \right),$$

where the energy flux constitutes of the standard parts, the convective part and flux due to the mechanical work exercised on the system, and of the new expected contribution in mixture, the flux due to the diffusion in the force field. The remaining flux is referred to as a heat flux $\boldsymbol{j}_q$ with the yet undetermined quantities P, $\boldsymbol{j}_{\mathrm{D}\alpha}$, and $\boldsymbol{j}_q$, which will be related to the state variables via constitutive relations.

Again we use Gibbs' definition of entropy for a mixture composed of n constituents $T\mathrm{d}s = \mathrm{d}\varepsilon + p\mathrm{d}(1/\rho) - \sum_{\alpha=1}^{n} \mu_\alpha \mathrm{d}(\rho_\alpha/\rho)$, where we defined the chemical potential μ_α of the component α (as a partial specific Gibbs function, i.e., energy per kilogram). When adopting the local equilibrium hypothesis, we may calculate the time derivative of entropy as above using material time derivative

$$T\frac{Ds}{Dt} = \frac{D\varepsilon}{Dt} + p\frac{D\partial 1/\rho}{Dt} - \sum_{\alpha} \mu_\alpha \frac{D\rho_\alpha/\rho}{Dt},$$

and hence, the flux and source in the balance of entropy

$$\frac{\partial \rho s}{\partial t} = \mathrm{div}(\rho \boldsymbol{v} s + \boldsymbol{j}_s) + \sigma_s$$

can be identified as in the single continuum case. Note that we employ the second law of thermodynamics in a local form (hence assuming its validity locally in every continuum point) and for the total entropy of the system. One entropy entails one temperature in the mixture, which might be a simplification too far in certain applications. In such a case, it is necessary to consider partial entropies as state variables.

When considering m chemical reactions with rates r_i (in $mol \times \mathrm{m}^{-3}\mathrm{s}^{-1}$) among constituents of the mixture, the mass source terms can be rewritten as $\widehat{\rho}_\alpha = \sum_{i=1}^{m} \nu_{i\alpha} M_\alpha r_i$, where $\nu_{i\alpha}$ is the stoichiometric coefficient of the constituent α in the i-th reaction and M_α is the molar mass. The entropy production then reads [12, 53]

$$\sigma_s = \boldsymbol{j}_q \cdot \nabla \frac{1}{T} - \sum_{\alpha=1}^{n} \boldsymbol{j}_{D\alpha} \cdot \left(\nabla \frac{\mu_\alpha}{T} - \frac{\boldsymbol{f}_\alpha}{T} - \frac{\widehat{p}_\alpha}{\rho_\alpha T} \right) - \frac{1}{T} \mathsf{P}_{\text{dis}} : \nabla \boldsymbol{v} + \frac{1}{T} \sum_{i=1}^{m} r_i \mathcal{A}_i \geq 0,$$

(2)

where the affinity of a chemical reaction is defined as $\mathcal{A}_i = -\sum_\alpha \nu_{i\alpha} M_\alpha \mu_\alpha$ and $\mathsf{P} = p\mathsf{I} + \mathsf{P}_{\text{dis}}$. Note that if the kinetic energy of diffusion is included in the internal energy, then an analogous derivation of entropy production entails an additional cause of reaction rates yielding a generalised affinity $\mathcal{A}_i^{\text{gen}} = \mathcal{A}_i - \sum_\alpha \nu_{i\alpha} M_\alpha \frac{u_\alpha^2}{2}$ with diffusion velocity $\boldsymbol{u}_\alpha$ [53].

If we employ linear constitutive theory, i.e., assuming linear flux–force relations, and neglect any coupling among forces, we recover Fourier's heat law (focusing on the first term), Fick's diffusion law with the gradient of electrochemical potential as a driving force, the constitutive relation for a Newtonian fluid, or the law of mass action (the last term).

As an illustration, we briefly mention how the law of mass action is included in the linear theory following from the obtained form of entropy production. Let us consider chemical reactions (phase transitions, interaction networks)

$$\sum_\alpha \bar{\nu}_{i\alpha} N_\alpha \overset{k_{\pm i}}{\rightleftarrows} \sum_\alpha \bar{\nu}'_{i\alpha} N_\alpha, \quad i = 1, .., s,$$

where N_α denotes the individual component, while noting that we explicitly denoted the stoichiometric coefficients of substrates only as $\bar{\nu}_{i\alpha}$ while those of products as $\bar{\nu}'_{i\alpha}$. The *law of mass action* follows from the linear flux–force relation $r_i = k_i \mathcal{A}_i$, where the explicit form of chemical potential for dilute (ideal) mixtures is used, while the proximity to a steady state is used to expand the thermodynamic force in terms of the differences in concentrations from their equilibrium value [24]:

$$r_i = k_{+i} \prod_{\alpha=1}^{n} \left(\frac{\rho_\alpha}{M_\alpha} \right)^{\bar{\nu}_{i\alpha}} - k_{-i} \prod_{\alpha=1}^{n} \left(\frac{\rho_\alpha}{M_\alpha} \right)^{\bar{\nu}'_{i\alpha}}.$$

(3)

Consequently, with this notation in mind, the change in the concentration of a substance in time is

$$\frac{\partial \rho_\alpha}{\partial t} = \sum_i (\bar{\nu}_{i\alpha} - \bar{\nu}'_{i\alpha}) M_\alpha r_i,$$

where we disregarded transport (including the diffusion) for simplicity as we focus on the law of mass action.

As a final note we remark that the typically mentioned drawback of Fick's and Fourier's law of diffusion is the "infinite speed of information propagation" meaning that a localised initial condition (or a perturbation) generates a non-zero response arbitrarily far away. However, one can either use the Extended Irreversible Thermodynamics framework [23] (where a flux is considered as an additional variable), where effectively a relaxation of the flux is introduced and the governing equation becomes a telegraph equation, or keep the intrinsic non-linearity of the force dependence on the state variables (and one can have a finite speed of front propagation due to the non-linearity, a difference in ad- and de-sorption rates or other "non-Fickian" phenomena, cf. [30]). Similarly, Souček et al. [61] showed that if a given mixture can be equivalently described by a single continuum model, certain restrictions of the structure of fluxes emerge. If this observation is applied to diffusive flux, a similar generalisation of Fick's law is achieved.

3.1 Coupling Phenomena, CIT

We saw how the classical physical laws of heat conduction or diffusion follow from the nonequilibrium thermodynamic framework. The key addition, however, lies in formulating more complex models involving the non-trivial interplays among various thermodynamical forces in a thermodynamically consistent way (complying with the second law). We can find requirements and motivation for such models across disciplines including biology and physics. For example, consider the influence a temperature has almost on any process, the Maxwell–Stefan diffusion, the Dufour and Soret effects [40], or the famous Turing instability [37, 46]. In biology in particular, there has been a recent growing interest in mechanotransduction and mechanical influence in general on biological and chemical processes [24, 34]. More generally, one can view all contributions to this book as motivations for studying coupling phenomena (how to add together Darcy's law and elasticity, putting together fluid dynamics and viscoelasticity, etc).

We first pause on a recent addition to understanding the description of coupling phenomena that comply with the linear nonequilibrium thermodynamic framework (including the discussion of the celebrated Onsager–Casimir reciprocal relations [9, 29, 47, 48]), while we draw an illustration at the end where we illustrate the coupling phenomena by including the effect of mechanical loading on chemical reactions.

3.1.1 Congruent Dependence of Phenomenological Coefficients on State Variables

If we revisit the entropy production (2), we can note its bilinear structure

$$\sigma_s = J_s X_s + \boldsymbol{J}_v \cdot \boldsymbol{X}_v + \boldsymbol{J}_a \cdot \boldsymbol{X}_a + \mathsf{J}_t : \mathsf{X}_t,$$

where each term is a product of two terms of the same tensorial order (scalars, vectors, antilinear vectors representing antisymmetric tensors, and symmetric tensors).

Linear nonequilibrium thermodynamics assumes linear flux–force relations as closures

$$J_i = \sum_j L_{ij} \mathbf{X}_j,$$

to satisfy the second law of thermodynamics, which is expressed as non-negativity of local entropy production

$$\sigma = \sum_i \mathbf{J}_i \mathbf{X}_i \geq 0.$$

Let us first focus on a one-dimensional stationary case (steady state outside of an equilibrium)

$$J_i = L_{ij} X_j,$$

where we further assume that the thermodynamic force is a gradient of a state variable $X_i = z_i'$ with z_i a state variable such that J_i are constant in space x. These requirements are further discussed below, but note that one can take this as a definition of state variables: conserved fluxes entail the existence of constant fluxes in one-dimensional space, and then, via the bilinear form of entropy production, we can assess the corresponding thermodynamic forces and, in turn, the state variables.

If we can argue that all but one force can vanish outside of equilibrium, we may show that there is a shared dependency in the phenomenological matrix [32]. We usually consider forces independent that correspond to this requirement, and it was shown to be possible via the choice of boundary conditions sufficiently near to the equilibrium [29]. In particular, for $X_k = 0$, $\forall k \neq j_0$, one has

$$J_1/L_{1j_0} = X_{j_0} = J_2/L_{2j_0} = \ldots = J_n/L_{nj_0},$$

where J_i are constants. Hence, we have

$$L_{ij}(\mathbf{z}) = f_j(\mathbf{z})g_{ij}(z_1, \ldots, z_{j-1}, z_{j+1}, \ldots, z_n), \quad \forall i, j,$$

i.e., there is a shared dependence on the state variable z_j in the whole j-th column of the phenomenological matrix L_{ij}.

On the other hand, one can identify exactly when the linear flux–force relations allow all but one force to vanish outside of equilibrium: exactly when L_{k,i_0} share the same functional dependency on z_{i_0} for all k.

In addition, there is an interesting connection to the monotonic profiles of state variables. If the state variables are analytic in the spatial variable, then they have to be monotonic in the 1D space. Finally, via the monotonicity of all state variables, one can show even stronger constraints on the phenomenological matrix: $\forall i$, L_{ij} share the same dependence on all z_k [29].

The above observations that the phenomenological matrix has to share the same functional dependence on the state variables in its columns if the forces are independent were independent of the Onsager–Casimir reciprocal relations. What happens if they apply in a given system? The congruent dependence in columns spreads throughout the matrix where the coupling is non-zero. Hence, the Onsager–Casimir reciprocal relations yield even stronger functional constraints, being a significant extension of the reciprocal relations: *the phenomenological matrix has to satisfy* $L_{ij}(\mathbf{z}) = f(\mathbf{z})\bar{L}_{ij}$ *with* $\bar{L}_{ij}$ *being a constant semi-definite matrix.*

To illustrate the above finding, we include a particular experimental example where these constrains were confirmed [32], water and proton transport in a Nafion ionomer membrane. Transport of water and protons is considered, while the linear force–flux relations were identified to be valid for the range $\varphi \in (0, 0.2)$ of the applied electric potential φ across a membrane electrode assembly. The measured data suggest the same functional dependence of diagonal (L_{pp}) and cross ($L_{pw} = L_{wp}$) coefficients on water concentration, see Fig. 2. In addition, Fig. 3 shows that the functional dependence of proton conductivity L_{pp} and water conductivity L_{ww} (the two diagonal components of the phenomenological matrix) on water concentration can be regarded equal. These observations together with conclusions from the preceding figure indeed suggest that the proposed hypothesis holds

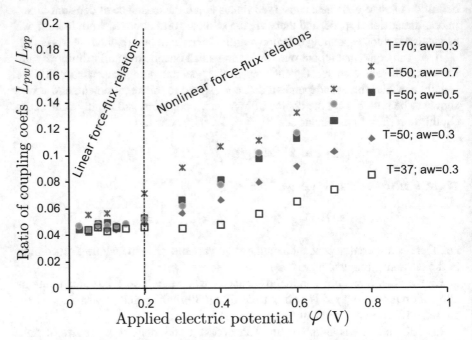

Fig. 2 Experimental verification of the proposed hypothesis of the congruent dependence of phenomenological coefficients on the state variables for a Nafion membrane electrode assembly. The linear regime of the force–flux can be identified for the range $\varphi \in (0, 0.2)$ of applied electric potential φ across a membrane electrode assembly. Temperature and water activity at open circuit is as noted on the graph. The measured data suggest the same functional dependence of diagonal (L_{pp}) and cross ($L_{pw} = L_{wp}$) coefficients on water concentration. Reused with permission from [32]

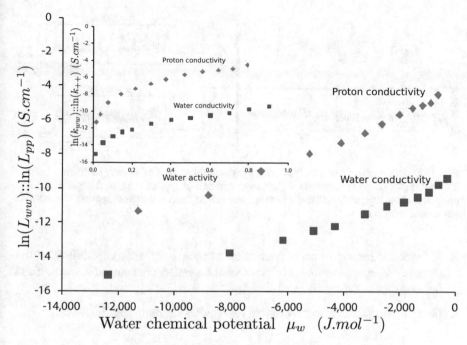

Fig. 3 Further measurement supporting the experimental verification of the proposed congruent dependence for a Nafion membrane electrode assembly. As data indicate (the double colon on the y-axis is used to separate the two plotted quantities), the functional dependency of proton conductivity L_{pp} and water conductivity L_{ww} on water concentration can be regarded equal (note that the shift in the logarithmic values corresponds to multiplication by a constant). These observations together with conclusions from Fig. 2 suggest that the proposed hypothesis holds true in the studied situation. Particularly, the whole matrix of phenomenological coefficients shares the same functional dependence on water concentration. The inset graph in the upper right shows the same data plotted as a function of water activity. Reused with permission from [32]

true in the studied situation. Particularly, the whole matrix of phenomenological coefficients shares the same functional dependence on water concentration.

Let us end this section by a note about the generality of the result. We argue that any model (not necessarily constrained to one dimension or the stationary state) should directly reduce to a description of a steady process in a single dimension without the need to redesign the model. In other words, the steady state in one dimension should be contained in any model as a special case. Hence, the revealed functional constraints should extend to the general model as well.

The significance of these observations can be summarised by the following two points:

1. Non-trivial constraints for models that are expected to be valid in a linear response theory

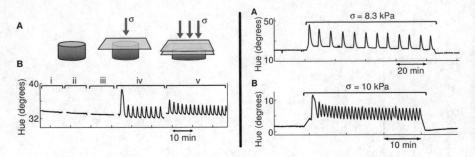

Fig. 4 Left: A schematic of a mechanically triggered Belousov–Zhabotinsky oscillations, where the bottom plot shows the response to increasing amounts of applied stress to the gel. Right: The properties of the triggered oscillations in concentrations depend on the stress applied. Reused with permission from [11]

2. A feedback for experiments (reduction of their number, compatibility of the obtained results, assessment of whether the studied phenomena lie within the linear response theory via monotonicity)

while we refer the reader to the original two studies for more details [29, 32].

3.1.2 Coupling Phenomena, an Example: Extended Law of Mass Action

We shall briefly illustrate the significance of coupling phenomena in a chosen situation and how to propose models for such situations using the above framework.

Experiments in gels capable of sustaining Belousov–Zhabotinsky reactions showed that a certain form of "revival" of chemical oscillations can be obtained using mechanical actuation [11], see the left figure in Fig. 4. In addition, it was observed that the frequency of the mechanical stimulus affects the frequency of oscillations and the duration of the revival, see the right figure in Fig. 4.

If we recall the bilinear structure of entropy production (2) and the force driving the chemical reactions is a scalar (affinity), we may write the general form of the (linear) constitutive relation for a scalar flux (rate of chemical reaction) as

$$J_s = L_{ss} X_s + \boldsymbol{L}_{sv} \cdot \boldsymbol{X}_v + \boldsymbol{L}_{sa} \cdot \boldsymbol{X}_a + \overset{\circ}{\mathsf{L}}_{\mathrm{st}(s)} : \overset{\circ}{\mathsf{X}}_{\mathrm{t}(s)},$$

where each pair consists of quantities of the same tensorial order (the first term corresponding to scalar quantities, the second to vectorial, the third to an antisymmetric second-order tensor, and $\overset{\circ}{\mathsf{L}}_{\mathrm{st}(s)}$ is a symmetric tensor with zero trace) [24].

The rate of deformation tensor D from the continuum mechanics [20] is equal to the symmetric velocity gradient

$$\mathsf{D}^{ij} = \frac{1}{2}\left(\frac{\partial v^i}{\partial x^j} + \frac{\partial v^j}{\partial x^i}\right) = \frac{1}{2}((\nabla v)^{ij} + (\nabla v)^{ji}),$$

and hence, its first invariant, the rate of volume variation $\mathsf{D}^{(1)}$, is linked to the local change in density

$$\mathsf{D}^{(1)} = \mathrm{tr}\,(\nabla v) = \mathrm{div}\,v = -\frac{1}{\rho}\frac{d\rho}{dt}.$$

The entropy production for a reactive system undergoing mechanical actuation reads

$$T\sigma_s = -\mathsf{P}_{\mathrm{dis}} : \nabla v + \sum_{i=1}^{s} r_i \mathscr{A}_i =$$

$$= \boxed{\left(\sum_{i=1}^{s} r_i \mathscr{A}_i - \frac{1}{3}\mathrm{tr}\left(\mathsf{P}_{\mathrm{dis}}\right)\mathrm{tr}\left(\nabla v\right)\right)} - (\nabla v)_{(a)} \cdot \left(\mathsf{P}_{\mathrm{dis}}\right)_{(a)} - \overset{\circ}{\mathsf{D}} : \left(\overset{\circ}{\mathsf{P}}_{\mathrm{dis}}\right)_{(s)} \geq 0,$$

where the gradient of velocity ∇v was decomposed into a scaled unit tensor and a symmetric and antisymmetric part with zero traces to separate the distinct tensorial contributions [24, 40].

Hence, when taking into account the Curie principle stating that only the forces of the same tensorial order can couple in an isotropic system [12], the rate of chemical reaction (flux) is influenced by both scalar forces

$$r_i = l_{ii}\mathscr{A}_i - l_{iv}\mathsf{D}^{(1)},$$

rather than just the chemical activities. Expressing the affinity and chemical potential for dilute mixtures leads to the classical law of mass action (3) when the coupling is neglected, while we obtain an extended law of mass action (reflecting the mechano–chemical coupling) [24, 31]

$$r_i = l_{ii}\mathscr{A}_i - l_{iv}\mathsf{D}^{(1)} = k_{+i}\prod_{\alpha=1}^{n}[\mathrm{N}_\alpha]^{\nu_{i\alpha}} - k_{-i}\prod_{\alpha=1}^{n}[\mathrm{N}_\alpha]^{\nu'_{i\alpha}} - \boxed{l_{iv}\mathsf{D}^{(1)}}. \tag{4}$$

One can see that not only the local changes in the concentrations caused by (dynamic, i.e., velocity type) mechanical actuation affect the reaction rates but also exactly in what form.

One can easily see the link to the motivation above where mechanical actuation caused a revival of depleted oscillations in Belousov–Zhabotinsky gels. In addition, such extensions have direct implications in the modelling of biomaterials where the mechanical properties of the materials are changing in time (typically due to the evolution of concentrations of certain constituents), and hence, via this mechano–chemical coupling, we have a feedback loop driving the material adaptation. Schematically, we may picture these ideas as $C = C\left(\rho_\alpha\right) = C\left(\rho_\alpha(r_i(\mathsf{D}^{(1)}))\right)$.

3.2 Other Approaches

The mixture theory framework [12, 41], which is summarised above, provides a robust way to propose a thermodynamically consistent model. However, this framework is not the only framework available. Why should we be interested in following other routes? We already mentioned that complex materials requiring state variables beyond the mentioned one (for which we are able to identify the sources and fluxes) are not well covered in this framework. In addition, the constitutive theory is the part where there are different approaches to closures and in particular for the theory of mixtures where fluid and solid constituents can appear at once. Our subjective opinion is that it is essential to keep an open mind to other approaches and techniques to overcome the difficulties and open problems in any of the frameworks.

We do not have an ambition to provide a review of the available continuum theories. We rather provide a very brief and crude guidepost for some of the flavours of the nonequilibrium thermodynamics framework, see [55] for a more detailed discussion in this manner. Those include generalisations of what was presented above, e.g., the extended irreversible thermodynamics [23], rational extended thermodynamics [44], thermodynamically constrained averaging theory [16].

Slightly alternative approaches include: Cosserat's kinematic description where each material point has its own microstructure [5]; a single balance, the energy balance, is considered, while from certain invariance arguments one can *as a sequel* obtain the remaining balance laws [17]; similarly one can use the variational approach to obtain the balance laws [14].

Finally, there are also conceptually different approaches. For example, the so-called symmetric hyperbolic thermodynamically consistent (SHTC) framework adds a focus on the mathematical properties of the final governing equations guaranteeing their local in-time well-posedness while introducing new state variables into the description and a different way of dissipation (but note that the theory of mixtures is still under development there) [15, 44, 56]. As a final example, we mention the GENERIC framework [19, 49, 52], which we briefly introduce below.

As its motivation is quite different conceptually, we motivate it first by the amount of information one can get from considering the effect of time-reversal transformation (velocity type variables are inverted, while time proceeds in the negative direction) on the evolution equations [51, 52]. First note that the reversible equation does not change under time-reversal transformation. Now, consider, for example, the Navier–Stokes equation and the action of the time-reversal transformation on it while noting the link between the reversible and irreversible parts:

$$\underbrace{\frac{\partial \rho \mathbf{v}}{\partial t}}_{even} = \underbrace{\overbrace{-\mathrm{div}(\rho \mathbf{v} \otimes \mathbf{v}) - \nabla p}^{even}}_{reversible} \underbrace{\overbrace{+\mu \Delta \mathbf{v}}^{odd}}_{irreversible}.$$

In one crude sentence, GENERIC framework provides systematic and more precise observations about and study of the structure of evolution equations, which is based on a collection of experimental observations and experiences about the evolution of various systems. In addition, it is a natural setting for studying reductions and extensions of models to different levels of description and in such a way that they are consistent with experimentally accessible levels of description (including the prominent example of equilibrium thermodynamics), see, e.g., [50, 52]. It is built on the notion that entropy is a potential always linked to two levels of description, which entails the possibility to identify new entropy on a different level of description while providing an estimate for evolution connecting the two levels [18, 33]. As a hallmark example, let us mention that kinetic theory can be naturally connected to continuum theories in this way [52].

The fundamental observation is that the time-evolution equations can be written in the form

$$\frac{\mathrm{d}x}{\mathrm{d}t} = L\frac{\delta E}{\delta x} + M\frac{\delta S}{\delta x},$$

where x are the independent state variables, E denotes the total energy, S entropy, L, M are the linear operators, and $\delta/\delta x$ is the variational derivative. The dissipative part is represented by the action of a linear operator on the entropy gradient (Ginzburg–Landau evolution equation), while the reversible part corresponds to the first term, the action of the operator L on the gradient of energy.

The observation that the two parts are characterised by reversibility and irreversibility is not immediate and was shown to be a consequence of the degeneracy requirements

$$L\frac{\delta S}{\delta x} = 0, \quad M\frac{\delta E}{\delta x} = 0,$$

by careful discussion of the time-reversal transformation [51].

Furthermore, if we introduce two brackets

$$\{A, B\} = \left\langle \frac{\delta A}{\delta x}, L\frac{\delta B}{\delta x} \right\rangle, \quad [A, B] = \left\langle \frac{\delta A}{\delta x}, M\frac{\delta B}{\delta x} \right\rangle,$$

then the general properties of L and M are

1. Antisymmetry and Jacobi identity for $\{,\}$:

$$\{A, B\} = -\{B, A\} \text{ and } \{A, \{B, C\}\} + \{B, \{C, A\}\} + \{C, \{A, B\}\} = 0$$

2. Symmetry and non-negativity for [,]:

$$[A, B] = [B, A] \text{ and } [A, A] \geq 0$$

The evolution equation for an arbitrary functional A of given state variables is

$$\frac{dA}{dt} = \{A, E\} + [A, S],$$

while it holds $dE/dt = 0$, $dS/dt \geq 0$.

Hence, there are four building blocks required for any GENERIC model: the potentials E, S and the operators L, M. Of course, one has to identify the state variables as a preceding step to these four building blocks.

Alternatively, one can use the dissipation potential formulation (which is a more general case of the above)

$$[A, B] = \left\langle \frac{\delta A}{\delta x}, \frac{\delta \Psi}{\delta(\delta B/\delta x)} \right\rangle,$$

where $\Psi(0) = 0$, Ψ reaches its minimum at 0, and Ψ is convex in the neighbourhood of 0.

Evolution equation takes the form

$$\frac{dx}{dt} = L\frac{\delta E}{\delta x} + \frac{\delta \Psi}{\delta(\delta S/\delta x)},$$

and a particular choice of $\psi(z) = (1/2)\langle z, Mz \rangle$ yields the previous GENERIC formulation.

Although the GENERIC framework might seem too abstract or not constrained enough, it is a nonequilibrium thermodynamic framework that significantly constrains the reversible part of the evolution equations [40]. We add an example below, where we demonstrate that this impression is only a semblance.

One can identify the possible extension of the abovementioned law of mass action within the GENERIC framework. Without going into the details here (and we refer the interested reader to [27]), a coupling between mechanical and chemical thermodynamic forces was considered and its consequences on the law of mass action identified. Surprisingly (using a very different framework to the classical linear nonequilibrium theory described above), the effect of dynamic mechanical state variables turned out to be the same as revealed before in Eq. (4), i.e., acting as sources of mass. In addition, the even-parity (with respect to the time-reversal transformation) state variables affect the reaction rate coefficients (in line with the known effect of pressure).

One can use such extensions to study the observed revival of oscillations in Belousov–Zhabotinsky reactions. One can show that this extended law of mass action (taking into account the mechano–chemical coupling) when used for a generalisation of the reaction kinetics model yields the following theoretical predictions [28]:

- Mechanotransduction is predicted to occur through volume change, but there are more possibilities that were identified.
- An increase in static loading, e.g., in the pressure, is predicted to have a stimulatory effect, whereas the increase in dynamic loading, e.g., in the rate of volume change, is predicted to be stimulatory only up to a certain threshold.
- A physically consistent explanation why some Belousov–Zhabotinsky gels require mechanical stimulation for a "revival" of oscillations.
- Indication of experimental setups enables both the validation of the non-linear coupling and the mechanical effects on the Belousov–Zhabotinsky reaction.

3.3 Single or Multiple Continua: Which One to Choose?

We saw that the formulation of single and multiple continua models is very similar in principle (balance laws, first principles, the first and second laws of thermodynamics, constitutive theory, thermodynamically consistent models).

On the other hand, the single continuum framework and mixture theory differ in many aspects including the number of state variables and parameters, and hence, there is a trade-off in the necessary number of experiments and data for parameter estimation (including the problem of overestimation).

Let us mention the famous Duncan–Toor experiment, see Fig. 5, where a surprising behaviour of diffusion was observed if the mixture is not seen in high enough detail. A rather complicated "explanation" is then offered using a classical diffusion model where the mixture is considered consisting of two diffusing components (and hence there is a single independent driving force of diffusion), see Fig. 6. In particular, one can observe various "regimes" of diffusion that are not found in the classical Fickian description (termed "normal" diffusion in the figure) where the diffusive flux occurs solely in the direction of the negative chemical potential gradient.

Now compare this single continuum description with a multiple continua description, the Maxwell–Stefan model of diffusion, which is just the classical model of diffusion but including coupling among all diffusion forces [38].

Imagine that the mixture consists of constituents that have very different heat conductivity or dielectric permeability. Similarly as in the Duncan–Toor experiment (or Fick vs. Maxwell–Stefan diffusion model), it is understandable that it would be much easier to model such a mixture using the theory of mixtures rather than trying to identify an implicit constitutive relation that would be able to capture the complex response of the whole mixture to temperature variation. With the more detailed description, it is straightforward to have insights into the cause and effect relationship while allowing natural descriptions of the problems at hand.

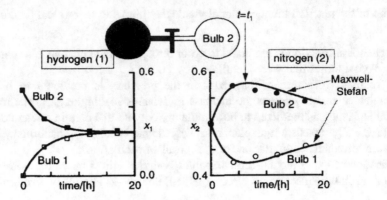

Fig. 5 Duncan–Toor experiment of a mixture of three constituents: hydrogen (1), nitrogen (2), carbon dioxide (3). Initial composition in Bulb 1 is: $x_1 = 0$, $x_2 = 0.509$, $x_3 = 0.491$, while in Bulb 2 is: $x_1 = 0.501$, $x_2 = 0.499$, $x_3 = 0$. The time evolution of concentrations is counter-intuitive as the concentration of nitrogen does not follow the Fick's law of diffusion. We shall argue that this behaviour is a consequence of multiple constituents (requiring mixture theory) and coupling phenomena. Reused with permission from [38]

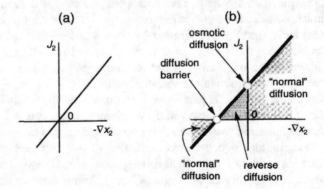

Fig. 6 If a mixture is not seen in a high enough detail, the corresponding model might struggle to offer understanding to the observed phenomena. For example, the Duncan-Toor experiment with three gasses is difficult to explain using a standard binary diffusion theory entailing a rather complicated explanation and requiring the introduction of "novel diffusive phenomena," as suggested in (**b**). On the other hand, if the mixture is recognised as a three-constituent mixture, a standard diffusion model (a direct extension of the binary diffusion model as depicted in (**a**)) offers a clear description and interpretation. Reused with permission from [38]

Let us conclude with an explicit listing of the positives and negatives of the mixture theory approach as opposed to the single continuum framework:

Advantage: Better insight into the cause and effect relationship (see, for example, Biot model and compare it to the biphasic [36] and triphasic theory [39] models of cartilage tissue)

Advantage: More realistic representation of a complex material

Advantage: Wider applicability

Disadvantage: Acquirement of data for parameter fitting; the number of necessary experiments

Disadvantage: Identification of plausible boundary conditions (a significant drawback for a wider application of mixture theory, see below).

4 Application: Biphasic Model

We shall illustrate the above theory by the famous application—a model of cartilage tissue [2, 26, 39, 43]. We shall introduce the biphasic model, which is the simplest extension of the single continuum model, illustrating both the pros and cons of the theory of mixtures. Note that this formulation follows [36] and see the references therein for other variants of biphasic models. The poroelasticity theory is a particular limit of the biphasic model.

We start with a crude overview of cartilage tissue to appreciate the suitability of a biphasic model as opposed to a single continuum description. Cartilage consists of an extracellular matrix and chondrocytes. The latter produce collagen and proteoglycans and respond to both mechanical and chemical stimuli. The extracellular matrix is mainly water (its content varies with depth 65–80%, age, disease), collagen (the sticky tape of animal kingdom; 10–20%), and proteoglycans responsible for the hydrophilic behaviour (and hence providing compressive strength), see Fig. 7.

Let us also mention the key observations (for our purposes) about the structure of the cartilage. In the superficial zone, collagen fibres are parallel to the surface, chondrocytes are flattened, and there is the highest concentration of collagen and the lowest concentration of proteoglycans. In the middle zone, the fibre orientation becomes random and chondrocytes are round. In the deep layer, fibres are perpendicular to the surface and cross the tidemark, while there is the highest concentration of proteoglycans. Finally, the tidemark separates the articular (uncalcified) cartilage from the deeper calcified cartilage and represents a border between nutritional

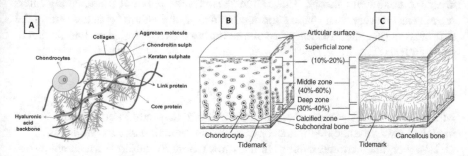

Fig. 7 A cartoon of the hyaline cartilage structure. Note the spatial variation of its structure. Reused with permission from [22]

sources for chondrocytes—cartilage is avascular and hence nourished from either synovial fluid or the subchondral bone.

As we shall employ certain simplifications below, we require quantitative data related to mechanical loading to justify the simplifications. During physiological loading of hyaline cartilage, its height is approximately 2 mm, while the stress is in the range 1–5 MPa with a typical frequency of 1 Hz and rates smaller than 10^3 MPa/s. Cartilage undergoes a very slow adaptation, e.g., too much or little exercise yields thinning and proteoglycan loss, an elastic deformation of the solid part of the cartilage occurs during its compression accompanied by weeping and lubrication.

From the modelling perspective, we shall view the cartilage structure as consisting of two phases: solid (cartilage collagen meshwork) and fluid (interstitial medium) with solid content cca 20%. The crucial effect of aggrecans (carrying charge) is modelled via a swelling pressure due to the fixed charge density attached to the solid phase rather than as a third phase.

We shall identify the bulk equations governing the cartilage behaviour. However, we shall assume the following:

- Solid phase defines the cartilage geometry (the Lagrange coordinates X are tracking solid).
- Incompressible phases.
- Saturation of the mixture, $\varphi_f + \varphi_s = 1$, where φ_α denotes the volume fraction of the α phase.
- A three-dimensional rotationally symmetric problem, hence effectively one-dimensional problem, as this complies with the experimental setup used for parameter estimation and allows qualitative analysis of the final model.
- The state variables are the volume fraction φ_s, φ_f (replacing the partial densities due to the assumption of incompressibility), solid displacement $\boldsymbol{u} = \boldsymbol{u}(X, t) = \boldsymbol{x}(X, t) - X$, fluid velocity $\boldsymbol{v}^f$ (hence tracking both partial velocities), and the Lagrange multiplier $\mathfrak{p}$ corresponding to pressure (see below).

Note that incompressibility is a kinematic constraint, a constraint in the "first layer" of continuum theory. This is the reason why p is not immediately interpretable as pressure. In a single continuum, an arbitrary volume changes during the deformation (i.e., deformation of a volume during the mapping from Lagrangian to Eulerian frame) as

$$|v| = J|v_0|,$$

where $J = \det \mathsf{F}$. Hence, the incompressibility in a single continuum can be modelled via Lagrange multipliers or directly via reduction of state variables.

However, the incompressibility in mixtures requires a different treatment as can be seen from the following argument. Let us introduce another partial density for a clear exposition of this point. In multiple continua, the bulk density of constituent α, ρ_α, is relative to the mixture volume (i.e., tissue volume in cartilage). The true density ρ_α^T of α is, on the other hand, expressed relative to its own volume,

$\rho_\alpha = \varphi_\alpha \rho_\alpha^T$. If we further introduce a superscript R to denote the quantities in the reference frame, we may express the requirement of incompressibility as ρ_α^R is constant in time.

With $F_{i,J}^\alpha = \partial \chi_i^\alpha / \partial X_J^\alpha$ with Jacobian $J^\alpha = \det F^\alpha$, we have from the balance of mass of components:

$$\varphi_\alpha \rho_\alpha^T J^\alpha = \varphi_\alpha^R \rho_\alpha^{TR} = \varphi_\alpha^R \rho_\alpha^T,$$

as $\rho_\alpha J^\alpha = \rho_\alpha^R$. Therefore, the Jacobian of the partial deformation gradient in the mixture

$$J^\alpha = \varphi_\alpha^R / \varphi_\alpha$$

is a measure of a change in the composition, not of incompressibility. However, as ρ_α^R are constants in an incompressible mixture, we may rewrite the mass balances in terms of volume fractions (which then justifies our choice of state variables), see [26] for more details.

Note that the saturation condition, $\varphi_f + \varphi_s = 1$, is natural in the cartilage setting as tracking the fluid and solid constituents accounts for everything significant volume-wise. However, this assumption might be quite restrictive in applications in general and might require the introduction of another component of the mixture.

The partial mass balance (assuming non-reactive mixture, hence no source of partial mass)

$$\frac{\partial \rho_\alpha}{\partial t} + \sum_{i=1}^{3} \frac{\partial}{\partial x_i} \left(v_\alpha^i \rho_\alpha \right) = 0$$

can be rewritten as a solid mass balance

$$J \varphi_s = \varphi_s^0,$$

and a conservation of volume (by rewriting the sum of partial mass balances)

$$\sum_{i=1}^{3} \frac{\partial}{\partial x_i} \left(v_f^i \varphi_f + v_s^i \varphi_s \right) = 0,$$

due to incompressibility and saturation.

By comparing the magnitude of all terms in the linear momentum balance (1) and using the physical dimensions of the problem including the typical fluid velocity and strain rates [36], one can estimate that the dominant balance in the momentum equation occurs within the divergence of the stress tensor, while the inertial terms can be neglected. Hence, the partial momentum balance equations read

$$0 = \sum_{i=1}^{3} \frac{\partial}{\partial x_j} \left(\tau_\alpha^{ij} \right) + \widehat{p}_\alpha,$$

where the partial sources satisfy $\widehat{p}_s = -\widehat{p}_f = \varphi_s \varphi_f \gamma (v_s - v_f) + \tilde{q}$ and γ is a drag coefficient. The drag coefficient γ is yet unspecified, it can depend on the state variables, but can be shown to satisfy $\gamma \geq 0$. Note that the structure of the momentum source terms, proportionality to φ_α, guarantees that the momentum exchange naturally vanishes if any of the constituents vanishes.

A constitutive theory plausible for mixtures of solids and fluids, the Coleman–Noll framework, yields [36] $\tilde{q} = p \nabla \varphi_f = -p \nabla \varphi_s$, where p having the dimensions of pressure is the Lagrange multiplier for the space-filling constraint, $\varphi_f + \varphi_s = 1$. The idea of this theory is the same as we discussed above: the entropy production inequality (or equivalently in terms of the free energy) is employed to obtain plausible constitutive relations satisfying the second law of thermodynamics in any time instant. The fluid contribution to entropy production is the same as we used above, i.e., via Gibbs' relation, while we would like to keep the entropy (or free energy) of the solid general (so that we can specify it a posteriori). Therefore, we do not use a particular form of Gibbs' relation from the start but rather use the second approach via a definition of temperature. This allows to specify the details of the solid (like inner structure, inclusions, fibres) at later stages of the modelling.

Solid and fluid constitutive relations are standard (and follow from the same Coleman–Noll framework):

$$\tau_f = -\varphi_f p \mathsf{I} + \underbrace{2\mu D_{dis}}_{\widehat{\tau}_f},$$

$$\tau_s = -\varphi_s p \mathsf{I} + \underbrace{\frac{2}{\det \mathsf{F}} \mathsf{F} \frac{\partial \psi(\mathsf{C})}{\partial \mathsf{C}} \mathsf{F}^T}_{\widehat{\tau}_s},$$

where we introduced the right Cauchy–Green tensor $\mathsf{C} = \mathsf{F}^T \mathsf{F}$.

Note that p is decomposed into both partial stresses τ_f and τ_s according to the volume fraction, and hence, we similarly redefine the partial stresses to be weighed by the volume fraction. Therefore, each constituent contribution to the total stress is weighed by its volume fraction. Hence, the partial momentum balance equations are

$$0 = \frac{\partial}{\partial x_j} \left(\varphi_\alpha \tau_\alpha^{ij} \right) + \widehat{p}_\alpha^i,$$

where

$$\widehat{p}_s = -\widehat{p}_f = \varphi_s \varphi_f \gamma (v_s - v_f) + \tilde{q}$$

$$= \varphi_s \varphi_f \gamma (\boldsymbol{v}_s - \boldsymbol{v}_f) - p\nabla\varphi_s$$

and

$$\varphi_\alpha \tau_\alpha = \varphi_\alpha \left(-p\mathsf{I} + \widehat{\tau}_\alpha \right).$$

In addition, from the scaling arguments, we may approximate $\widehat{\tau}_f \approx 0$.

It is instructive to rewrite the partial momentum equations into another form, using a relation for the whole tissue (where the momentum is conserved)

$$0 = \nabla \cdot \tau_{tot} = \nabla \cdot \left(\varphi_f \tau_f + \varphi_s \tau_s \right),$$

i.e.,

$$0 = -\frac{\partial p}{\partial x_i} + \frac{\partial}{\partial x_j} \left(\varphi_s \widehat{\tau}_s^{ij} \right) = \frac{\partial \tau_{tot}^{ij}}{\partial x_j},$$

while the momentum balance for fluid acquires Darcy's form

$$0 = -\varphi_f \frac{\partial p}{\partial x_i} - \varphi_s \varphi_f \gamma (v_f^i - v_s^i) = -\gamma \varphi_s \left[\frac{\kappa}{\mu} \frac{\partial p}{\partial x_i} + \varphi_f (v_f^i - v_s^i), \right],$$

μ is the fluid viscosity, and $\kappa = \mu\varphi_f/(\gamma\varphi_s)$ permeability (note $\gamma \geq 0$). Note that permeability can be related to state variables, most frequently to volume fractions reflecting the closing of pores.

4.1 Swelling Pressure and the Effect of Fixed Charge

The presence of ions in cartilage has a profound effect on cartilage tissue response to loading. There are two types of charges—one is fixed to the solid phase, while the other is dissolved in the joint synovial fluid. This setup allows for a gradual increase of pressure within cartilage balancing the applied loading as the cartilage is deformed. We shall now address this effect of charge in the cartilage model.

As aggrecans are fixed in space to the solid phase, we use solution volume-based concentrations to express the concentration of fixed charge $\tilde{c}^F$:

$$\varphi_f \tilde{c}^F J = \varphi_f^0 \tilde{c}_0^F, \quad \tilde{c}^F = \tilde{c}_0^F \frac{\varphi_f^0}{J\varphi_f} = \tilde{c}_0^F \frac{1 - \varphi_s^0}{J - \varphi_s^0}.$$

Donnan approximation of a swelling pressure for external bath ion concentration $\tilde{c}_b$ reads

$$p^{swell} = RT\left(\sqrt{(\tilde{c}^F)^2 + \tilde{c}_b^2} - \tilde{c}_b\right)$$

and hence relates the additional pressure to the fixed charge concentration and ion concentration in the external bath. Note that

$$p^{swell} \to \infty \quad \text{as } \varphi_f \to 0,$$

and hence, the compaction (closing of pores) is implicitly included in this treatment.

The question remains how to add the swelling pressure to the balance equations. Swelling pressure is a bulk phenomenon, not just at the interface; hence, it should appear in the bulk force balance, i.e., in the balance of linear momentum. Considering the swelling pressure as an external force (which exactly corresponds to the addition of electrical forces to the chemical potential to obtain the electrochemical potential via the bulk forces), it does not affect the Coleman–Noll procedure.

Swelling pressure induces an external force per unit volume $-\nabla p^{swell}$ in the *total* momentum balance. In compliance with the definition of total mixture, we distribute the corresponding force among phases via volume fractions φ_α:

$$0 = -\frac{\partial \mathsf{p}}{\partial x_i} - \frac{\partial}{\partial x_i}\left(p^{swell}\right) + \frac{\partial}{\partial x_j}\left(\varphi_s \widehat{\tau}_s^{ij}\right) = \frac{\partial \tau_{tot}^{ij}}{\partial x_j},$$

$$0 = \frac{\kappa}{\mu}\left(\frac{\partial \mathsf{p}}{\partial x_i} + \frac{1}{\varphi_f}\frac{\partial}{\partial x_i}\left(\varphi_f p^{swell}\right)\right) + \varphi_f(v_f^i - v_s^i).$$

4.2   Initial and Boundary Conditions in 1D

It is instructive to write the one-dimensional model in Lagrange coordinates of the solid to discuss the number and significance of the initial and boundary conditions. In Lagrange coordinates, we have

$$0 = \frac{\kappa}{\mu}\frac{1}{J}\left[\frac{\partial}{\partial X}\left(\mathsf{p} + p^{swell}\right) + p^{swell}\frac{\partial}{\partial X}\ln\left(1 - \varphi_s^0/J\right)\right] + (1 - \varphi_s^0/J)(v_f - v_s),$$

$$0 = \frac{\partial}{\partial X}\left(\frac{\varphi_s^0}{J}\widehat{\tau}_s - (\mathsf{p} + p^{swell})\right),$$

$$0 = \frac{\partial}{\partial X}\left(v_s + (1 - \varphi_s^0/J)(v_f - v_s)\right),$$

for three unknowns u ($v_s = \frac{\partial u}{\partial t}$), v_f, and p.

With an impermeable base ($v_f = 0 = v_s$) at $X = 0$, we have

$$0 = v_s + (1 - \varphi_s^0/J)(v_f - v_s),$$

and hence, $\mathfrak{p}$ and u decouple where the latter satisfies a second-order, non-linear parabolic equation for $u(X, t)$,

$$\frac{\partial u}{\partial t} = \frac{\kappa}{\mu} \frac{1}{J} \left[\frac{\partial}{\partial X} \left(\varphi_0^s J^{-1} \widehat{\tau}^s(J) \right) + p^{swell} \frac{\partial}{\partial X} \ln\left(1 - \varphi_0^s / J\right) \right],$$

where $J = 1 + \partial u / \partial X$.

Now the number of required boundary conditions becomes clear. Two boundary conditions for the displacement u are required, where one, $u(X = 0, t) = 0$, has already been used. At the top, one can introduce a kinematic boundary condition $u(X = h_0, t) = \int v^*$, where v^* denotes the specified velocity at the top of the specimen, at $X = h_0$. If we have instead stress boundary condition (as in a creep experiment), we have to apply a force balance at the interface instead noting that it involves the state variable $\mathfrak{p}$.

Hence, a further boundary equation is required. Typically, this is the fluid free draining yielding $\mathfrak{p}|_{X=h_0} = p^{ext}$, where p^{ext} is the external pressure. Now the state variable $\mathfrak{p}$ can indeed be interpreted as a pressure.

Note that this biphasic model was derived without any assumption on the spatial homogeneity of the initial profiles of volume fractions nor the fixed charge density. Therefore, it is suitable to study the effects of this kind, which is exactly what has been recently done for the cartilage tissue [36]. One can show that if there is any heterogeneity present in the model, one cannot choose the initial solid deformation at will but rather have to be computed. Such a situation is analogous to the presence of a prestress in a material that prevents the material to relax to a deformation free state. This observation, together with the details of the inclusion of the swelling pressure, has a significant impact on the tissue response. Note that the redistribution of fluid within the tissue is crucially dependent on the choice of permeability coefficient and as a result affects the sharpness and movement of the draining front.

As a final remark, let us mention that there is no viscoelasticity in the solid, and we considered the simplest model of a fluid, Newtonian fluid. Yet, the response of the mixture as a whole shows complex viscoelastic responses that would be very difficult, if not impossible, to obtain with a single continuum model (the biphasic model captures a wide range of responses including the effect of ion concentrations in the external bath or the effect of heterogeneity in the tissue). We refer the interested reader to the original article [36] for further details including the comparison of the model behaviour to experimental data.

5   Note on Boundary Conditions

We have seen in the previous sections that there is a well-developed framework (actually several of them) for obtaining thermodynamically consistent bulk governing equations. However, the description of the problem at hand is not complete without the specification of initial and boundary conditions as we saw in the biphasic model of cartilage. The increase of the level of detail in the bulk modelling entails concomitant refinement in the boundary conditions. Although we see in the cartilage example that one can come up with reasonable boundary conditions in certain applications, there is no systematic way of obtaining them (one could, for example, question the free draining boundary condition). This issue has been recognised as the most significant drawback of the theory of mixtures preventing its wide application.

First, let us mention that there are two different issues connected to the boundaries but which are quite distinct. There are *boundary phenomena* that represent processes occurring at the boundary, for example, chemical reactions at the interface, apical growth, surface tension, and its dependence on state variables, active surfaces. By the *problem of boundary conditions*, we mean to have a "well-posed system," the correct number, and type of boundary conditions for a given problem.

For the description of the *boundary phenomena*, one can use the same framework that we presented for the derivation of the bulk equations above. Consider, for illustration, two single continua separated by an interface (a reactive boundary for example). The idea is [1, 4, 62] to characterise the interface with a level set function. Then, the balance laws (with a surface of discontinuity) yield balance laws in the two parts of the domain and a jump (compatibility) condition as we saw above. Finally, entropy production and constitutive theory proceed analogously as above, and only the introduced interfacial phenomena will appear as additional terms. Similarly, one can proceed with complex fluids, although the complexity of derivation quickly increases, see, for example, the boundary phenomena in Korteweg type fluids [59]. One can quite straightforwardly extend this idea to reactive mixtures with a *single* velocity, where bulk variables may act as source terms in the surface balance equations, see [60].

Boundary conditions depend on the complexity of the model (level of description) and hence become a more pronounced problem in the theory of mixtures. Consider, for example, the no-slip boundary condition vs. various models of slip. At the microscale, the boundary condition is no-slip, but with the upscaling to macroscale continuum, it might become more appropriate to replace the no-slip condition with a slip (although still at a finer length scale we would observe that the velocity does not slip at the boundary). As a typical problem with boundary conditions, consider a boundary separating two different mixtures, for example, a single and biphasic one. Are the boundary conditions derivable from some principles?

5.1   Are BCs Derivable?

Despite the claims in the literature, e.g. [21], one cannot derive boundary conditions in general for a mixture model. We shall illustrate this point with the simplest case, a biphasic mixture of solid and fluid being in contact with a fluid or another biphasic

Fig. 8 A cartoon of the experimental findings about the tangential component of the velocity. Reused with permission from [3]

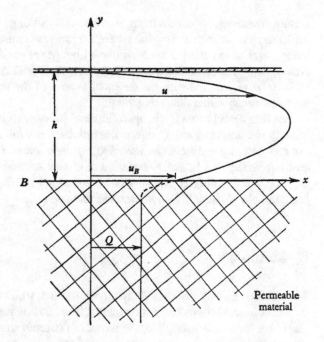

mixture. As we considered above, solid defines the mixture boundary, and hence, the continuity of solid velocity across the interface corresponds to the requirement that the interface moves with the solid velocity at the interface. However, one can show that, despite to what is claimed in the mentioned text, one cannot infer the fluid velocity condition at the interface, in particular in the tangential direction. The tangential velocity component is not constrained by the jump conditions at the surface of the discontinuity. Similarly, one can show that the continuity of pressure across the interface is not a plausible requirement [35].

This observation can be traced back in a way to the Beavers–Joseph experiment who realised that the tangential flow above a porous medium is unrelated to the tangential flow within the porous medium on the macroscale (while there is a continuity on the microscale, cf. the discussion of no-slip boundary condition above), see Fig. 8. As a result, they suggested an additional degree of freedom, a constitutive relation for the tangential velocity at the boundary, which has to be determined by an experiment. This condition bears their name, in the notation from the figure reads [3]

$$\frac{du}{dy}\Big|_{y=0^+} = \beta(u_B - Q),$$

and was subsequently derived [58].

The problem of boundary conditions is indeed a significant one and has been pointed out for quite some time [25, 57]. Although one can write down reasonable boundary conditions (as we saw in the application to cartilage modelling above) in certain situations, a systematic approach is still missing. We are about to finish an initial attempt in such a direction where we propose to use the "inverse" projection using MaxEnt to estimate the least biased boundary conditions in a given system. Note that this concept is based within the GENERIC framework illustrating the need to be acquainted with the various flavours of the mixture theory as they are complementary rather than competing.

Another useful route is via upscaling and homogenisation, which enables us to identify the boundary conditions as the link between microscale and macroscale is not lost. See, for example, the upscaled biphasic model [63] where the method of multiple scales and boundary layer method was successfully used to upscale not only the bulk but also the Nernst–Planck equations and the boundary conditions are interpretable via their link to microscale variables.

6   Summary

We provided an overview of the theory of mixtures while building upon similarities with the classical single continuum theory. The mixture theory can be formulated on "different levels of description," in terms of different state variables. In particular, the choice whether one considers all partial velocities or a single averaged mixture velocity is a key step in the formulation. The second law of thermodynamics is used as a fundamental constraint for obtaining the constitutive relations, the closures. To invoke the second law, one has to relate the entropy to the considered state variables. To this end, one can either use directly a definition of entropy (e.g., Gibbs' relation together with the local equilibrium assumption), a definition of some of the related potentials [8], or even a definition of temperature, which all allow a connection between the entropy and the other state variables for which balance equations are available. Via this relation, we may identify the entropy production that has to be positive. This inequality can be satisfied by several different approaches, called constitutive theories, and we commented on the most widely used, the linear nonequilibrium thermodynamics. Finally, we discussed the significance and role of coupling in the model formulation and illustrated it with several examples.

The superiority of mixture theory over the single continuum framework is self-evident (and can be appreciated, for example, in the Duncan–Toor experiment), but there is a trade-off in terms of more parameters that need to be estimated and the number of boundary conditions. In the latter, difficulties are inherent to the theory as no successful general resolution of this problem has been provided. They are not derivable from first principles and require further (constitutive) modelling, although there are situations where boundary conditions can be assessed. Upscaling methods linking the microscale to macroscale parameters might be a partial way forward, but this approach has its own limitations. Preliminary results suggest that there might be a systematic approach via MaxEnt principle within the GENERIC framework.

Acknowledgments Václav Klika is grateful for support from the Czech Grant Agency, project number 20-22092S.

References

1. Abarbanel, H.D., Brown, R., Yang, Y.M.: Hamiltonian formulation of inviscid flows with free boundaries. The Physics of Fluids **31**(10), 2802–2809 (1988)
2. Ateshian, G.A.: On the theory of reactive mixtures for modeling biological growth. Biomechanics and Modeling in Mechanobiology **6**(6), 423–445 (2007)
3. Beavers, G.S., Joseph, D.D.: Boundary conditions at a naturally permeable wall. Journal of Fluid Mechanics **30**(1), 197–207 (1967)
4. Bedeaux, D., Albano, A., Mazur, P.: Boundary conditions and non-equilibrium thermodynamics. Physica A: Statistical Mechanics and its Applications **82**(3), 438–462 (1976)
5. Bedford, A., Drumheller, D.S.: Theories of immiscible and structured mixtures. International Journal of Engineering Science **21**(8), 863–960 (1983)
6. Bowen, R.M.: Theory of mixtures. In: A. Eringen (ed.) Continuum Physics, vol. 3. Academic Press, New York (1976)
7. Bulíček, M., Málek, J., Průša, V.: Thermodynamics and stability of non-equilibrium steady states in open systems. Entropy **21**(7), 704 (2019)
8. Callen, H.B.: Thermodynamics and an Introduction to Thermostatistics. John Wiley & Sons (1985)
9. Casimir, H.B.G.: On Onsager's principle of microscopic reversibility. Rev. Mod. Phys. **17**, 343–350 (1945). https://doi.org/10.1103/RevModPhys.17.343
10. Chadwick, P.: Continuum mechanics: concise theory and problems. Courier Corporation (2012)
11. Chen, I.C., Kuksenok, O., Yashin, V.V., Balazs, A.C., Van Vliet, K.J.: Mechanical resuscitation of chemical oscillations in Belousov–Zhabotinsky gels. Advanced Functional Materials **22**(12), 2535–2541 (2012)
12. De Groot, S.R., Mazur, P.: Non-equilibrium thermodynamics. Courier Corporation (2013)
13. Drew, D.A., Passman, S.L.: Theory of multicomponent fluids, vol. 135. Springer Science & Business Media (2006)
14. Drumheller, D.: On theories for reacting immiscible mixtures. International Journal of Engineering Science **38**(3), 347–382 (2000)
15. Godunov, S.K., Romenskii, E.: Elements of continuum mechanics and conservation laws. Springer Science & Business Media (2003)
16. Gray, W.G., Miller, C.T.: Introduction to the thermodynamically constrained averaging theory for porous medium systems. Springer (2014)
17. Green, A.E., Naghdi, P.: A unified procedure for construction of theories of deformable media. iii. Mixtures of interacting continua. Proceedings of the Royal Society of London. Series A: Mathematical and Physical Sciences **448**(1934), 379–388 (1995)
18. Grmela, M., Klika, V., Pavelka, M.: Reductions and extensions in mesoscopic dynamics. Physical Review E **92**(3), 032111 (2015)
19. Grmela, M., Öttinger, H.C.: Dynamics and thermodynamics of complex fluids. I. Development of a general formalism. Physical Review E **56**(6), 6620 (1997)
20. Gurtin, M.E., Fried, E., Anand, L.: The mechanics and thermodynamics of continua. Cambridge University Press (2010)
21. Hou, J., Holmes, M., Lai, W., Mow, V.: Boundary conditions at the cartilage-synovial fluid interface for joint lubrication and theoretical verifications. Journal of Biomechanical Engineering **111**(1), 78–87 (1989)
22. Izadifar, Z., Chen, X., Kulyk, W.: Strategic design and fabrication of engineered scaffolds for articular cartilage repair. Journal of Functional Biomaterials **3**(4), 799–838 (2012)

23. Jou, D., Casas-Vázquez, J., Lebon, G.: Extended irreversible thermodynamics. Springer (1996)
24. Klika, V.: Comparison of the effects of possible mechanical stimuli on the rate of biochemical reactions. The Journal of Physical Chemistry B **114**(32), 10567–10572 (2010)
25. Klika, V.: A guide through available mixture theories for applications. Critical reviews in solid state and materials sciences **39**(2), 154–174 (2014)
26. Klika, V., Gaffney, E.A., Chen, Y.C., Brown, C.P.: An overview of multiphase cartilage mechanical modelling and its role in understanding function and pathology. Journal of the Mechanical Behavior of Biomedical Materials **62**, 139–157 (2016)
27. Klika, V., Grmela, M.: Coupling between chemical kinetics and mechanics that is both nonlinear and compatible with thermodynamics. Physical Review E **87**(1), 012141 (2013)
28. Klika, V., Grmela, M.: Mechano-chemical coupling in Belousov-Zhabotinskii reactions. The Journal of Chemical Physics **140**(12), 124110 (2014)
29. Klika, V., Krause, A.L.: Beyond Onsager–Casimir relations: shared dependence of phenomenological coefficients on state variables. The Journal of Physical Chemistry Letters **9**(24), 7021–7025 (2018)
30. Klika, V., Kubant, J., Pavelka, M., Benziger, J.B.: Non-equilibrium thermodynamic model of water sorption in Nafion membranes. Journal of Membrane Science **540**, 35–49 (2017)
31. Klika, V., Maršík, F.: Coupling effect between mechanical loading and chemical reactions. The Journal of Physical Chemistry B **113**(44), 14689–14697 (2009)
32. Klika, V., Pavelka, M., Benziger, J.B.: Functional constraints on phenomenological coefficients. Physical Review E **95**(2), 022125 (2017)
33. Klika, V., Pavelka, M., Vágner, P., Grmela, M.: Dynamic maximum entropy reduction. Entropy **21**(7), 715 (2019)
34. Klika, V., Pérez, M.A., García-Aznar, J.M., Maršík, F., Doblaré, M.: A coupled mechano-biochemical model for bone adaptation. Journal of Mathematical Biology **69**(6–7), 1383–1429 (2014)
35. Klika, V., Votinská, B.: Towards systematic approach to boundary conditions in multiphasic and mixture models: Maximum entropy principle estimate. International Journal of Engineering Science (2021). Submitted
36. Klika, V., Whiteley, J.P., Brown, C.P., Gaffney, E.A.: The combined impact of tissue heterogeneity and fixed charge for models of cartilage: the one-dimensional biphasic swelling model revisited. Biomechanics and Modeling in Mechanobiology **18**(4), 953–968 (2019)
37. Krause, A.L., Klika, V., Woolley, T.E., Gaffney, E.A.: From one pattern into another: analysis of Turing patterns in heterogeneous domains via WKBJ. Journal of the Royal Society Interface **17**(162), 20190621 (2020)
38. Krishna, R., Wesselingh, J.: The Maxwell-Stefan approach to mass transfer. Chemical Engineering Science **52**(6), 861–911 (1997)
39. Lai, W., Hou, J., Mow, V.: A triphasic theory for the swelling and deformation behaviors of articular cartilage. Journal of Biomechanical Engineering **113**(3), 245–258 (1991)
40. Lebon, G., Jou, D., Casas-Vázquez, J.: Understanding non-equilibrium thermodynamics, vol. 295. Springer (2008)
41. Málek, J., Souček, O.: Theory of mixtures. Lecture notes (2019). http://geo.mff.cuni.cz/~soucek/vyuka/materials/theory-of-mixtures/theory_of_mixtures-lecture-notes.pdf. Accessed on 23 Oct,2020
42. Massoudi, M.: On the importance of material frame-indifference and lift forces in multiphase flows. Chemical Engineering Science **57**(17), 3687–3701 (2002)
43. Mow, V.C., Kuei, S., Lai, W.M., Armstrong, C.G.: Biphasic creep and stress relaxation of articular cartilage in compression: theory and experiments. Journal of Biomechanical Engineering **102**(1), 73–84 (1980)
44. Müller, I., Ruggeri, T.: Rational extended thermodynamics, vol. 37. Springer Science & Business Media (2013)
45. Murdoch, A.: On material frame-indifference, intrinsic spin, and certain constitutive relations motivated by the kinetic theory of gases. Arch. Ration. Mech. Anal **83**(2), 183 (1983)

46. Murray, J.D.: Mathematical biology: I. An introduction, vol. 17. Springer Science & Business Media (2007)
47. Onsager, L.: Reciprocal relations in irreversible processes. I. Phys. Rev. **37**, 405–426 (1931). https://doi.org/10.1103/PhysRev.38.2265
48. Onsager, L.: Reciprocal relations in irreversible processes. II. Phys. Rev. **38**, 2265–2279 (1931). https://doi.org/10.1103/PhysRev.38.2265
49. Öttinger, H.C., Grmela, M.: Dynamics and thermodynamics of complex fluids. II. Illustrations of a general formalism. Physical Review E **56**(6), 6633 (1997)
50. Pavelka, M., Klika, V., Esen, O., Grmela, M.: A hierarchy of Poisson brackets in non-equilibrium thermodynamics. Physica D: Nonlinear Phenomena **335**, 54–69 (2016)
51. Pavelka, M., Klika, V., Grmela, M.: Time reversal in nonequilibrium thermodynamics. Physical Review E **90**(6), 062131 (2014)
52. Pavelka, M., Klika, V., Grmela, M.: Multiscale thermo-dynamics: introduction to GENERIC. Walter de Gruyter GmbH & Co KG (2018)
53. Pavelka, M., Maršík, F., Klika, V.: Consistent theory of mixtures on different levels of description. International Journal of Engineering Science **78**, 192–217 (2014)
54. Pavelka, M., Peshkov, I., Klika, V.: On Hamiltonian continuum mechanics. Physica D: Nonlinear Phenomena **408**, 132510 (2020)
55. Pavelka, M., Peshkov, I., Sỳkora, M.: A note on construction of continuum mechanics and thermodynamics. In: Continuum Mechanics, Applied Mathematics and Scientific Computing: Godunov's Legacy, pp. 283–289. Springer (2020)
56. Peshkov, I., Romenski, E.: A hyperbolic model for viscous Newtonian flows. Continuum Mechanics and Thermodynamics **28**(1–2), 85–104 (2016)
57. Rajagopal, K.: On boundary conditions for fluids of the differential type. In: Navier Stokes Equations and Related Nonlinear Problems, pp. 273–278. Springer (1995)
58. Saffman, P.G.: On the boundary condition at the surface of a porous medium. Studies in Applied Mathematics **50**(2), 93–101 (1971)
59. Souček, O., Heida, M., Málek, J.: On a thermodynamic framework for developing boundary conditions for Korteweg-type fluids. International Journal of Engineering Science **154**, 103316 (2020)
60. Souček, O., Orava, V., Málek, J., Bothe, D.: A continuum model of heterogeneous catalysis: Thermodynamic framework for multicomponent bulk and surface phenomena coupled by sorption. International Journal of Engineering Science **138**, 82–117 (2019)
61. Souček, O., Průša, V., Málek, J., Rajagopal, K.: On the natural structure of thermodynamic potentials and fluxes in the theory of chemically non-reacting binary mixtures. Acta Mechanica **225**(11), 3157–3186 (2014)
62. Waldmann, L.: Reciprocity and boundary conditions for transport-relaxation equations. Zeitschrift für Naturforschung A **31**(12), 1439–1450 (1976)
63. Whiteley, J.P., Gaffney, E.A.: Modelling the inclusion of swelling pressure in a tissue level poroviscoelastic model of cartilage deformation. Mathematical Medicine and Biology: A Journal of the IMA (2020)

Modeling Biomechanics in the Healthy and Diseased Heart

Renee Miller, David Marlevi, Will Zhang, Marc Hirschvogel,
Myrianthi Hadjicharalambous, Adela Capilnasiu, Maximilian Balmus,
Sandra Hager, Javiera Jilberto, Mia Bonini, Anna Wittgenstein,
Yunus Ahmed, and David Nordsletten

Abstract The hierarchical construction of the myocardium plays a pivotal role in the biomechanics of the heart muscle and the resulting flow of blood. In disease, the construction of the heart remodels, altering the structure of the tissue from the subcellular level all the way to the whole organ. Elucidating the impact of these fundamental alterations on the biomechanics of the heart presents challenges to diagnosis, therapy planning, and treatment. Computational modeling provides an innovative tool, enabling the simulation of complex biomechanics that capture the complexity of tissue, its growth and remodeling, and the resulting blood flow. In this chapter, we review the key ways that computational models can address challenging biomechanical questions in the heart and how these tools can change the way treatment is approached across a range of heart diseases.

R. Miller · M. Hirschvogel · A. Capilnasiu · M. Balmus · S. Hager · A. Wittgenstein
Kings College London, London, UK
e-mail: renee.miller@kcl.ac.uk; marc.hirschvogel@kcl.ac.uk; adela.capilnasiu@kcl.ac.uk;
maximilian.balmus@kcl.ac.uk; sandra.hager@kcl.ac.uk; anna.wittgenstein@kcl.ac.uk

D. Marlevi
Massachusetts Institute of Technology, Cambridge, MA, USA
e-mail: marlevi@mit.edu

W. Zhang · J. Jilberto · M. Bonini · Y. Ahmed
University of Michigan, Ann Arbor, MI, USA
e-mail: willwz@umich.edu; jilberto@umich.edu; mbonini@umich.edu; yuah@med.umich.edu

M. Hadjicharalambous
University of Cyprus, Nicosia, Cyprus
e-mail: hadjicharalambous.myrianthi@ucy.ac.cy

D. Nordsletten (✉)
University of Michigan, Ann Arbor, MI, USA
Kings College London, London, UK
e-mail: david.nordsletten@kcl.ac.uk; nordslet@umich.edu

1 Introduction

The prevalence of cardiac and cardiovascular diseases has continuously increased worldwide [27], making research into heart physiology and pathophysiology a continued area of focus. Normal cardiac physiology depends on the anatomy, structure, and function of the heart from the fundamental cellular building blocks to the whole organ itself. Diseases of the heart manifest into changes that occur across spatial scales, often altering the cells, extracellular matrix, local tissue structure, and whole organ anatomy. These phenomena, working to remodel the heart, often occur in an effort to maintain blood flow to the body—but can, in turn, result in further exacerbation of disease. Understanding the influence of these changes on biomechanical function of tissue and the evolution of remodeling itself are critical goals that are important for understanding and treating heart disease.

Current research in diseases of the heart is diverse, with efforts in tissue engineering, medical imaging, devices, pharmaceuticals, and computational models looking to provide new approaches for early and accurate diagnosis, development of new therapies, and improvement of long-term treatment strategies. Across all targets, computational models of cardiac biomechanics provide a unique tool. Accurate cardiac models allow for rapid design and testing of potential cardiac therapies *in silico*, avoiding or reducing animal experimentation. They allow for a controlled setting in which to perturb individual factors (e.g., geometry, collagen content, tissue properties, etc.) to investigate their impact on cardiac function. Additionally, cardiac modeling provides quantitative measures, unavailable through imaging alone, of cardiac function and behavior such as localized stresses. Computation in combination with novel imaging methods provides the ability to create patient-specific biomechanical models describing the mechanics of myocardium and blood flow and providing a significant step toward personalized treatment planning.

In this chapter, we explore the biomechanical action of the heart, focusing on how mathematics and computational models can be used to capture the essence of cardiac function. Section 2 begins by providing an overview of the anatomy and structure of the human heart and how these components influence physiological function. Section 3 introduces the continuum mechanics framework used for developing constitutive equations to describe the behavior of myocardial tissue, incorporating tissue composition and structure from the preceding section. As an adaptive organ under constant change, Sect. 4 introduces mathematical approaches to describe growth and remodeling (G&R) of cardiac tissue. In Sect. 5, governing principles of blood flow, hemodynamics, and modeling are reviewed. Finally, Sect. 6 presents four diverse pathologies and discusses how previously presented modeling techniques have been applied to study disease and investigate treatments.

2   Structure and Mechanical Function in the Heart

To provide a foundation for future sections detailing mechanical modeling approaches of the tissue and blood flow in the heart, this section will explain the anatomy and structure beginning at the organ level. Explanations of the primary cell types and constituents that make up the extracellular matrix give further understanding of the building blocks of heart tissue. The hierarchical structure of muscle as well as collagen fibers will be described in order to better understand their role in tissue behavior. Finally, this section will present the phases of the cardiac cycle, relevant indices for measuring cardiac function, and mechanisms for how the heart adapts to hemodynamic changes.

2.1   Organ Structure

A labelled diagram of the bisected heart is presented in Fig. 1, showing cardiac anatomy from apex (lowermost point) to base (uppermost point). There are four chambers in the heart—two muscular ventricles (left ventricle: LV, right ventricle: RV) and two pliant atria (left atrium: LA, right atrium: RA). The heart can also be divided into low-pressure (right) and high-pressure (left) units, separated by the septum. The atria work primarily as reservoirs that pool blood and facilitate ventricular filling, while the right and left ventricles pump blood out of the heart

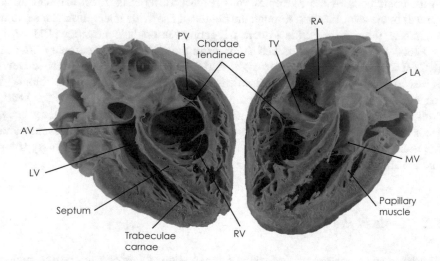

Fig. 1 Image of a plastinated human heart (adapted with permission from Visible Heart Laboratories, University of Minnesota/©Medtronic [1]) cut into two halves along the long axis illustrating the primary structures, including the ventricles, atria, valves, and endocardial muscular structures (e.g., trabeculae carneae). LV: left ventricle, RV: right ventricle, AV: aortic valve, PV: pulmonary valve, TV: tricuspid valve, MV: mitral valve, LA: left atrium, RA: right atrium

into the pulmonary and systemic vasculature, respectively. The tissue characteristics of these chambers are tailored to facilitate their respective functions. For example, because the left side of the heart needs to generate higher pressures for driving blood through the entire body, the LV walls tend to be thicker [217]. In contrast, the RV contains thinner walls with less compact, more trabeculated myocardium, capable of filling under low pressures while providing sufficient pulmonary pressures. Similarly, the walls of the LA are thicker than the RA due to higher pressures generated, whereas the RA holds a greater volume of blood [460].

Separating the ventricles from atria and arterial outflows are four valves, promoting unidirectional flow through the heart (mitral valve: MV, tricuspid valve: TV, aortic valve: AV, pulmonary valve: PV). The atrioventricular valves (TV and MV) are designed to open under low differential pressures and are reinforced by papillary muscles (via the chordae tendineae) to avoid reverse flow when ventricles contract [191]. Ventricular outflow tracts are opened and closed by semilunar valves (PV and AV), which are thick, open under larger pressures and can withstand significant pressure differences. All valves integrate into the heart through stiff, thin annular rings that resist dilation.

The heart also has structures that control the electrical activation and, consequently, the complex mechanical pattern of muscle contraction and relaxation. The electrical signal starts at the sinoatrial (SA) node located in the RA and spreads through both atrial chambers, triggering atrial contraction. During this process, the signal travels to the atrioventricular (AV) node, located in the center of the four chambers. The cells in this node—the only physiological electrical conduit for atrioventricular coupling—have a slower conduction velocity that causes a delay in the contraction of the ventricles with respect to the atria. Electrical activation spreads in the ventricles first through the bundle of His in the interventricular septum and further through the Purkinje fibers, triggering ventricular contraction [158, 177].

Blood flow to the heart tissues stems from the right and left coronary arteries (RCA/LCA), supplying blood to right and left sides of the heart, respectively. A vascular network, originating from these two main arteries, supplies blood to the myocardium. The large arteries are located in the outer surface of the heart, while smaller arteries penetrate the heart walls feeding coronary capillary beds. Deoxygenated blood is returned to the RA through a network of veins that surround the heart. The main path for venous blood return is by way of the coronary sinus and the anterior cardiac veins.

2.2   Cells in the Heart

The heart is comprised of various cell types—depending on the specific region of tissue—that maintain its structure and function, see [6, 286, 287, 389, 422] for reviews. Myocardial muscle, the main structural component of the heart, is comprised of cardiac myocytes, fibroblasts, and pacemaker cells that contribute to muscle contraction, creation of an extracellular matrix, and electrical stimulation, respectively. While these constituents make up the majority of the tissue in the heart, other cell types, serving specialized functions, will also be discussed briefly.

Myocytes are the primary source of contraction in the heart, providing the force to eject blood through the heart chambers. The most abundant cell types (by volume) in the myocardium, myocytes, are approximately cylindrical in shape, with a length of $80\,\mu\text{m}$–$100\,\mu\text{m}$ and a diameter of $10\,\mu\text{m}$–$20\,\mu\text{m}$. Most cardiomyocytes have more than one nucleus [378] and, apart from the usual organelles of a cell (e.g., sacroplasmic reticulum, etc. [6]), contain numerous mitochondria (sources of energy through adenosine triphosphate, ATP) and myofibrils (sources of contraction). Myofibrils, a structural component unique to muscle cells, run along the length of the myocyte, enable contraction, and give myocytes their striated appearance [341]. Myofibrils are comprised of sarcomeres arranged in series (see Fig. 2f). Within the sarcomeres, the interaction between actin (thin) and myosin (thick) filaments sliding relative to one another through crossbridge cycling is the primary mechanism underpinning cardiac muscle contraction. Z disks delineate the beginning and ending of each sarcomere and act as attachment points for thin actin filaments. Thick filaments have heads that attach to thin filaments as illustrated in Fig. 2g.

Fibroblasts cells are the primary producer of connective tissue in the heart. In contrast to myocytes, fibroblasts are flat and elongated cells and are the most common cell type (by number) in the heart [47, 473]. Connective tissue, composed of collagen fibers and other proteins, makes up the myocardial extracellular matrix (ECM) [59, 455]. Fibroblasts not only express proteins for collagen assembly, but are also involved in the degradation of the ECM [135]. These regulatory processes can be triggered through a sensitivity to mechanical stresses and strains [59, 111] as well as by gene expression [422]. In cardiovascular diseases, which result in increased collagen deposition, changes in the phenotype of fibroblasts can be observed, consistent with their role in remodeling processes [462].

As mentioned in Sect. 2.1, the electrical activation of the heart, which induces contraction, is triggered in the SA node. More specifically, it is triggered by specialized cardiomyocytes called pacemaker cells that automatically generate electrical impulses [214]. Morphologically, pacemaker cells have fewer myofibrils, which influence the striated organization as well as its contraction pattern. They also have a larger surface area than typical myocytes. Other structures that make up the electrical conduction system (e.g., Purkinje fibers, bundle of His) are formed of secondary pacemaker cells, which have even fewer myofibrils when compared to pacemaker cells found in the SA node [253]. More detailed information on the mechanisms controlling the behavior of pacemaker cells can be found in [243, 275, 338, 388].

Other cell types exist that carry out specialized functions in the heart. For example, endothelial cells (EC) line the chambers of the heart as well as the major vessels, creating a barrier between the blood and the chamber walls. They are also constituents of the ECM [212, 422] and are associated with collagen synthesis and remodeling processes [199]. ECs in the valves act as mechanotransducers, triggering structural changes in response to mechanical strains [422]. Immune cells (e.g., macrophages, T-cells) are also present and respond to antigens/pathogens and are a part of the inflammatory processes. The heart is also innervated by different nerve cells that act to modulate cardiac function through autonomic pathways [178, 286]. Efforts in cardiac tissue engineering have shown the importance of understanding the structural and functional role of each cell type [339].

2.3 Myocardial Tissue Structure

The myocardium exhibits a hierarchical order, as shown in Fig. 2, ensuring the orchestrated contraction of the heart [473]. Cardiomyocytes connect in series via intercalated discs, which enable synchronized contraction, forming supracellular myofibers. Furthermore, these fibers form fiber bundles that are interwoven with capillaries and are embedded in the ECM (see Fig. 2c). The main structural component of the ECM in cardiac tissue is collagen, playing an important role in cardiac mechanics as a scaffolding for cells [135]. The most common types of collagen molecules are high-strength type I (75 − 80%) and highly deformable type III (10 − 20%) with varying amounts depending on the location and function within in the heart [273].

Collagen fibers form the scaffold required to support the myofibers and capillaries, maintaining their arrangement and distributing loads across the tissue. Cardiac myocytes are embedded in a fine and flexible network of collagen fibers connecting neighboring cells. This, dominantly type III collagen network, is called endomysial collagen [135, 212, 273]. Perimysial collagen, on the other hand, groups myocytes and capillaries into bundles called laminar sheets, which typically consist of between four to six layers of myocytes. Perimysial collagen, interleaving laminar sheets, forms the so-called cleavage planes (see Fig. 2b). Epimysial collagen, consisting predominantly of type I collagen, forms a thick collagen layer around the myocardium and is a connecting layer to the most superficial layers of the endo- and epicardium [212].

The laminar structure of the myocardium confers a local orientation within the tissue. Elongated myocytes are aligned end to end, forming muscle fibers. Muscle fibers are then tethered together forming sheet structures that are bound together through perimysial collagen layers. These local directions led to the definition of three material orientations in the myocardium: fiber, sheet, and the resulting orthogonal normal direction [99]. These local microstructural directions are often assumed to be orthonormal and used to define a coordinate transformation from global coordinates into the local microstructural coordinate frame.

On the tissue scale, the local orthotropy of myocardium is leveraged to create global structures that enable efficient muscle contraction of the heart. Cardiac fiber orientations can be described by a helical angle, measured with respect to the circumferential direction. Early studies [464] revealed a transmural change in helical angle from approximately $-60°$ at the subepicardium to $+60°$ at the subendocardium. Detailed imaging of the microstructure, obtained using scanning electron microscopy [274] and confocal microscopy [273], showed regular patterns

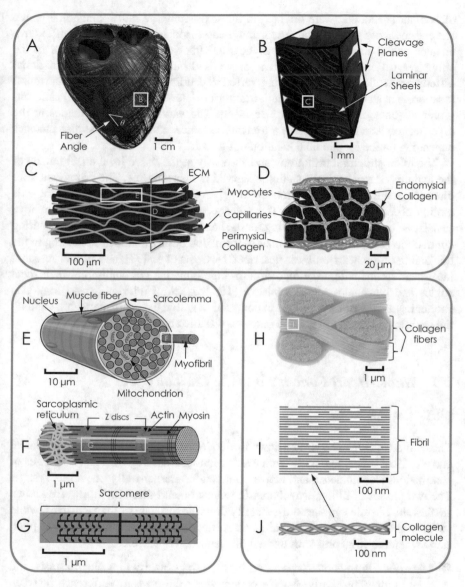

Fig. 2 Illustration of multiscale cardiac structure. (**a**) Biventricular anatomy with fiber orientation, (**b**) tissue block illustrating laminar structure, (**c**) local structural arrangement of myocytes, capillaries, and ECM, (**d**) myocytes and capillaries in endomysial and perimysial collagen, (**e**) myofiber cut to view internal structures, (**f**) myofibril illustrating individual components, (**g**) sarcomere made of actin (thin) and myosin (thick) filaments, (**h**) bonds between multiple collagen fibrils form collagen fibers, (**i**) collagen fibril formed from staggered collagen molecules, (**j**) three amino polypeptide strands form a single collagen molecule. ((**e**) and (**f**) reproduced from Anatomy and Physiology under Creative Commons Attribution License 4.0 from OpenStax. ©2016)

of laminar sheets that were oriented in approximately the radial–tangential plane. Recent advances in diffusion tensor magnetic resonance imaging (DTMRI) have enabled for quantification of the myocardial fiber and sheet orientations in the entire heart. Results from DTMRI in an excised human heart illustrated regional differences in fiber orientations that varied depending on not only the transmural location, but also the circumferential (anterior, lateral, posterior, septal) and longitudinal (base, mid, apex) positions [408]. The variation of sheet angles in the myocardium was shown to have a transmural concave curvature that also changes depending on the longitudinal location in the heart.

The microstructural orthotropy and assembly at the tissue level enable many of the features of mechanical function observed in heart muscle. The prominent fiber direction and transmural variation are largely responsible for torsion and long-axis motion observed in the muscle. The presence of cleavage planes enables relative translation between sheets, allowing for the heart wall to thicken significantly during contraction [273, 274, 393]. Laminar deformation during contraction leading to the thickening of the heart wall was described by Costa et al. [77], and further research into shear strains within the myocardium highlights the correlation between sheet angles and mechanics of the ventricle [10, 427]. Wall thickening and long-axis shortening are the primary drivers of contraction within the heart chambers, yielding the significant changes in volume between diastole and systole.

2.4   Whole-Heart Function and the Cardiac Cycle

2.4.1   Cardiac Cycle

The cardiac cycle can be broken down into four phases that describe the mechanical and electrical events that occur with every heartbeat (see Fig. 3). The two phases of contraction, isovolumetric and active contraction, are generally known as systole. The two phases of filling, isovolumetric relaxation and diastolic filling, are known as diastole. Systole and diastole generally describe the state of the ventricles, while the phases of contraction for the atria are shifted in time. The following description starts when the atrial pressures exceed the ventricular pressures:

■ **Diastolic filling** is often described in three stages: early diastole, diastasis, and atrial contraction. Early diastole occurs when the pressure difference between the atria (high) and ventricles (low) causes the mitral and tricuspid valves to open and blood to flow into the ventricles. As blood flows into the ventricles, the volume and pressure in the ventricles begin to increase. Diastasis occurs during mid diastole when the atrial blood flow in the ventricles decreases as a result of the rising ventricular pressures. The final phase in diastolic filling is atrial contraction, stimulated by the electrical impulse sent through the atria from the SA node. Atrial systole increases intra-atrial pressure, forcing more blood into the ventricles. As atrial contraction ends, decreasing atrial pressures cause both the MV and TV to close, ending diastolic filling. The volume and pressure in the ventricles at the end of filling are known as end-diastolic volume (EDV) and end-diastolic pressure (EDP), respectively.

■ **Isovolumetric contraction** begins as signals from the atrioventricular node spread through the Purkinje network and begin stimulating muscle contraction. As activation spreads through the muscle, mechanical contraction of myocytes across the heart yields continually increasing pressures in both ventricular chambers. Both AV and PV remain closed owing to the higher systemic and pulmonary pressures of the aorta and pulmonary trunk, respectively. The TV and MV also remain closed, retaining their competence in part due to the papillary muscles and chordae tendineae that act to prevent prolapse. As the name implies, the volume of blood in the ventricles remains constant, while contraction initiates.

■ **Active contraction** initiates as the chamber pressure of both ventricles begins to exceed systemic and pulmonary pressures, resulting in the opening of the AV and PV. As ventricular contraction continues, myocytes shorten, walls thicken, the long axis of the heart shortens, and the heart twists around the long axis. During active contraction, the pressure in the ventricles is at their highest. While active contraction is marked by the ejection of blood from ventricular chambers, blood flow from the left and right pulmonary veins and venae cavae begins to fill both atrial chambers. In mid systole, ventricular flow peaks and then slowly declines as the heart reaches its maximal contractile state. The pressure in the ventricles drops as blood continues to be ejected. Systole ends with a decline in ventricular pressure and eventual closure of both aortic and pulmonary valves. The volume and pressure in the ventricles at the end of contraction are known as end-systolic volume (ESV) and end-systolic pressure (ESP), respectively.

■ **Isovolumetric relaxation** denotes the period of relaxation in the ventricular chambers, while chamber pressures fall below systemic and pulmonary pressures yet remain larger than those in the atria. The volume of blood in the ventricles remains constant throughout this phase, but the pressure in the ventricles continually decreases as the myocardium relaxes. Atrial chambers continually fill, increasing volume and chamber pressures due to passive stretch of the atrial walls. Once the pressure in the ventricles drops below the pressure in the atria, the MV and TV valves open, ending IVR and re-starting the cardiac cycle [373].

2.4.2 Cardiac Functional Metrics

There are various metrics used to describe and quantify cardiac function. One measurement, the stroke volume (SV), is defined as the volume of blood pumped by the ventricle with each contraction (SV = EDV - ESV) [447]. The ejection fraction (EF) is the percentage of how much blood the ventricle pumps out with each contraction as a fraction of the EDV, e.g., EF = SV/EDV x 100%. The EF is an important marker of systolic function (healthy 55% > EF < 65%) [447]. In some pathologies, such as hypertrophic cardiomyopathy, EF is normal but the SV is depressed.

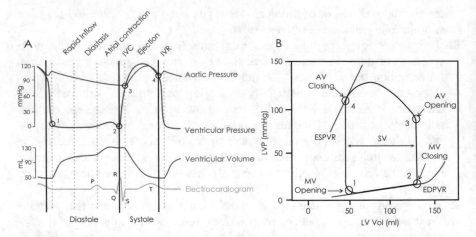

Fig. 3 (**a**) The Wiggers diagram illustrates the LV and aortic pressures along with the ventricular volume and a representative electrocardiogram. Each phase discussed in the text is delineated by dashed lines, whereas diastole and systole are shown by solid lines. (**b**) An example pressure–volume loop for the LV is shown, highlighting three functional metrics: SV, EDPVR, and ESPVR. Valve opening and closing times are drawn in both figures as open circles. It should be noted that these traces represent pressures and volumes in the LV. Analogous traces and curves can be drawn for the RV and both atrial chambers

Stroke work (SW) is a metric that quantifies the work done by the ventricles to eject a volume of blood for each contraction. It is represented as the area within the pressure–volume loop (see Fig. 3b) but is often estimated as the product of SV and mean systolic pressure during ejection. Cardiac output (CO) quantifies the volume of blood being pumped by the heart per unit of time and is the product of heart rate (HR) and SV. The slopes of the end-diastolic pressure–volume relationship (EDPVR) and the end-systolic pressure–volume relationship (ESPVR) represent the global compliance and contractility of the heart, respectively (see Fig. 3b) [239].

2.4.3 Cardiac Adaptation

The Frank–Starling mechanism is the ability of the heart to change its force of contraction in response to changes in pressure and volume in order to preserve the EF [217]. An increase in preload (e.g., elevated left ventricular EDV) results in a greater force of contraction and an increased SV. An increase in afterload (e.g., high systemic or pulmonary pressure) results in an increase in the force generated by the heart to overcome downstream pressure [29]. This ability of heart to adapt to alterations in pressure and volume is advantageous for exercise and short-term hemodynamic changes but can be detrimental in chronic pathological states of elevated pressure and/or volume. It can result in sustained alterations in force generation, leading to tissue remodeling (hypertrophy) and ultimately heart failure [447].

3   Modeling Passive Cardiac Tissue Mechanics

Measuring the complex underlying structures of the heart at the micro- and macro-level (Sect. 2) enables exploration of cardiac biomechanics through modeling. While closed-form approaches for understanding muscular function of the heart (e.g. Yin et al. [526]) can provide insights, modeling biomechanics through continuum approaches such as using finite elasticity can provide a framework for numerical predictions that incorporate many of the hierarchical components of myocardial tissue and structural components of the heart itself. Much of this complexity is embedded through constitutive models, which provide a bridge between experimental data on stress–strain responses and appropriate continuum mechanics representations. While cardiac biomechanics has traditionally been described using hyperelastic constitutive models, viscoelastic modeling approaches for myocardial tissue have gained more recent attention. Combining insights from cardiac structure, continuum approaches, constitutive models, and computational methods, it becomes possible to explore the biomechanical world of the heart through modeling.

3.1   Continuum Mechanics

Continuum mechanics is founded on the concept that a material (be it fluid, solid, or mixture) and its properties can be approximated by fields that are well defined almost everywhere [139, 170, 299]. This is referred to as the continuum assumption, which considers the scale at which discrete microstructural variations appear to much smaller than the smallest scales of interest. This enables the averaging of material characteristics, transforming the discrete nature of materials into well-defined point-wise fields. Quantities such as mass, momentum, and angular momentum can all be moved from the individual constituent molecules to continuous fields over which we may apply limits and differential calculus. Conservation laws can be derived to enable computational simulations of heart tissue, detailing its movement, deformation (kinematics), and stress responses (kinetics). In this section, these fundamental concepts will be briefly reviewed. For more detailed overviews, refer to [45, 201, 364, 458].

3.1.1   Kinematics

To describe the motion of a continuum body in d-dimensional space, we define the region occupied by the body at rest (the reference domain) as $\Omega_0 \subset \mathbb{R}^d$ (see Fig. 4). Reference coordinates $\mathbf{X} \in \Omega_0$ describe the undeformed position of material particles within the body. In response to deformations, the material particles move to the coordinates $\mathbf{x} \in \Omega(t)$ (the physical or deformed domain) at some time t. Assuming that the reference coordinates, $\mathbf{X}$, and physical coordinates, $\mathbf{x}$, are related

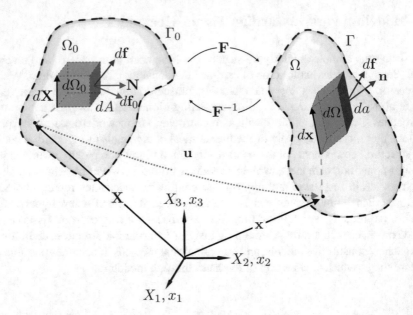

Fig. 4 The mapping of local coordinates and applied forces from an initial reference configuration (Ω_0, with boundary Γ_0, and the reference material coordinates X_i), to a deformed physical configuration (Ω, with boundary Γ, and the physical coordinates, x_i). The deformation gradient tensor, **F**, describes the local change of any text vector d**X** to d**x** as the result of the applied deformation. Illustrated in the domain is a differential volume, illustrating the mapping of forces (reference, d**f**$_0$, and physical, d**f**) along with normals (reference, **N**, and physical, **n**) and areas (reference, dA, and physical, da)

by a continuous displacement field $\mathbf{u}(\mathbf{X}, t)$, i.e., $\mathbf{x}(\mathbf{X}, t) = \mathbf{u}(\mathbf{X}, t) + \mathbf{X}$, we can define the Jacobian of the mapping, also known as the deformation gradient tensor, as

$$\mathbf{F} = \frac{\partial \mathbf{x}}{\partial \mathbf{X}} = \nabla_{\mathbf{X}} \mathbf{u} + \mathbf{I}, \quad F_{ij} = \frac{\partial u_i}{\partial X_j} + \delta_{ij}. \tag{1}$$

F characterizes the local deformation at a material point, **X**, i.e., relating every infinitesimal line segment emanating from **X** to its counterpart in Ω by a linear transformation, d**x** = **F**d**X** [45, 201].[1] The magnitude of d**x** can be computed as $\|\mathrm{d}\mathbf{x}\|^2 = \mathrm{d}\mathbf{X}^T \mathbf{C} \mathrm{d}\mathbf{X}$, where $\mathbf{C} = \mathbf{F}^T \mathbf{F}$ is the right Cauchy–Green tensor. In a similar way, we can express the magnitude of d**X** as $\|\mathrm{d}\mathbf{X}\|^2 = \mathrm{d}\mathbf{x}^T \mathbf{B}^{-1} \mathrm{d}\mathbf{x}$, where $\mathbf{B} = \mathbf{F}\mathbf{F}^T$ is the left Cauchy–Green tensor. Both **C** and **B** are tensors providing information about the local and directionally dependent stretch behavior of the material. Another commonly used kinematic quantity is the Green–Lagrange strain $\mathbf{E} = \frac{1}{2}(\mathbf{C} - \mathbf{I})$.

---

[1] $\nabla = \sum_i \mathbf{e}_i \partial/\partial x_i$ is the gradient operator in physical coordinates, $\nabla_{\mathbf{X}} = \sum_i \mathbf{e}_i \partial/\partial X_i$ is the gradient operator in the reference coordinates, and $\mathbf{e}_i$ is the set of rectangular Cartesian unit vectors.

The determinant $J = \det(\mathbf{F})$ quantifies the local change in volume, while the local change in surface area is given by Nanson's formula [364],

$$\mathrm{d}a\,\mathbf{n} = J\mathbf{F}^{-T}\mathrm{d}A\,\mathbf{N},$$

where $\mathbf{N}$ and $\mathbf{n}$ are the surface normal vectors in the reference and current configuration, respectively (see Fig. 4).

For developing constitutive models, coordinate independence and rigid body invariance are essential [364, 458]. The use of tensor invariants of the right Cauchy–Green tensor, $\mathbf{C}$, is often considered as these ensure objectivity. The principal invariants are common, but derivatives that span the same space may also be used. For example, Bonet et al. [45] introduced the following three invariants of the right Cauchy–Green strain tensor for modeling isotropic materials:

$$I_1 = \sum_{i=1}^{d} \lambda_i = \mathrm{tr}(\mathbf{C}) = \mathbf{C} : \mathbf{I},$$

$$I_2 = \sum_{i=1}^{d} \lambda_i^2 = \mathrm{tr}(\mathbf{C}\mathbf{C}^T) = \mathbf{C} : \mathbf{C},$$

$$I_3 = \prod_{i=1}^{d} \lambda_i = \det(\mathbf{C}),$$

where $\lambda_i, i = 1\ldots d$, are the eigenvalues of $\mathbf{C}$.[2,3,4] To model anisotropy, additional pseudo-invariants are introduced for directional dependence [200],

$$I_4 = \mathcal{I}(\mathbf{C}, \mathbf{a}, \mathbf{a}), \quad I_5 = \mathcal{I}(\mathbf{C}^\mathsf{T}\mathbf{C}, \mathbf{a}, \mathbf{a}), \quad I_8 = \mathcal{I}(\mathbf{C}, \mathbf{a}, \mathbf{b}), \tag{2}$$

where $\mathbf{a}$ and $\mathbf{b}$ are the orientation vectors (e.g., the local direction of fibers or sheets) and $\mathcal{I}(\mathbf{A}, \mathbf{a}, \mathbf{b}) = \mathbf{A} : \mathrm{sym}(\mathbf{a} \otimes \mathbf{b})$.[5,6]

Many models consider splitting volumetric changes from distortional changes within the material. In this case, isochoric definitions of the deformation gradient (e.g., $\tilde{\mathbf{F}} = J^{-1/d}\mathbf{F}$) are commonly used with corresponding changes to stretch tensors. In this case, the isochoric invariants become

---

[2] I_2 is related to the standard second invariant, $II_\mathbf{C} = \lambda_1\lambda_2 + \lambda_1\lambda_3 + \lambda_2\lambda_3$, by $I_2 = I_1^2 - 2II_\mathbf{C}$.

[3] The Frobenius product between tensors $\mathbf{A}, \mathbf{B} \in \mathbb{R}^{d \times d}$ is defined as $\mathbf{A} : \mathbf{B} = \sum_i \sum_j A_{ij} B_{ij}$.

[4] The trace operator on $\mathbf{A} \in \mathbb{R}^{d \times d}$ is defined as $\mathrm{tr}(\mathbf{A}) = \sum_i A_{ii}$.

[5] The dyadic product of vectors $\mathbf{a}, \mathbf{b} \in \mathbb{R}^d$ produces a tensor $\mathbf{A} \in \mathbb{R}^{d \times d}$ with components $A_{ij} = [\mathbf{a} \otimes \mathbf{b}]_{ij} = a_j b_i$.

[6] The symmetric part of a tensor $\mathbf{A} \in \mathbb{R}^{d \times d}$ is defined as $\mathrm{sym}(\mathbf{A}) \in \mathbb{R}^{d \times d}$ with components $[\mathrm{sym}(\mathbf{A})]_{ij} = \frac{1}{2}(A_{ij} + A_{ji})$.

$$\tilde{I}_1 = J^{-2/d} I_1, \quad \tilde{I}_2 = J^{-4/d} I_2, \quad \tilde{I}_3 = 1. \tag{3}$$

3.1.2 Kinetics

Stress is a tensor variable that enables quantification of the internal tractions (forces per unit area). Cauchy postulates that a force on any surface that passes through a point depends only on its unit normal $\mathbf{n}$. Considering an infinitesimally small representative volume element at a point with volume $d\Omega$ and surface area of da, the traction on each face is given by $\mathbf{t} = d\mathbf{f}/da$, where $d\mathbf{f}$ is the part of the force vector $\mathbf{f}$ that acts on the surface da (see Fig. 4c). Cauchy theorizes that the traction on all planes passing through the point can be expressed as a linear combination of three mutually orthogonal traction vectors. In other words, he proposed that there exists a stress tensor, σ, such that the traction on a plane with normal $\mathbf{n}$ is given by

$$\frac{d\mathbf{f}}{da} = \mathbf{t} = \sigma\mathbf{n}.$$

This is also known as Cauchy's stress theorem [359]. By definition, the Cauchy stress, also known as the true stress, is defined on the current deformed configuration (see Fig. 4).

While the Cauchy stress is defined relative to the physical domain, equivalent forms can also be constructed on the reference domain (see Fig. 4). These are known as the first and second Piola–Kirchhoff stresses (denoted PK1 and PK2, respectively). PK1, denoted by $\mathbf{P}$, satisfies that $\frac{d\mathbf{f}}{dA} = \mathbf{P}\mathbf{N}$ for all surfaces passing through the point on the reference configuration with normal $\mathbf{N}$ and surface area dA. This form of stress can be more convenient since $\mathbf{N}$ and dA are often more easily measured at the beginning of experiments. Mathematically, PK1 can be related to the Cauchy stress by

$$\mathbf{P} = J\sigma\mathbf{F}^{-\mathsf{T}}, \tag{4}$$

which can also be understood as the pullback of the Cauchy stress. If another pullback operation is applied, we get the definition for PK2, i.e.,

$$\mathbf{S} = J\mathbf{F}^{-1}\sigma\mathbf{F}^{-T}. \tag{5}$$

$\mathbf{S}$, given by a force vector $d\mathbf{f}_0 = \mathbf{F}^{-1}d\mathbf{f}$, is convenient to use for constitutive modeling since it is entirely defined on the reference configuration $d\Omega_0$ (Fig. 4).

3.1.3 Conservation Laws

The conservation laws describe the physical variation of quantities (e.g., mass, momentum, energy, charge) for which processes for transport, production, or destruction of the quantity are well defined. Typically, conservation laws are formulated through control volume analysis, whereby the alterations in the amount of a quantity over a unit of space-time are quantified and examined in the limit as the unit of space-time tends to zero. Here, we present the conservation laws associated with mass, linear and angular momentum:

■ **The Conservation of Mass** states that the mass of a body in a closed system can only be created or destroyed by known sources (or sinks). Thus, the mass balance equation is given by

$$\frac{D\rho J}{Dt} = \hat{\rho} J, \tag{6}$$

which states that the rate of change in mass, ρJ, is equal to the rate of mass being generated by a source $\hat{\rho}$. Here, $\frac{D}{Dt} = \frac{\partial}{\partial t} + \mathbf{v} \cdot \nabla$ defines the Lagrangian time derivative. For incompressible materials with constant density and no source of growth, the remaining mass balance equation is equivalent to $J - 1 = 0$, $\forall \mathbf{X} \in \Omega_0$.

■ **The Conservation of Linear Momentum** states that the momentum of an isolated body remains constant in the absence of external forces. The momentum balance equation is given by

$$\frac{D\rho J\mathbf{v}}{Dt} = J\nabla \cdot (\sigma) + \rho J\mathbf{b}, \tag{7}$$

where the rate of change in momentum is dependent only on the divergence of the Cauchy stress and the body force, **b**. The momentum balance can also be projected to the reference domain, resulting in the form

$$\frac{D\rho_0\mathbf{v}}{Dt} = \nabla_{\mathbf{X}} \cdot \mathbf{P} + \rho_0\mathbf{b}, \tag{8}$$

where ρ_0 is the initial density at the material point.[7,8]

■ **The Conservation of Angular Momentum** states that the angular momentum of an isolated body remains constant in the absence of external forces. The angular momentum balance equation is given by

$$\mathbf{x} \times \left[\frac{D}{Dt} (\rho J\mathbf{v}) - J\nabla \cdot (\sigma) - \rho J\mathbf{b} \right] = J \left(\mathcal{E} : \sigma^\mathsf{T} \right),$$

where $\mathcal{E}$ is the rank three Levi–Civita symbol, which is 1 if the indices are even, -1 if the indices are odd, and 0 if the any index is repeated. The extra term $\mathcal{E} : \sigma^\mathsf{T}$

---

[7] Here, we assume $\hat{\rho} = 0$, $\rho_0 = \rho J$ by the conservation of mass.
[8] ∇ and $\nabla_{\mathbf{X}}$ are related by $\nabla = \mathbf{F}^{-\mathsf{T}}\nabla_{\mathbf{X}}$.

is a consequence of the divergence theorem on $\int_{\Gamma} \mathbf{x} \times (\sigma \mathbf{n}) d\Gamma$, where Γ is the boundary of $d\Omega$. Given the conservation of linear momentum,

$$\mathcal{E} : \sigma^{\mathsf{T}} = \left(\sigma_{32} - \sigma_{23}; \; \sigma_{13} - \sigma_{31}; \; \sigma_{21} - \sigma_{12}; \; \right)^{\mathsf{T}} = \mathbf{0},$$

that is, the Cauchy stress tensor is *symmetric*, i.e., $\sigma = \sigma^{\mathsf{T}}$. Note from (5) and (4) that $\mathbf{S}$ is also symmetric but $\mathbf{FP}^{\mathsf{T}} \neq \mathbf{PF}^{\mathsf{T}}$ in general.

3.2 Stress–Strain Behavior of Myocardial Tissues

The biomechanical rheology of the human heart is a critical component of cardiac function. While its active properties lead to the contracture of the myocardium, its passive properties play a key role in cardiac pathophysiology, such as diastolic heart failure [539], heart failure with preserved ejection fraction (HFpEF) [438], and myocardial infarction [72]. This has driven over 1.5 centuries of research into characterizing the behavior of myocardium [43, 518].

Experimental studies in animals have shown that myocardial tissue exhibits complex stress–strain behavior. Early studies noted the nonlinear stress–strain response [390] of heart tissue is thought to reflect the progressive recruitment of extracellular matrix collagen and large intracellular structural proteins that resist strain. While initial studies were performed predominantly in papillary muscle under uniaxial tension, testing on portions of the heart wall under biaxial stretch [92, 465, 527] and shear [98, 100] highlighted the inherent anisotropic response of tissue. Reflecting the hierarchical nature of heart tissue, stress response to strain depends significantly on the orientation of the local microstructure. Many studies examined effects in a variety of animals, including recent work in bovine tissue [277]. However, recent works have also demonstrated the nonlinear anisotropic response of myocardial in human tissues [452, 453] (see Fig. 5). Efforts at developing constitutive relations largely paralleled available data, with early descriptions focusing on transversely isotropic approximations [167, 209, 211, 465], followed by orthotropic descriptions [76, 203, 427, 428].

Beyond nonlinearity and anisotropy, the viscoelasticity is also a common feature of the stress–strain response of the heart (see Fig. 5). Early evidence of hysteresis and nonlinear stress–strain relations was demonstrated in the canine papillary muscle by Walker [495], with evidence later shown in whole organ experiments in both feline [263] and tortoise [363] hearts. A comprehensive mechanical assessment was later performed in the rabbit papillary muscle by Pinto and Fung [390], showing stress relaxation, creep, hysteresis, and the frequency dependence of apparent stiffness. While observed in nearly all data collected in the heart, fewer works have been published that directly address these characteristics of passive myocardium.

While not the focus of this chapter, a key aspect to the biomechanics of heart tissue is its active contractile response. Unlike the passive behavior of the heart that undergoes slow variation in time due to growth and remodeling, the active

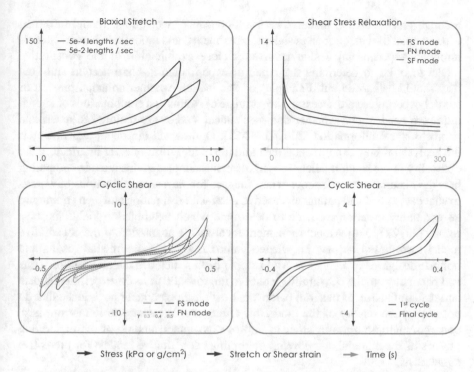

Fig. 5 Illustration of rheological tests in myocardial tissues. (Top left) Recreation of biaxial test data performed in canine myocardium under different frequencies [93]. (Top right) Recreation of shear stress relaxation in human myocardium for different microstructural orientations [453]. (Bottom left) Recreation of cyclic shear stress tests in human myocardium for different microstructural orientations [453]. (Bottom right) Recreation of cyclic shear stress tests in porcine myocardium [100]

contraction of the heart is highly dynamic and dependent upon electrochemical reactions taking place with each beating of the heart. Force generation in human myocardium, due to contraction, has shown a strong dependence on strain [482] and available calcium and ATP [172, 488]. However, while these effects are commonly discussed, they remain only a prelude to the complex interdependent reaction processes that drive contraction—including genetics, metabolic state, pH, protein isoforms, and expression, among numerous others. Hunter, McCulloch and Ter Keurs [213] provide an excellent overview of some of these key factors influencing active contraction, and Chabiniok et al. [61] provide a review of modeling efforts.

3.3 Data-Model Integration

Experimental data acquisition, constitutive modeling, and parameter estimation are closely linked, and their tight integration is crucial for the development of an accurate and predictive model. The main concepts involved are model-informed experimental design, optimal design of experiments, and model refinement. Model-informed experimental design involves a close examination of the constitutive model in order to determine the types of experiments that are needed and how they should be carried out. Examples of this include whether uniaxial, biaxial, or triaxial experiments are necessary, what range of strains and combinations of stretch ratios are needed, and whether non-mechanical studies are required to determine microstructural information [256, 417, 533]. Optimal design of an experiment is the analysis of how experimental data and the model form interact in affecting the optimality of the resulting material parameters. Examples of this include computing how the material parameter covariance changes for various combinations of experimental data and model parameters used in parameter estimation and then computing an optimality criterion for each to determine which combination yields the best results [23, 277, 536]. Model refinement involves the adaptation of the constitutive model form based on new knowledge gained from the experimental results and optimal design of experiments. An example of this includes the addition of terms and parameters to the constitutive model to improve the fit to the experimental data and the elimination of material parameters with high parameter covariance that do not impact the quality of the model fit. Often, it is necessary to iterate on these concepts multiple times in order to improve our understanding of the underlying physics of the material and develop the optimal constitutive models for predictive simulations.

3.4 Hyperelastic Modeling Approaches

The relationship between the experimentally observed stresses and strains is expressed through constitutive equations. Typically, a scalar-valued strain-energy density, Ψ, is used to describe the stored internal energy resulting from material deformations. The energy of the system must obey the second law of thermodynamics that is expressed through the Clausius–Duhem inequality (21). For an isothermal hyperelastic body, considering that all processes are reversible and thus there is no energy dissipation,

$$\mathbf{S} = 2\frac{\partial \Psi}{\partial \mathbf{C}}, \qquad S_{ij} = 2\frac{\partial \Psi}{\partial C_{ij}}.$$

Therefore, the stress is, by definition, the gradient of the strain-energy function.

In order to use hyperelasticity to model the constitutive behavior of the myocardium, some simplifying assumptions may be utilized for either experimental or computational purposes. First, compressibility of the myocardium was observed [24] as a result of intravascular blood flow [17]. In this case, isochoric invariants (3) are needed in the definition of the constitutive equation, with a bulk term to model the response to changes in volume [232]. However, most often, the incompressibility assumption is drawn, as it is considered that the myocardium is made primarily of water. Second, as presented in Sect. 3.2, myocardium has been shown to be organized into sheets with clear fiber (f), sheet (s), and normal (n) directions. As a result, constitutive models of myocardium generally describe the material as anisotropic, utilizing the pseudo-invariants for anisotropy (2). Although numerous studies have developed orthotropic models, another common assumption is to model the material as transversely isotropic. There are three general approaches—phenomenological, structural, and hybrid—for determining the form of Ψ that are discussed in the following sections.

3.4.1 Phenomenological Models

Phenomenological models (e.g., [167, 210]) seek to establish a simple mathematical relationship necessary to accurately predict stress within a reasonable range of kinematic strains. The mathematical relationships established typically utilize measures of strain (e.g., invariants) along with material parameters to fit the phenomenological model to some given data. Usually the logic of the model form is not structurally based; rather it is determined by experience or trial and error. Due to their relative simplicity, they are often computationally fast to solve and efficient to work with. Numerous phenomenological constitutive equations exist that describe the passive strain energy, a review of which can be found in [203].

One phenomenological model by Guccione et al. [166] describes the passive behavior using a transversely isotropic Fung-type law,

$$\Psi = \frac{c}{2}\left[\exp\{Q\} - 1\right], \quad Q = b_f E_{ff}^2 + 2b_{fs}\left(E_{fs}^2 + E_{fn}^2\right) + b_n\left(E_{ss}^2 + E_{nn}^2 + 2E_{sn}^2\right),$$

where C, b_f, b_{fs}, and b_n are the material parameters and E_{ff}, ..., E_{sn} are the components of the Green–Lagrange strain tensor in each respective direction, i.e., $E_{ff} = \mathcal{I}(\mathbf{E}, \mathbf{n}_f, \mathbf{n}_f)$, $E_{fs} = \mathcal{I}(\mathbf{E}, \mathbf{n}_f, \mathbf{n}_s)$, and so on.

The generalized Fung model is well known for its ability to match a wide range of soft tissue responses. Although there are parameter restrictions for transverse isotropy, the Guccione model is still flexible in the range of responses it is able to model. In addition, with $b_f, b_{fs}, b_n > 0$, strong ellipticity is established for Q and thus Ψ for numerical stability. Having only four parameters and utilizing a phenomenological form, this model is computationally efficient. However, the myocardium is known to be orthotropic, and thus, this transversely isotropic restriction may not be sufficient. Additionally, due to the removal of coupling terms from the generalized Fung model form, the Guccione model cannot match the shearing behavior of fibers in soft tissues [536].

Addressing some of these issues is the orthotropic model by Holzapfel et al. [203], which takes into account the interaction between the fiber and sheet directions observed in Dokos et al. [99],

$$\Psi = \frac{a}{2b}\{e^{[b(I_1-3)]} - 1\} + \frac{a_f}{2b_f}\{e^{[b_f(I_{4f}-1)^2_+]} - 1\}+$$

$$\frac{a_s}{2b_s}\{e^{[b_s(I_{4s}-1)^2_+]} - 1\} + \frac{a_{fs}}{2b_{fs}}\{e^{[b_{fs}I^2_{8fs}]} - 1\},$$

where $I_{4f} = \mathcal{I}(\mathbf{C}, \mathbf{n}_f, \mathbf{n}_f)$, $I_{4s} = \mathcal{I}(\mathbf{C}, \mathbf{n}_s, \mathbf{n}_s)$, $I_{8fs} = \mathcal{I}(\mathbf{C}, \mathbf{n}_f, \mathbf{n}_s)$, see (2). The I_4 terms are responsible for the response of the fibers in the fiber (f) and sheet (s) directions, respectively, and I_8 term is responsible for the interaction between the f and s directions. The I_C term is responsible for the isotropic behavior of the extracellular matrix and the normal ($\mathbf{n}$) direction. In general, since myocardial fibers and sheets exhibit stiffening behavior in tension but not in compression, the model relies on $(\cdot)_+$ of the invariants, returning 0 if the bracketed term is less than 0.

Phenomenological models are a popular choice due to their ease of implementation, efficiency in large-scale applications, and tunable material response. However, phenomenological models are often loosely tied to the underlying tissue microstructure, making the response to alterations within the material challenging to predict. Additionally, phenomenological models tend to predict stresses poorly when extrapolating to deformation outside of the range used to determine the model parameters [469, 536]. In these instances, structural models provide an alternative approach for describing the constitutive response of myocardium.

3.4.2 Structural Models

Unlike phenomenological models, structural models directly integrate structural information into the constitutive model forms [256, 417, 533], e.g., distribution of fiber orientations and distribution of collagen fiber crimping. These models draw on the link between the biomechanical behavior and the tissue's microstructure to formulate the constitutive model form. One benefit of this type of constitutive model is that it enables one to better understand the underlying mechanisms relating to structure and function. They also exploit knowledge and experimental data of the microstructure of materials to enhance predictive capabilities outside of typical values of measured strain [533].

Structural models first divide the strain-energy function into parts corresponding to the major microstructural components of the extracellular matrix, for example, the ground matrix (Ψ_g), myofibers (Ψ_f), and collagen (Ψ_c),

$$\Psi = \Psi_g(I_1) + \Psi_f(\mathbf{C}) + \Psi_c(\mathbf{C}).$$

The ground matrix accounts for the response after removing the contributions of the collagen and myofibers. The ground matrix is often assumed to be isotropic (modeled with the first invariant I_1) due to its low stiffness relative to the collagen fibers. The response of single collagen fibers can be examined experimentally [282, 283, 420, 421, 533] and modeled. Similarly, the overall collagen fiber architecture, which includes the fiber orientations and fiber crimps, can also be examined using imaging techniques such as second harmonic imaging [124, 188, 533] or polarized spatial frequency-domain imaging [522]. Collagen fibers are known to be undulated in their unstressed state and only provide stiffness when straightened. The mechanical properties of individual myofibers are not easily determined, but groups of myofibers sharing the same directions (fiber ensembles) can be examined through modeling [22]. The orientations of myofibers, like collagen fibers, can also be determined through imaging techniques such as diffusion tensor imaging [30, 148, 399, 408, 431].

To homogenize the mechanical response from the fiber level to the tissue level, the structure and behavioral information of collagen and myofibers are integrated stochastically using probability density distribution functions extracted from imaging data. From images, the resulting fiber orientation distribution, $R(\mathbf{n}(\theta))$, for the collagen and myofibers, and the recruitment distribution, $D(\lambda_s)$, corresponding to the stretch, λ_s, needed to straighten a crimped collagen fiber are determined. Affine deformation is typically assumed and has been shown to be accurate for collagen fibers in common soft tissues [533]. The stretch of a fiber ensemble can be computed using I_4 (2), and as an example, the strain energy of myofibers can be given by

$$\Psi_f = \int_{-\pi/2}^{\pi/2} R(\mathbf{n}_f(\theta))w(I_{4f})d\theta,$$

where $w(I_{4f})$ is the constitutive model for the response of an ensemble of fibers, and $\mathbf{n}_f$ is the orientation of the fiber ensemble. Here, the mechanical response of myocardium due to the myofibers is accounted for by all myofibers that are under stretch. $w(I_{4f})$ is typically assumed to be 0 when I_{4f} is negative. As myofibers are stretched and rotated into frame, the stress of the tissue along different directions increases and decreases. Essentially, this model form uses the structural reorientation of myofibers as the tissue deforms to compute the mechanical response of the tissue. As long as the assumptions for the mechanisms of fiber-level deformations and the fiber ensemble models are accurate, this structural model form is very robust in its predictive capability.

The structural model for collagen can be extended further due to available fiber crimp information. Thus, a fiber-level model, w_f, can be used instead of a fiber ensemble-level one. Fiber-level models are typically linear in behavior [282, 283, 420, 421, 533] and thus are not prone to error from extrapolation. The response of a collagen ensemble is similarly representative of all the collagen fibers under stretch. This can be computed by integrating over $D(\lambda_s)$. The stretch of a fiber ensemble gradually recruits collagen fibers to bear stress; hence, it is called the recruitment distribution. Thus, a typical collagen model typically has the form

$$\Psi_c = \int_\theta R(\mathbf{n}_c) \int_{\lambda_s=1}^{\lambda_c} D(x)w_f(\lambda_t)d\lambda_s d\theta,$$

where $\lambda_c = \sqrt{\mathcal{I}(\mathbf{C}, \mathbf{n}_c, \mathbf{n}_c)}$ is the stretch of the collagen fiber ensemble, $\mathbf{n}_c$ is the direction of fiber ensemble, and $\lambda_t = \lambda_c/\lambda_s$ is the true stretch of the collagen fibers. As long as the mechanisms behind the reorientation and straightening of the collagen fibers are accurate, the extrapolation of this structural model is equivalent to the extrapolation of the linear collagen fiber model, which has been shown to accurately predict behavior of dense collagenous soft tissues [533]. One recent structural model of myocardium, developed using this full approach, integrated full three-dimensional collagen and myofiber information and accounted for their interactions [24].

While structural models provide a clearer representation of tissue with respect to underlying tissue structure, making them easier to adapt with microstructural changes, they can present challenges. The required experimental data, especially microstructural data for non-homogenous specimens, may be difficult or impossible to obtain with current technology. In addition, the integration of such data in the model, by means of numerical integration over fiber distributions, often makes these models computationally intractable in large simulations. Further, other complexities, such as cross-linking or other interactions, are often neglected or difficult to determine. As a consequence, hybrid approaches aim to bridge between the phenomenological and structural worlds.

3.4.3 Hybrid Models

Hybrid approaches using structural tensors are one way to bridge the gap between the computational efficiency of phenomenological models and structural data integration [202]. One way to do this is by the introduction of a structural tensor ($\mathbf{H}$) describing the fiber orientation distribution ($\Gamma(\mathbf{N})$) in three dimensions:

$$\mathbf{H} = \frac{1}{4\pi} \int_S \Gamma(\mathbf{N})\mathbf{N} \otimes \mathbf{N} \mathrm{d}S, \qquad \mathrm{tr}(\mathbf{H}) = 1,$$

where S denotes the surface of a unit sphere. This allows a new invariant to be defined from fiber orientation data as $I_H = \mathbf{H} : \mathbf{C}$. Constitutive models using this new invariant can have anisotropic characteristics that are closely correlated with its structural model equivalent but avoid the use of numerical integration.

3.5 Viscoelastic Modeling Approaches

Classically, viscoelastic models are built on two fundamental elements, springs and dashpots, which represent the elastic and viscous components, respectively. For linear materials, these are the strain and strain-rate terms. For nonlinear materials, it is often more convenient to use a hyperelastic PK2 model, $\mathbf{S}_e$, for the springs, and the time derivative $\dot{\mathbf{S}}_e$ for the dashpots. In cases where the form of the viscoelastic

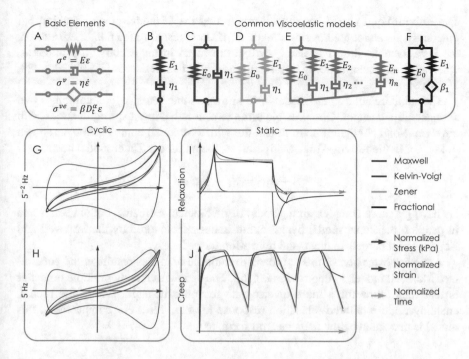

Fig. 6 (Top) Common viscoelastic models based on arrangements of spring, dashpot, and spring-pot elements: (**a**) Maxwell, (**b**) Kelvin–Voigt, (**c**) Standard linear (Zener), (**d**) Generalized Maxwell, (**e**) Fractional Kelvin–Voigt, (**f**) Fractional Maxwell. The linear spring, dashpot, and spring-pot are defined in the lower row. (Bottom) Comparing the behavior of classical viscoelastic models (Maxwell (red), Kelvin–Voigt (blue), and Zener (green)) and the fractional approach (black) (**g**) at low frequency, (**h**) at high frequency, (**i**) in relaxation tests, and (**j**) in creep tests

behavior is significantly different from the hyperelastic part, the time derivative of a separate model form $\dot{\mathbf{S}}_Q$ can be used to represent the dashpots. The simplest combinations of these two elements form the Maxwell and Kelvin–Voigt models. The Maxwell model is composed of a spring and dashpot in series (Fig. 6a),

$$\mathbf{S} + a_1\dot{\mathbf{S}} = b_1\dot{\mathbf{S}}_Q,$$

where a_1 and b_1 are the constants derived from the stiffness of the spring and dashpot. The Maxwell model is particularly useful in modeling relaxation behavior but is known to be unable to predict typical creep behavior [329]. Under constant stress, materials usually exhibit a strain increase whose rate is decreasing. However, this model displays a linear strain increase (constant rate).

The Kelvin–Voigt model is described by a spring and dashpot arranged in parallel (see Fig. 6b),

$$\mathbf{S} = \mathbf{S}_e + b_1\dot{\mathbf{S}}_Q. \tag{9}$$

The Kelvin–Voigt model has the advantage that it can model creep behavior, but inherently, it cannot predict stress relaxation. Upon reaching the maximum stretch, the instantaneous drop in velocity will result in an instantaneous drop in stress, whereas an exponential decay is normally observed in soft tissues [52, 97, 225, 228, 395, 498].

The standard solid linear model, also known as the Zener model, combines two springs and a dashpot. One possible arrangement is having a spring in series with a Kelvin–Voigt configuration or in parallel with a Maxwell configuration, as shown in Fig. 6c. In the Kelvin–Voigt configuration, the nonlinear Zener model becomes

$$\mathbf{S} + a_1 \dot{\mathbf{S}} = \mathbf{S}_e + b_1 \dot{\mathbf{S}}_Q. \tag{10}$$

By having a more complex arrangement of rheological elements, both in series and in parallel, the Zener model bypasses the issues encountered by the Maxwell and Kelvin–Voigt models in creep and relaxation, respectively.

A further extension of this approach to multiple branches results in the generalized Maxwell model, as depicted in Fig. 6d. The benefit of having multiple branches is that they allow for a more diverse rate response behavior within the model, enabling a more tailored relaxation response [3, 425, 426]. For compactness, this model is presented below in its integral form, as

$$\mathbf{S}(t) = \mathbf{S}_e(t) + \int_0^t \sum_{k=1}^n b_k \exp\{(s - t)/\tau_k\} \dot{\mathbf{S}}_Q(s) \, \mathrm{d}s.$$

Then, each branch is defined by the relaxation time τ_k and relaxation weight b_k. The generalized Maxwell model is the most comprehensive classical viscoelastic model, providing flexibility in accounting for a spectrum of relaxation rates. However, this generality comes at the added cost of an increased number of parameters that are difficult to determine.

3.5.1 Fractional Viscoelasticity

One approach to extend classical viscoelastic models is to use fractional derivatives in place of first-order derivatives. Fractional calculus began with the generalization of integral and differential operators from the set of integers to the set of real numbers (for reviews, see [296, 410]). Many different definitions for fractional differential and integral operators of arbitrary order have been introduced [392], with the Caputo definition being a popular choice for physics-based applications. For an n-times differentiable function f and $0 < \alpha \leq n$, the Caputo derivative is defined as

$$D_t^\alpha f = \frac{1}{\Gamma(\lceil \alpha \rceil - \alpha)} \int_0^t \frac{f^{\lceil \alpha \rceil}(s)}{(t - s)^{1 + \alpha - \lceil \alpha \rceil}} \, \mathrm{d}s,$$

where $\lceil \alpha \rceil$ denotes the ceiling of α. It is common to consider only $0 \leq \alpha \leq 1$, since the composition of the fractional derivative has the property that $D_t^a D_t^b f = D_t^{a+b} f$. Mathematically, if $\alpha = 0$, then the behavior of a fractional derivative term converges to that of a spring, whereas if $\alpha = 1$, its behavior converges to that of a dashpot. The intermediate values $0 < \alpha < 1$ represent the gradual transition of the mechanical response from elastic to viscous. Due to having both the characteristics of a spring and a dashpot, the rheological element it represents is sometimes also called a spring-pot. The spectrum of relaxation behaviors possible from a fractional derivative is advantageous for describing the multiscale structure of the myocardium, as the spectrum of relaxation times from the different components can be conveyed through a single parameter, α. Being able to reduce the complex behavior of a viscoelastic model to a single parameter is both robust for parameter estimation and more physically intuitive. Furthermore, this also allows the moduli and characteristic time scales of the Prony terms to be easily pre-computed for computational efficiency.

A fractional viscoelastic model can be assembled by starting with a classical form, such as the Zener model, and replacing the time derivative with a fractional derivative, i.e.,

$$\mathbf{S}_{ve} + a_1 D_t^{\alpha_a} \mathbf{S}_{ve} = b_0 \mathbf{S}_e + b_1 D_t^{\alpha_b} \mathbf{S}_Q.$$

Setting the fractional orders α_a and α_b to 1 recovers (10). Conversely, disregarding the derivative on the left-hand side (i.e., setting $a_1 = 0$) gives the fractional adaption of the classical Kelvin–Voigt form (9), which is equivalent to an approximation to the generalized Maxwell model [535]. The behavior of this model for an isotropic nonlinear material is examined in Zhang et al. [534]. In comparison with classical approaches, the fractional approach is able to match the response across a more complete range of experiments. For example, the classical Maxwell (Red) is unable to match the experimental behavior at low frequencies of loading (see Fig. 6g), while the Kelvin–Voigt model (Blue) is unable to match the experimental behavior at high frequencies of loading (see Fig. 6h). The fractional models have no problem in this regard. In addition, none of the classical Maxwell, Kelvin–Voigt, and Zener models can match the relaxation and creep behavior exhibited when constant strain and stress are cycled on and off repeatedly (see Fig. 6i,j). These features make fractional viscoelastic models appealing for use in complex biological tissues.

3.5.2 Cardiac Viscoelastic Models in the Literature

While viscoelastic response is clearly observed in cardiac tissue, the use of viscoelastic models remains less common than the use of hyperelastic approaches. This is due, in part, to the increased difficulty with implementation and practical application. Loeffler and Sagawa [288] developed a 1D model based on the transient response of cat papillary muscle. They found that the passive part needed at least five linear elements in order to capture the behavior, and therefore, used the generalized

Maxwell configuration with two branches. For a complete 3D definition, a passive cardiac model was proposed by Huyghe et al. [215], who used an orthotropic, exponential formulation for the elastic part and an isotropic quasi-linear formulation for the viscoelastic portion. The 11-parameter model was fit to uniaxial, biaxial, and relaxation data obtained from three previous studies. However, large variations were observed in the viscoelastic parameters, potentially due to the quasi-linear assumption.

In computational cardiac mechanics, viscosity was integrated into elastic and poroelastic models through the addition of a linear dashpot element in either the passive part alone [64] or both the passive and active components [65]. This captures energy dissipation for simulation purposes. But, a single viscous element is unlikely to represent typical relaxation data observed in rheological testing.

More recently, two groups devised a similar 3D orthotropic model for the passive myocardium, considering multiple Maxwell branches. Both Cansız et al. [56] and Gültekin et al. [169] modeled the elastic part as a sum of the underlying matrix, fiber, sheet, and relative fiber-sheet shear contributions using exponential functions. In the model proposed by Gültekin et al., the viscous contributions were defined for the matrix, fiber, sheet, and fiber sheet as well using a first-order linear differential equation, leading to 16 model parameters. By comparison, Cansız et al. defined the viscous contributions in the fiber, sheet, and normal directions, employing a quadratic function of the strain logarithm (resulting in 14 parameters). Both models were suitable for capturing shear data in six deformation modes. Additionally, the Gültekin et al. model was also validated against biaxial data. However, large parameter variability was observed when the model was fit to shear data.

To date, fractional approaches have not yet been employed in modeling cardiac mechanics in the literature. Therefore, the suitability of these models remains to be investigated.

3.6   Applications of Cardiac Constitutive Modeling

Studying the biomechanical function of myocardium can enable a better understanding of healthy and pathological states. Constitutive models presented in preceding sections can be used to not only predict behavior, but also elucidate properties and functional metrics. As discussed previously, there is a close correspondence between the tissue microstructure and biomechanical behavior [483]. Studies investigating how collagen orientation, angular dispersion, fiber stiffness, crimp angle, and collagen density affect mechanical response [297, 491] have found that the collagen structure and orientation are important determinants of passive tissue response. A material model incorporating collagen shape parameters (i.e., elongation, flatness and anisotropy) was shown to accurately predict behavior of cardiac tissue in a healthy and hypertensive heart [181]. Compressibility of myocardium is another important mechanical property to be explored. Studies [320] have shown experimental evidence for volume change in passive, excised porcine myocardium. Constitutive modeling can also be used to interrogate the source of viscoelastic behavior in myocardium [216, 242, 332, 333, 439, 449, 450, 513, 529]. For example, a poroelastic description can be used to model the viscous properties due to the flow of interstitial fluid through the extracellular matrix as a result of mechanical deformation. However, Yang et al. [524], using an incompressible, isotropic, poroelastic model, found that poroelasticity alone is insufficient to explain hysteresis seen in passive cardiac muscle. On the other hand, Emery et al. [114] found through modeling that non-elastic behavior of the myocardium is better explained by murine (or plastic) effects than viscoelasticity, which could be due to damage to elastic components.

The combination of appropriate constitutive models with accurate structure and geometry can also lead to better insights into organ-level properties and function. Non-invasive imaging, such as ultrasound, CT, and MRI, can be used to determine patient-specific cardiac geometry, microstructural orientation [323, 463], and deformation [126, 430]. Non-invasive imaging modalities that provide temporal geometric information can also be used to obtain deformation in three dimensions by utilizing geometric meshing and image-to-mesh information mapping through rasterization [106, 134, 437, 442, 454, 478], which can be further improved in accuracy through the addition of a constitutive model [377] and machine learning [517]. Patient-specific parameters of the constitutive model can be estimated by minimizing the difference between modeled versus measured deformation [19, 20, 174, 497, 499]. These and other studies have enabled the development of modeling pipelines for assessment of patient-specific parameters in myocardium [134, 144, 161, 330, 398, 481, 507, 537].

4 Tissue Growth and Remodeling

The heart muscle is constantly subjected to alterations in loading conditions: at rest, during different degrees of exercise, or as a consequence of cardiovascular disease. Sustained deviations from physiological conditions, e.g., chronically elevated blood pressure or endurance exercise, can induce structural changes in myocardial tissue and mass in order to retain homeostasis and meet the increased demand. These changes are referred to as growth and remodeling (G&R).

Depending on the type of supraphysiological loading, the myocardium may undergo different distinct patterns of G&R. Eccentric or concentric shape adaptations are prominent patterns of ventricular hypertrophy and often occur as a consequence of volume or pressure overload [412]. In general, myocardial hypertrophy is considered to be an adaptive response and hence, in the beginning, can be beneficial for myocyte regeneration and survival (e.g., "athlete's heart" [120, 331]). A persisting overload, however, favors apoptosis and fibrotic mass production, both of which are maladaptive processes that impair the heart's function and can eventually lead to systolic or diastolic heart failure.

Models that account for hypertrophic and tissue constituent changes are essential when describing the mechanical function of the myocardium and the interplay of tissue mechanics with changes in circulatory system hemodynamics. G&R models can help to address the following key questions: What degree of deviation of physiological loading conditions induces hypertrophy? When is hypertrophy beneficial and at which stage does it become maladaptive? How does G&R contribute to hemodynamics, blood pressure, tissue oxygenation, and myocardial regeneration? What are some treatment strategies to trigger reverse G&R, or at least to reduce maladaptive G&R progression?

In Sect. 4.1, common approaches for modeling of G&R in a continuum mechanical setting are presented to give an overview of ideas, methodologies, advantages, and possible disadvantages of these methods. Since the theory of kinematic growth is widely used and builds upon early and fundamental continuum mechanical approaches for inelastic material behavior, this theory is expanded upon in Sect. 4.2. Thereafter, an overview of how G&R models are applied in the context of modeling the diseased heart is given in Sect. 4.3 in order to demonstrate their capabilities as well as highlight future potential avenues for research.

4.1 Modeling Approaches in G & R

Growth and remodeling (G&R) can be approached through a variety of theoretical approaches. For more comprehensive reviews, the reader may refer to Ambrosi et al. [9], Niestrawska et al. [358], or Lee et al. [268].

■ **The Theory of Kinematic Growth** relies on a classical approach in continuum mechanics for inelastic material behavior that was first introduced by Lee [265] for modeling finite strain plasticity. The deformation gradient (1) at a time instance $t = s$, $\mathbf{F}(s)$, that describes the kinematic pathway from a reference to a deformed configuration is multiplicatively split into one part associated with the inelastic deformation, $\mathbf{F}^{\mathrm{g}}(s)$ (e.g., growth, plasticity, or damage), and one part that describes the purely elastic behavior, $\mathbf{F}^{\mathrm{e}}(s)$, cf. Fig. 7b. In the context of growth in soft tissues, this methodology was first described by Rodriguez et al. [406]. Due to its prominence and widespread application, especially in modeling cardiac growth, we present the theory in detail in Sect. 4.2.

The theory benefits from its relative simplicity and straightforward incorporation into existing continuum mechanical analysis frameworks (also from a computational perspective). However, the underlying assumptions are purely kinematic and do not include effects of remodeling; hence, a change of tissue constituents and their mechanical properties would have to be addressed by making additional assumptions or alteration of material parameters.

■ **The Constrained Mixture Theory** is a specialization of the general mixture theory and makes assumptions as to how different tissues or cell constituents evolve over time. These may be fibrillar collagen, muscle cells, or other

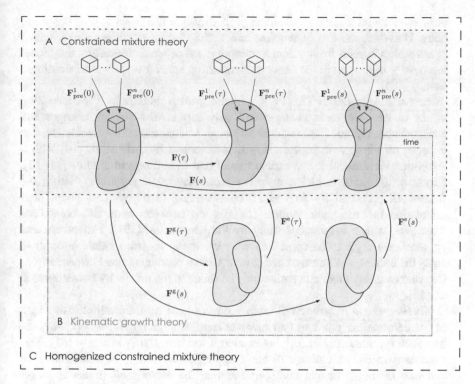

Fig. 7 (**a**) In *constrained mixture theory*, a body is composed by n individual constituents, each consisting of multiple mass increments that were deposited with a pre-stretch $\mathbf{F}_{\text{pre}}^n(t)$ at different times. The elastic pre-stretch depends on the individual stress-free natural configuration of each constituent. All constituents undergo the same elastic deformation together, despite having been deposited with different pre-stretches at different times. (**b**) In kinematic growth theory, the deformation gradient $\mathbf{F}$ is multiplicatively split into an elastic part $\mathbf{F}^e$ and a part governing the growth deformation $\mathbf{F}^g$. The intermediate configuration may be incompatible if growth violates restrictions imposed by boundary conditions or is inhomogeneous in general. (**c**) The *homogenized constrained mixture theory* combines the kinematic growth and the constrained mixture approach in such a way that the benefits from both are inherited [83]

constituents. Growth and remodeling processes are attributed to the replacement or reorganization of existing constituents with new ones in order to maintain a constant preferred state of (biomechanical) homeostasis. Changes in mechanical or chemical stimuli with respect to a baseline state dictate the rates of constituent turnover. New constituents are deposited at preferred stretches and are mechanically constrained to comply with the other tissue constituents, see Fig. 7a.

The underlying mathematical theory was first introduced by Humphrey and Rajagopal [208]. Figure 7a depicts the evolution of a continuous body over time and sketches the deposition of new constituents with a defined pre-deformed state at a certain time t characterized by $\mathbf{F}_{\text{pre}}^n(t)$. The constrained mixture theory can be viewed as a "true" G&R model due to its ability to incorporate new constituents with distinct mechanical properties into the growing (and adapting) body. However, the theory requires accounting for past configurations, which poses a high burden from a computational point of view. Therefore, the model has mostly been used in the context of modeling arterial vessels with simplified shapes [484].

■ **Structural Adaptation Theory** Kinematic growth models lack a description of the tissue constituent changes and only capture the volume changes and consequences thereof (i.e., residual stresses). Microstructural reorientation that depends on the growth process was first described by Arts et al. [14] with a simplified mechanical left ventricle model and in subsequent studies [11, 13]. Kroon et al. [250] modeled adaptive reorientation of myofibers within the framework proposed by Arts et al. by assuming that myofibers adapt their orientation in order to minimize shear strain between fiber and cross-fiber directions, with a subsequent study by Pluijmert et al. [391]. However, these approaches are yet to be combined with kinematic growth models in order to study the influence of volume change on myofiber reorientations. Further insights into studies of myofiber adaptation may be found in the review by Bovendeerd et al. [49].

■ **Fully Structural Approach** Growth models based on the multiplicative split of the deformation gradient that involves mass change require an open-system approach for which the underlying motion is not necessarily bijective [81]. This essentially means that no one-to-one mapping between a point in the reference and one in the deformed configuration may be established (material points may be absorbed or added). Lanir [257, 258] aimed at overcoming this issue by proposing a structural approach in terms of a mechanistic micromechanical G&R theory for fibrous tissues. Therein, tissue growth was modeled by fiber turnover. Two features of this approach are shared with constrained mixture models, that is, the evolving natural configuration of tissue constituents and the rule of mixtures, in order to evaluate the global tissue response. However, the main difference is that, in classical mixture theories, only remodeling of different tissue constituents, but not the 3D structural network itself, is considered [258].

■ **Hybrid Approaches** were recently developed, which inherit the benefits of both kinematic growth and constrained mixture models. A model that combined volumetric growth approaches (Sect. 4.1) with the constrained mixture model was applied in the context of aortic remodeling [7]. A *homogenized constrained mixture* approach was developed by Cyron et al. [83], which bypassed the need to keep track of past configurations while still being able to account for influences on a microstructural level, see Fig. 7c. It was successfully applied to model anisotropic volumetric growth and remodeling of arteries in Braeu et al. [50].

4.2 Kinematic Growth

Here, the theory of kinematic growth is discussed further due to its prevalence, especially in modeling growth in the heart. A detailed discussion of continuum mechanics of finite strain kinematic growth may be found in Menzel and Kuhl [325] or Himpel et al. [190].

4.2.1 Finite Strain Kinematics of Growth

The deformation of a continuous body from a reference to a current configuration is characterized by the deformation gradient tensor (1). Without loss of generality, this deformation may be split into infinitely many parts associated with distinct kinematic patterns that then can be superimposed in order to recover the full deformation. An example is the well-known polar decomposition of the deformation gradient [200]. Within the theory of finite strain plasticity, a multiplicative decomposition of the deformation gradient was first employed by Lee [265], which was adopted in various different approaches for modeling inelastic material behavior in a finite strain continuum mechanical setting. Kinematic growth, first applied in the context of soft elastic tissues by Rodriguez et al. [406], assumes that the deformation gradient at a time instance t can be decomposed into an elastic part, denoted as $\mathbf{F}^e(t)$, and an inelastic growth part, $\mathbf{F}^g(t)$:

$$\mathbf{F}(t) = \mathbf{F}^e(t)\mathbf{F}^g(t). \tag{11}$$

The geometric interpretation of (11) is depicted in Fig. 7b. A fictitious stress-free intermediate configuration that solely entails the deformation due to growth arises as a consequence of $\mathbf{F}^g$ acting on the reference state. This configuration, in general, can be incompatible, which is indicated by the portion of grown matter disjoint from the underlying body in Fig. 7b. Incompatibility can occur if growth is inhomogeneous or if boundary conditions on the domain inhibit growth to occur. This incompatibility is "restored" by $\mathbf{F}^e$ acting on the intermediate configuration in such a way that balance equations of linear and angular momentum are fulfilled.

4.2.2 Balance Equations of Growth

In addition to classical continuum mechanics of elasticity presented in Sect. 3, the balance equations of growth have to be considered in an open-system thermodynamic framework due to the fact that the body's mass may change during the growth process. For an in-depth treatment of thermodynamics with focus on volumetric growth, the reader is referred to Epstein and Maugin [116].

■ **Balance of Mass** Consider a growth deformation gradient $\mathbf{F}^g$ acting on the reference configuration yielding the incompatible intermediate state (Fig. 7b).

Under the assumption that any mass increase due to growth happens entirely throughout volume change, hence under density preservation, the change of mass due to growth requires the mass density in the reference configuration, denoted as ρ_0, to change:

$$\dot{\rho}_0 = M_0 + \nabla_\mathbf{X} \cdot \mathbf{M}, \tag{12}$$

where M_0 is the mass source and $\mathbf{M}$ represents a mass flux with respect to the reference configuration. Microscopically viewed, the mass source may be associated to hypertrophy or proliferation of cells, while the flux term would refer to cell migration.

Without loss of generality, we make the reasonable assumption that the mass flux is constant in space, and assume $\nabla_\mathbf{X} \cdot \mathbf{M} = 0$ in the following. If density is preserved, the mass source can be expressed kinematically as

$$M_0 = \rho_0 \, \mathrm{tr}\mathbf{L}^\mathrm{g}, \tag{13}$$

where the growth velocity gradient is

$$\mathbf{L}^\mathrm{g} = \dot{\mathbf{F}}^\mathrm{g}\mathbf{F}^{\mathrm{g}^{-1}}. \tag{14}$$

■ **Balance of Linear Momentum** The volume-specific balance of linear momentum has to account for the mass change and hence the change of the reference density due to growth. Therefore, (8) becomes

$$\nabla_\mathbf{X} \cdot \mathbf{P} + \hat{\mathbf{b}}_\mathbf{X} = \rho_0 \dot{\mathbf{v}} + \dot{\rho}_0 \mathbf{v}. \tag{15}$$

Under the assumption that growth happens on a significantly larger time scale than the dynamics due to elastic deformation, the additional inertia term $\dot{\rho}_0 \mathbf{v}$ due to mass change may be neglected. Furthermore, if growth is analyzed in a quasi-static setting, where any inertial forces vanish, (15) reduces to the standard quasi-static balance of linear momentum:

$$\nabla_\mathbf{X} \cdot \mathbf{P} + \hat{\mathbf{b}}_\mathbf{X} = \mathbf{0}.$$

■ **Balance of Entropy** For an open-system setting, the local form of the Clausius–Duhem inequality for a constant absolute temperature θ can be stated as

$$\mathcal{D} = \mathbf{P} : \dot{\mathbf{F}} - \dot{\Psi} - \theta H \geq 0. \tag{16}$$

Therein, the term H represents entropy sources or fluxes in order to satisfy the second law of thermodynamics.[9] Here, we assume the free energy Ψ to depend on not only the deformation gradient, but also on the growth internal variables, i.e., the growth tensor $\mathbf{F}^g$, as well as the mass density in the reference configuration:

$$\Psi = \Psi(\mathbf{F}, \mathbf{F}^g, \rho_0). \tag{17}$$

The time derivative of the free energy per unit volume (17) then reads

$$\dot{\Psi} = \frac{\partial \Psi}{\partial \mathbf{F}} : \dot{\mathbf{F}} + \frac{\partial \Psi}{\partial \mathbf{F}^g} : \dot{\mathbf{F}}^g + \frac{\partial \Psi}{\partial \rho_0} \dot{\rho}_0. \tag{18}$$

The first term in (18) is the change in the free energy due to elastic deformation, which will constitute the PK1 stress tensor. The second term in (18) represents a change of the free energy owing to inelastic processes, i.e., growth. The third term in (18) may be viewed as a change of the energetic state due to mass changes as a consequence of growth or atrophy. By introducing the free energy per unit mass, $\psi = \frac{1}{\rho_0} \Psi$, and making use of the mass source (13), this term can be expressed as

$$\frac{\partial \Psi}{\partial \rho_0} \dot{\rho}_0 = \psi \dot{\rho}_0 = \psi \rho_0 \, \mathrm{tr} \mathbf{L}^g = \Psi \, \mathrm{tr} \mathbf{L}^g. \tag{19}$$

Inserting (19) into (16) yields the Clausius–Duhem inequality in the form

$$\mathcal{D} = \left(\mathbf{P} - \frac{\partial \Psi(\mathbf{F})}{\partial \mathbf{F}} \right) : \dot{\mathbf{F}} + \frac{\partial \Psi}{\partial \mathbf{F}^g} : \dot{\mathbf{F}}^g - \Psi \, \mathrm{tr} \mathbf{L}^g - \theta H \geq 0. \tag{20}$$

Note that for a reversible (purely hyperelastic) mechanical problem without growth, (12) vanishes and (17) becomes $\Psi = \Psi(\mathbf{F})$. Hence, (20) reduces to the equality

$$\mathcal{D} = \left(\mathbf{P} - \frac{\partial \Psi(\mathbf{F})}{\partial \mathbf{F}} \right) : \dot{\mathbf{F}} = 0. \tag{21}$$

■ **Residual Stress States** Residual stresses play a ubiquitous role in many contexts of biological soft tissue and often are a consequence of spatially varying changes in tissue constituents over time. In the framework of finite deformation kinematic growth, residual stresses arise for any $\mathbf{F}^g$ that produces an incompatible fictitious intermediate configuration Ω_g (cf. Fig. 7). Therefore, the body in the current configuration $\Omega(t)$ may be in mechanical equilibrium in the absence of external

---

[9] In the most general from, H may consist of an entropy source H_0 and entropy flux $\mathbf{H}$ of the form $H = H_0 - \nabla_0 \cdot \mathbf{H}$.

forces and hence possess a reference configuration Ω_0^{res} that is not stress-free, i.e., with residual stress $\mathbf{P}^{res} \neq \mathbf{0}$. This residual stress distribution has to fulfill mechanical equilibrium, which can be stated with respect to the reference configuration as

$$\nabla_{\mathbf{X}} \cdot \mathbf{P}^{res} = \mathbf{0} \text{ in } \Omega_0^{res}, \quad \mathbf{P}^{res}\mathbf{N} = \mathbf{0} \text{ on } \Gamma_0^{res}. \tag{22}$$

A direct consequence of (22) is that either $\mathbf{P}^{res} = \mathbf{0}$ or that $\mathbf{P}^{res}$ is distributed inhomogeneously within Ω_0^{res}. These inhomogeneous residual stress distributions arise in many soft tissues, predominantly in arteries, where it is agreed upon that they are primarily caused by mechanisms of growth and remodeling [138, 446].

4.2.3 Constitutive Equations for Growth

In order to close the system of equations, we have to specify constitutive equations for the free energy and the evolution of growth internal variables. We assume (17) to be of a hyperelastic form; hence, the constitutive relation for PK1 can be deduced from the first term in (20). We want to restrict ourselves to the microstructural concept of characterizing the evolution of growth and define a growth tensor $\mathbf{F}^g$ that may take the general form

$$\mathbf{F}^g = \vartheta^{iso}\mathbf{I} + (1 - \vartheta^{ani})\mathbf{a}_0 \otimes \mathbf{a}_0, \tag{23}$$

where ϑ^{iso} and ϑ^{ani} are the isotropic and anisotropic growth stretches, respectively, and $\mathbf{a}_0$ describes a microstructural orientation unit vector in the reference configuration. For simplicity, we want to exemplify the theory for purely isotropic growth ($\vartheta^{ani} = 0$, $\vartheta^{iso} \equiv \vartheta$). Hence, (23) reduces to

$$\mathbf{F}^g = \vartheta\mathbf{I}. \tag{24}$$

For this case, the growth velocity gradient (14) reads

$$\mathbf{L}^g = \frac{\dot{\vartheta}}{\vartheta}\mathbf{I},$$

and the mass source (13) becomes

$$M_0 = 3\rho_0\frac{\dot{\vartheta}}{\vartheta}.$$

Conveniently, a growth evolution equation for the stretch ϑ is introduced of the form

$$\dot{\vartheta} = f(\vartheta, \mathbf{a}_0, \mathbf{\Sigma}^e, \mathbf{F}^e, \ldots), \tag{25}$$

which may depend on mechanical quantities like stresses (e.g., the Mandel stress, $\mathbf{\Sigma}^e$) or strains, microstructural orientations, or other factors. A thermodynamically motivated form of (25) for an isotropic growth tensor (24) may take the form

$$\dot{\vartheta} = k(\vartheta)\,\mathrm{tr}\mathbf{\Sigma}^e,$$

where

$$k(\vartheta) = \begin{cases} \frac{1}{\tau_+}\left(\frac{\vartheta_{max}-\vartheta}{\vartheta_{max}-1}\right)^{\gamma_+}, & \mathrm{tr}\mathbf{\Sigma}^e \geq 0, \\ \frac{1}{\tau_-}\left(\frac{\vartheta-\vartheta_{min}}{1-\vartheta_{min}}\right)^{\gamma_-}, & \mathrm{tr}\mathbf{\Sigma}^e < 0, \end{cases} \tag{26}$$

with $\vartheta_{max} > 1$ and $\vartheta_{min} < 1$ to being limiting values that can be reached by growth or atrophy, respectively. The parameters τ_+, τ_- denote time constants and γ_+, γ_- nonlinearities controlling growth or atrophy, respectively [190].

4.3 Applications of G & R in the Heart

Patterns of growth and remodeling in the heart can drastically depend on the disease and its manifestation over time. The atrophy and ballooning observed in ventricular aneurysms vary significantly from the thickening of the heart muscle observed in ventricular hypertrophy. Two common growth forms observed in the ventricles are eccentric dilation and concentric wall thickening. Processes that drive G&R are complex, be they neural, hormonal, genetic, or mechanical. Eccentric enlargement (Fig. 8, left) is often associated with an overstretching of the myocytes as a consequence of increased venous return (volume overload), while concentric hypertrophy (Fig. 8, right) is predominantly caused by sustained end-systolic ventricular pressures (pressure overload) [369]. Sarcomerogenesis is the accepted mechanism by which both of these adaptations occur [152]. Sarcomeres are added in series or in parallel to the pre-aligned ones in such a way that the supraphysiological state of loading is reduced [324, 326]. Approaches are presented for modeling eccentric and concentric growth based on the framework of kinematic growth presented in Sect. 4.2. Subsequently, cardiac remodeling approaches are introduced, which are based on structural adaptation theories that take microstructural reorganization due to mechanical stimuli into consideration. For an in-depth review of cardiac G&R, the reader is referred to [269, 358].

■ **Eccentric growth** can be observed in cardiovascular diseases where the heart is exposed to a chronic increase in end-diastolic volume load. Hence, the myocytes are elongated above a critical physiological stretch. Within the theory of kinematic growth, the general form of the growth tensor (23) thus may be adapted to account for growth only in the myofiber direction $\mathbf{f}_0$ ($\vartheta^{iso} = 1$, $\vartheta^{ani} \equiv \vartheta$):

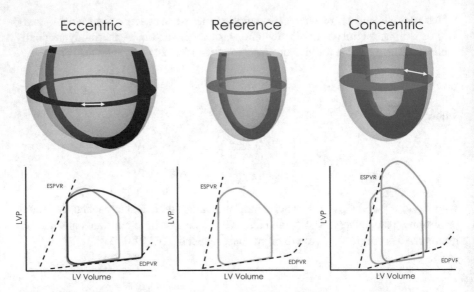

Fig. 8 Eccentric growth (left) applied to an idealized left ventricular geometry (center) results in cavity dilation along with increased stroke volume and decreased contractility, as shown by the changes in the PV loop and ESPVR curve. Concentric growth (right), however, results in wall thickening and is often accompanied by an increase in ESP as well as global elastance, illustrated by the steeper EDPVR curve

$$\mathbf{F}^{\mathrm{g}} = \mathbf{I} + (\vartheta - 1)\mathbf{f}_0 \otimes \mathbf{f}_0.$$

Here, ϑ denotes the anisotropic growth of fibers, which is determined based on a constitutive relationship governing the response of growth. Following the example in (26), ϑ could be defined based on growth to a critical stretch, $\hat{\lambda}_{\mathrm{myo}}^{\mathrm{crit}}$, i.e.,

$$\dot{\vartheta} = k(\vartheta)(\lambda_{\mathrm{myo}}^{\mathrm{e}} - \hat{\lambda}_{\mathrm{myo}}^{\mathrm{crit}}), \quad \text{where} \quad \lambda_{\mathrm{myo}}^{\mathrm{e}} = \frac{1}{\vartheta}\sqrt{\mathbf{f}_0 \cdot \mathbf{C}\mathbf{f}_0}.$$

Examples of cardiovascular diseases that can cause a state of volume overload are mitral or aortic valve regurgitation, meaning increased blood return to the left ventricle [345]. Furthermore, tissue inflammation, viral infections, or myocardial infarction can lead to increased end-diastolic volumes due to systolic dysfunction. Eccentric hypertrophy is often accompanied by functional changes including increased stroke volume and decreased contractility [31], as seen in the PV loop and ESPVR curves in Fig. 8 (bottom left).

■ **Concentric growth** is typically associated with pressure overload in the heart and results in addition of sarcomeres in parallel to sustain higher pressures. The general form of the tensor in (23) can be adapted to model growth in the cross-fiber direction ($\vartheta^{\mathrm{iso}} \equiv \vartheta$, $\vartheta^{\mathrm{ani}} \equiv 1 - \vartheta$):

$$\mathbf{F}^{g} = \vartheta\mathbf{I} + (1 - \vartheta)\mathbf{f}_0 \otimes \mathbf{f}_0.$$

As before, the growth driven by ϑ requires a constitutive form. Göktepe et al. [159] used a hydrostatic stress measure to drive the material toward a stress-driven equilibrium, $\hat{\Sigma}^{\mathrm{crit}}$, i.e.,

$$\dot{\vartheta} = k(\vartheta)(\mathbf{S}^{\mathrm{e}} : \mathbf{C}^{\mathrm{e}} - \hat{\Sigma}^{\mathrm{crit}}).$$

Strain-driven growth laws that describe volume change in the transmural direction were also used, cf. Kerckhoffs et al. [233].

Concentric growth often results from higher afterload such as the case in hearts with aortic valve stenosis. Due to the reduced orifice size, higher pressures are needed, above a critical threshold, to eject blood during systole [240]. Concentric growth within the heart results in increased wall thickness, decreased cavity volume as well as reduced global tissue compliance, as shown by the steeper EDPVR curve (Fig. 8 bottom right).

Cardiomyopathies are often associated with both a change in volume as well as a change in the myocardial microstructure. Building upon previous theories of structural remodeling (presented in Sect. 4.1), Kuhl et al. [252] suggested that a reorientation of the fibers is a natural consequence of altered loading states in the tissue. By introducing a biomechanical stimulus, it was possible to model the gradual realignment of fibers along the maximum principal strain direction by minimizing the free energy stored in the system [251, 324, 326].

While current computational models of G&R have great capabilities of providing a mechanistic description of both tissue constituent and ventricular shape adaptations in consequence of supraphysiological loading conditions, the gap to tools serving as a basis for therapy and intervention planning is yet to be closed. Advances in scientific models for treatment customization, however, are progressing continuously, and model-data integration is key to their success. Data on volume-overloaded porcine hearts, gathered over a period of 8 weeks, allowed model-based correlation of alterations on cellular and organ scales, finding that only roughly half of cardiac dilation is associated with cardiomyocyte lengthening [78, 384]. Therefore, other morphological changes play a crucial role in understanding heart failure progression [385].

Growth laws integrated with models of the electrophysiological activity of the heart offer further insights into mechanisms of cardiac hypertrophy due to failure in the heart's electrical conduction system, e.g., left bundle branch block [234], demonstrating how these models may guide cardiac resynchronization therapy in the future [15, 264]. Preliminary studies with growth models applied to infarcted myocardium conceptually agree with clinical observations, indicating that surgical intervention does not necessarily reduce G&R driving forces [241]. Further model studies that may pave the way to clinical applications use reversible strain-driven laws with a homeostatic equilibrium zone, showing how mass reduction (atrophy) as a consequence of ventricular unloading occurs [267]. Multiscale models of G&R, accounting for both mechanical and hormonal drivers, indicate that hormones are the predominant driver for growth in the case of transverse aortic constriction [119], suggesting how surgical and drug interventional treatments may be guided.

5 Hemodynamics and Blood Flow Modeling in the Heart

Controlling and maintaining blood flow is a central feature of the heart, and assessing the flow of blood—both inside the chambers as well as within the coronary vessel tree—is paramount to understanding cardiac behavior and performance. In fact, structural cardiopathologies will directly impact blood flow, and similarly, modifications in flow can also trigger structural changes. Measurable hemodynamic changes have even been postulated to occur before significant morphological changes can be detected [382], highlighting the importance of cardiac hemodynamics in clinical decision-making.

Cardiovascular diagnostics are also driven by hemodynamic assessment [402]. Peak flows and pressure drops are routinely evaluated by medical imaging [33, 154], and catheter-based measurements of fractional flow reserves (FFR) or regional pressure variations are key in percutaneous coronary intervention [276]. However, these tools are merely observational and have limited abilities in providing causal explanations of measured behavior, or predicting future outcomes. Measurement tools are also limited in spatiotemporal coverage, and fine-scale hemodynamic features cannot be appreciated in detailed resolution. In these instances, cardiac flow modeling can help elucidate observed behavior, be used to guide diagnostic and interventional strategies, and even be utilized to predict how proposed modifications will impact hemodynamic behavior. The following section outlines the fundamental principles of cardiac flow modeling, provides examples of approaches and requirements, and outlines specifics on fluid–structure interaction (FSI) modeling.

5.1 Kinetics of Blood Flow

The equations of motion governing a fluid are the same as those presented for tissues in Sect. 3.1.3. However, fluids are predominantly posed in the Eulerian or arbitrary Lagrangian–Eulerian (ALE) reference frame on a domain that does not necessarily correspond to the Lagrangian motion of the fluid itself (see Fig. 9). An Eulerian reference frame requires that all fields are described with respect to a fixed spatial position, i.e., $\mathbf{x}$, rather than tracking the particle motion with respect to the reference configuration (e.g., Lagrangian reference frame). The conservation laws in the two frames are related through the well-known material time derivative expansion. For some scalar or vector material property, denoted by $\mathbf{m}$, this expansion takes the form:

Tracking points | Eulerian Frame   Lagrangian Frame   ALE Frame

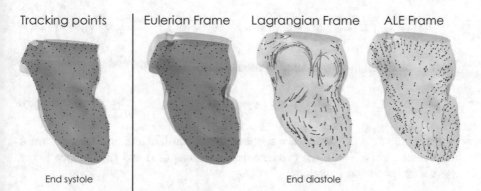

End systole | End diastole

Fig. 9 Representation of the Eulerian, Lagrangian, and ALE reference frames. (Left) A set of points in a plane cutting through the LV chamber are selected to represent the reference space. (Right) The same set of points are tracked based on the definition of spatial mapping that characterizes each frame. Thus in the case of the Eulerian description the points remain fixed and can exit the problem domain depending on the motion of boundary. In the Lagrangian description, the points follow the motion of fluid as clearly shown by the two vortexes in the upper part of the chamber. In the case of ALE, the mapping is based on the results of an artificial motion propagation problem

$$\frac{D\mathbf{m}}{Dt} = \frac{\partial \mathbf{m}}{\partial t} + \frac{\partial \mathbf{x}(\mathbf{X}, t)}{\partial t} \cdot \nabla \mathbf{m} = \frac{\partial \mathbf{m}}{\partial t} + \mathbf{v} \cdot \nabla \mathbf{m}, \qquad (27)$$

where $\mathbf{v} = \partial \mathbf{x}/\partial t$ is the material velocity, and the spatial conservation law

$$\frac{DJ}{Dt} = J\nabla \cdot \mathbf{v}, \qquad (28)$$

which holds for any mapping of space (where J and $\mathbf{F}$ are well defined). Arriving at equations for a fluid requires transforming temporal derivatives to their appropriate Eulerian equivalents. Considering mass in (6) as well as (27) and (28), the temporal derivative can be transformed into

$$\frac{D\rho J}{Dt} = J\frac{D\rho}{Dt} + \rho \frac{DJ}{Dt} = J\left(\frac{\partial \rho}{\partial t} + \mathbf{v} \cdot \nabla \rho\right) + \rho J\nabla \cdot \mathbf{v} = J\left(\frac{\partial \rho}{\partial t} + \nabla \cdot (\rho \mathbf{v})\right),$$

reducing (6) to the Eulerian conservation of mass equation

$$\frac{\partial \rho}{\partial t} + \nabla \cdot (\rho \mathbf{v}) = \hat{\rho}. \qquad (29)$$

Similarly, considering the momentum balance in (7) as well as (27) and (28), the temporal derivative can be transformed using procedure as before, i.e.,

$$\frac{D\rho \mathbf{v} J}{Dt} = J \left(\frac{\partial \rho \mathbf{v}}{\partial t} + \mathbf{v} \cdot \nabla(\rho \mathbf{v}) \right) + \rho \mathbf{v} J \nabla \cdot \mathbf{v} = J \left(\frac{\partial \rho \mathbf{v}}{\partial t} + \nabla \cdot (\rho \mathbf{v} \mathbf{v}) \right),$$

yielding the Eulerian conservation of momentum

$$\frac{\partial \rho \mathbf{v}}{\partial t} + \nabla \cdot (\rho \mathbf{v} \mathbf{v}) = \nabla \cdot \boldsymbol{\sigma} + \rho \mathbf{b}. \tag{30}$$

Assuming the fluid density, ρ, is a spatiotemporal constant and no sources of mass are present (i.e., $\hat{\rho} = 0$), the conservation equations (29) and (30) reduce to the typical Navier–Stokes equations,

$$\rho \frac{\partial \mathbf{v}}{\partial t} + \rho \mathbf{v} \cdot \nabla \mathbf{v} - \nabla \cdot \boldsymbol{\sigma} = \rho \mathbf{b}, \tag{31a}$$

$$\nabla \cdot \mathbf{v} = 0. \tag{31b}$$

While appropriate for many cardiovascular applications, the Eulerian form is less commonly applied in the heart where the moving cardiac chambers favor the ALE reference frame. ALE can be seen as a generalization of the Lagrangian and Eulerian frames, in the sense that it allows us to pose the problem on a dynamic domain without having to explicitly follow the material deformation. In order to arrive at the ALE form of the Navier–Stokes equations, let Λ be a reference domain, not necessarily coinciding to the material reference domain (which may be unknown), and let $\boldsymbol{\lambda} : \Lambda \times [0, T] \rightarrow \Omega$ be a bijective mapping that relates the reference domain to the spatiotemporal domain of the fluid problem. Similar to the material time derivative expansion (27), we can now define an ALE time derivative with respect to Λ in the form:

$$\left. \frac{\partial \mathbf{m}}{\partial t} \right|_{\hat{x}} = \frac{\partial \mathbf{m}}{\partial t} + \frac{\partial \mathbf{x}(\hat{\mathbf{x}}, t)}{\partial t} \cdot \nabla \mathbf{m} = \frac{\partial \mathbf{m}}{\partial t} + \frac{\partial \boldsymbol{\lambda}(\hat{\mathbf{x}}, t)}{\partial t} \cdot \nabla \mathbf{m}$$

$$= \frac{\partial \mathbf{m}}{\partial t} + \hat{\mathbf{v}} \cdot \nabla \mathbf{m}, \tag{32}$$

where $\hat{\mathbf{x}} \in \Lambda$ is a fixed coordinate in the reference domain and $\hat{\mathbf{v}}$ denotes the domain velocity. Using (32), we can construct the non-conservative form of the ALE Navier–Stokes equations [132],

$$\left. \rho \frac{\partial \mathbf{v}}{\partial t} \right|_{\hat{x}} + \rho (\mathbf{v} - \hat{\mathbf{v}}) \cdot \nabla \mathbf{v} - \nabla \cdot \boldsymbol{\sigma} = \rho \mathbf{b}, \tag{33a}$$

$$\nabla \cdot \mathbf{v} = 0. \tag{33b}$$

5.1.1 Constitutive Behavior of Blood

As with tissues, the equations governing fluid flow require some constitutive equation defining the relationship between stress, σ, and kinematics. To understand how blood constitutive models are defined, a brief review of both blood constitutive properties and circulatory physiology is necessary. Human blood is composed of plasma and cells in suspension. Plasma is an aqueous solution that is formed of approximately 90% water, as well as proteins and other solutes. Red blood cells account for the majority of the cell content of blood and approximately 37–49% of the blood volume (also known as the hematocrit concentration). This colloidal suspension of red blood cells introduces complex rheology, particularly throughout the cardiovascular system where flow moves through domains ranging from a few centimeters in the heart and great vessels to a few micrometers in the capillary beds. Similarly, flow characteristics change drastically, from velocities of up to 150–175 cm/s in the aorta, down to 0.5–1 mm/s in the capillaries.

Similar to tissue mechanics, continuum approximations can be employed to describe the molecular and particulate behavior of blood by means of continuous fields representing bulk properties. However, as the scale at which blood flow occurs varies by several orders of magnitude, the scale of material property homogenization must be adjusted accordingly. In practice, this translates to different continuum assumptions depending on the application. At the large-scale level, such as flow through the heart, a single-phase assumption is generally employed. At the level of capillaries, however, the interactions between individual red blood cells as well as between red blood cells and vessel walls may need to be accounted for through a two-phase continuum approach.

The rheological properties of blood have been studied extensively [73, 140, 456], driven both by evidence connecting property changes to diseases (e.g., myocardial infarction [222, 301] and arterial hypertension [394]) and by the need to predict and assess the performance of new technologies (e.g., implantable devices and numerical models). The devices used in rheological studies can be classified based on the scale at which properties are being measured as either macro- or microscale device. Of the former, we mention mechanical devices (e.g., tube viscometers, capillary viscometers, falling ball viscometers for the measurement of shear viscosity, and Couette-type rheometers for the quantification of yield stress) and optical devices [175, 343]. Alternatively, microfluidic [371] and micro-rheology [84] devices have been developed as miniaturized alternatives to macroscale devices and for characterization of red blood cell mechanical properties, respectively. Much of our understanding of the rheological properties of blood is owed to the seminal work of Chien et al. [68, 69] on shear thinning and the influence of hematocrit on blood viscosity; Cokelet et al. [74] and Merrill et al. [328] on identification and measurement of yield stress; as well as Fahraeus and Lindqvist [121, 122] on the dependence of hematocrit and apparent viscosity on the size of capillary diameter. A list of some of the main properties of blood can be found in Table 1.

In light of the studies mentioned above, the macroscopic continuum approximation used to describe the mechanics of the blood is that of a non-Newtonian fluid,

Table 1 List of main phenomena and rheological properties displayed by blood

Name	Description
Fåhraeus effect	Decrease in hematocrit in conjunction with a decrease in capillary vessel radius.
Fåhraeus–Lindqvist effect	Migration of red blood cells away from the capillary wall.
	Apparent viscosity decreases with smaller capillary radii.
Shear thinning	Apparent viscosity decreases when increasing shear rate.
	Viscosity plateaus for shear rates exceeding 100 s^{-1}, such that blood behaves as a Newtonian fluid.
Hematocrit dependence	Blood viscosity increases with the concentration of red blood cells.
Yield stress	For a vanishing shear rate, blood behaves like an elastic material.
Thixotropy	Viscosity reduction with time when subjected to constant stress.
	Linked to the aggregation and disruption of erythrocyte structures.
Viscoelasticity	Storage modulus is measurable in small amplitude oscillatory flow experiments.

which displays a nonlinear relationship between shear rate and strain. Consequently, several different constitutive laws have been proposed to characterize blood's stress–strain relationship with limited success at fitting existing experimental data: Carreau [70], Casson [140], Walburn–Schneck [493], power law [70], and generalized power law [28], to mention a few. Despite this, the size and Reynolds number of cardiac blood flow [512] make the Newtonian approximation a common approach for modeling blood flow in the heart, i.e.,

$$\boldsymbol{\sigma} = \mu(\nabla \mathbf{v} + \nabla \mathbf{v}^T) - p\mathbf{I},$$

where μ is the viscosity constant.

5.1.2 Artificial Domain Problem

The introduction of an arbitrary domain motion raises the question of how to define such motion. For most cardiac applications, the answer to this question is provided by defining an artificial domain problem. In short, this is an inverse problem based on a constitutive law, which takes as an input the prescribed boundary motion and gives as an output a mapping of a domain that satisfies the criteria defined in the preceding section. While the choice of a constitutive law is not based on any physical consideration, in practice, one must consider both solution existence and potential violation of the bijectivity requirement in the discrete setting. For this reason, different approaches have been proposed to try to mitigate this, including diffusion [187] and elastic solid problems [223, 459].

5.2   Hemodynamic Modeling in the Heart

Based on the fundamental principles outlined in Sect. 5.1, cardiac hemodynamics can be modeled in a variety of different ways, with modeling decisions influenced by available input data, computational resources, and by the clinical problem at hand. For cardiovascular flow, a variety of techniques have been utilized ranging from reduced-order 0D and 1D models to 3D or multiscale models. The following section reviews a few of these different approaches and requirements for cardiac hemodynamic modeling.

5.2.1   Cardiovascular Modeling Approaches

The earliest attempts to model cardiac blood flow were done using *reduced-order models*, where either the entirety or a part of the cardiovascular system was modeled using electrical circuit descriptors, known as *0D models* [443]. In *0D models*, global cardiac hemodynamics can be modeled in direct conjunction with the rest of the circulatory system, without inclusion of spatial dimensionality (Fig. 10a). Instead, 0D models assume uniform distribution of volume, pressure, and flow in any regional model compartment. As such, they provide direct feedback of global change as a function of regional modifications [37, 235]. The Windkessel model [477] is a simple, but effective example where pressure and flow are related to resistance and capacitance of a vessel. Extensions abound [54, 255, 409, 511] with the most complex being the Guyton model [171, 177] where a larger network is used to describe the equilibrium state between hemodynamic output and electrolyte regulation (among other regulatory mechanisms). More recently, the CircAdapt [12] model has been used in a wide range of clinical applications. Through 0D models, the global behavior of the complex cardiovascular system can be reduced to an efficient, easily modifiable model that can capture flow and pressure behavior through cardiovascular compartments. However, 0D models do not provide information on regional behavior within a compartment, which may be critical for assessing disease or the effect of devices.

Increasing model complexity while limiting computational costs, *1D models* represent cardiovascular arterial segments using vessel centerlines and the 1D Navier–Stokes equations (Fig. 10b) [131, 401]. 1D models can reveal the pressure and flow changes along the full length of a vessel, being used to understand phenomena such as pulse wave transmissions through the arterial network [492]. However, in the case of cardiac modeling, 1D models are seldom used in isolation. Instead, they are more often used in conjunction with either a *regional* 0D or 3D

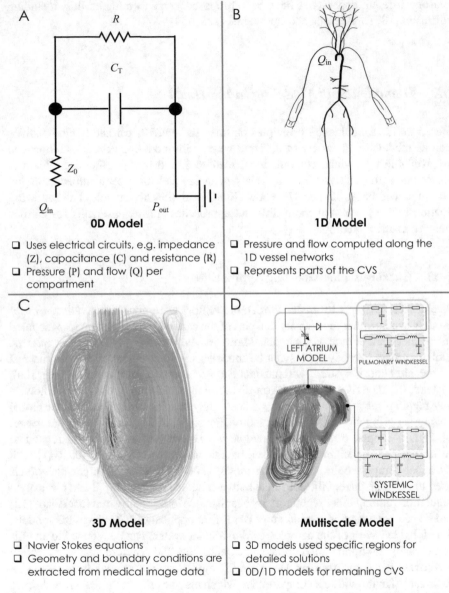

0D Model

❑ Uses electrical circuits, e.g. impedance (Z), capacitance (C) and resistance (R)
❑ Pressure (P) and flow (Q) per compartment

1D Model

❑ Pressure and flow computed along the 1D vessel networks
❑ Represents parts of the CVS

3D Model

❑ Navier Stokes equations
❑ Geometry and boundary conditions are extracted from medical image data

Multiscale Model

❑ 3D models used specific regions for detailed solutions
❑ 0D/1D models for remaining CVS

Fig. 10 Illustration of different modeling approaches in cardiac flow modeling. (**a**) 0D modeling, using electrical circuits as an analogue for the cardiovascular system. (**b**) 1D modeling, extending the reduced-order 0D model by allowing change in flow and pressure over the length of pre-described segments (adapted with permission [179]). (**c**) 3D modeling, isolating a portion of the heart and solving the Navier–Stokes equations to generate detailed insights into regional hemodynamic behavior. (**d**) Multiscale modeling, combining the aforementioned approaches, iteratively updating the boundary conditions of a 3D model by means of coupled reduced-order models

model. Specifically, the 1D model is used to represent the *remaining* parts of the cardiovascular system, not included in the regional model [133, 221]. In principle, 1D models are equally effective as 0D models when it comes to computational efficiency; however, they exhibit the same drawbacks associated with their difficulty in capturing more detailed hemodynamics.

In some instances, situations exist where resolution of local flow in fine detail is needed in order to understand physiological behavior: elucidating regional flow patterns in and around a stenosed valve, quantifying local flow changes in a stented coronary artery, etc. [67, 194]. In these cases, detailed hemodynamics can be simulated using *3D models* and the full Navier–Stokes equations (31a)–(31b), see Fig. 10c. For cardiac computational fluid dynamics (CFD), significant efforts are put into defining the problem at hand, with geometries extracted from medical image data and boundary conditions derived from invasive or non-invasive measurements. The computational demands associated with 3D modeling however mean that isolated cardiac entities are usually modeled, either in the form of single-chamber analysis [88, 414, 424], or using combined atrioventricular models [259]. Cardiac CFD has also become increasingly effective in investigating interventional outcomes, where optimal usage of left ventricular assist devices [316, 317], transcatheter valve implants [506] and coronary stents [344] have all been assessed in a 3D virtual environment.

A combination of any of the aforementioned approaches can also be used to create a multiscale model (see Fig. 10d). Here, the efficiency and computational simplicity of reduced-order models are used to describe general cardiovascular behavior, with the output of these subsequently fed and coupled into a higher-fidelity 3D model in areas where a detailed solution is required. This has been successfully employed specifically for cardiac modeling, where detailed intraventricular flow behavior can be mapped while maintaining accuracy in describing surrounding circulatory behavior [61, 192]. These models provide an attractive tool for evaluating highly resolved hemodynamics; however, it is important to understand how these models depend on the defined domains and chosen boundary conditions [219, 289, 528].

5.2.2 Requirements for Cardiovascular Modeling

Often the various approaches with which to model cardiac flow are selected by balancing the scientific question with the needs of the mathematical model. Modeling choices depend on appropriate data—including geometry, motion, and boundary conditions (see Fig. 11a). The following section reviews some fundamental considerations in cardiac flow modeling:

■ **Kinetics of Blood Flow in the Heart** are largely influenced by the type of flows commonly observed. In health, the majority of blood flow will include complex, non-turbulent flow, with Reynolds numbers ranging up to 2000 under normal healthy conditions [386]. In combination with the comparably large chambers

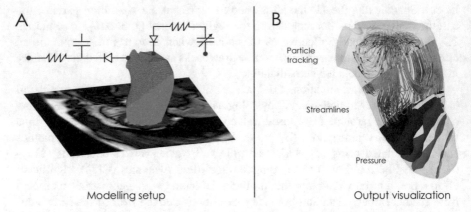

Fig. 11 Illustration of modeling setup (left) and output visualization (right) of a multiscale model. The model setup consists of a domain extracted from medical image data, isolating a specific cardiac entity of interest (here the left ventricle), with boundary conditions defined as per a reduced-order model. Output visualization then follows after completed simulation, where velocity and pressure data—available at each nodal point—can be represented in the form of traced particles, visualized streamlines, or pressure iso-contours, just to mention a few

experiencing flow at fairly high shear rates, a Newtonian description of blood flow kinetics is typically valid. However, if assessing isovolumic or highly regional flows—e.g., in the atrial appendage [404], around papillary muscles [487], or in distal coronary sections [226]—non-Newtonian constitutive models may be required (as per Sect. 5.1.1). Likewise, if attempting near-wall analysis, the choice of rheological model can impact estimated wall shear behavior [103]. At even finer resolutions, microscale flow simulations may necessitate inclusion of multiphase flow definitions, where different blood components are modeled as separate entities within a transportable mass fluid [238, 525]. However, the use of these simulations at large scales significantly increases computational demand. Turbulent flow may be induced in the heart pathologies, such as valvular stenosis [108, 461]. Turbulent flow models may be invoked if modeling high-turbulence phenomena, using e.g., the Reynolds-averaged Navier–Stokes equations or large eddy simulation formulations [129, 357].

■ **Defining the Dynamic Domain** of the human heart is essential for cardiac flow, providing the internal movement that drives blood flow. Early attempts were made using simplified domains [25, 101], and more recent models now often use non-invasive imaging data to define domain boundaries and segment cardiac entities of interest. A variety of modalities have also been used to extract cardiac geometries, each with its specific benefit. Echo-based imaging provides excellent temporal information but lacks image quality [35], computed tomography offers high spatial resolution but lacks temporal mapping [259, 487], and magnetic resonance imaging [414] offers competitive soft tissue differentiation and direct coupling to in vivo flow measurements yet lacks the ability to capture highly transient anatomical events. In all instances, domain extraction is achieved by image-based segmentation, where a multitude of methods have been developed ranging from threshold-based segmentation [266] to trained neural networks identifying isolated cardiac chambers [471].

■ **Boundary Conditions** are essential in order to describe and capture physiological behavior not explicitly included in the model itself. For flow modeling, boundary conditions usually involve a combination of prescribed wall motion and inflow and outflow conditions. In the case of *open-loop* 3D models, boundary conditions typically take the form of Dirichlet or traction conditions at defined boundaries [311]. In its most basic form, zero pressure or zero traction is commonly employed for open-loop 3D cardiac modeling; however, their simplistic nature limits their applicability [311]. The employment of Windkessel circuits is instead typically used in order to capture the interactive relationship between pressure and flow at defined boundaries. In different forms, the Windkessel circuit has been extensively employed in 3D cardiac modeling [245, 502], yet instances remain where even these approaches fail to appropriately capture physiological behavior [38, 298].

Boundary conditions are also required at model walls, with motion derived from imaging data. Commonly, no-slip (zero velocity) or no-penetration (normal velocity equal to that of the wall) conditions are prescribed [354]. Since the heart experiences fairly high degrees of geometrical change, the motion of the fluid and the heart wall must be coupled. This can be achieved either by geometry-prescribed methods (assuming that near-wall flow is exclusively governed by forces imposed by the wall on the fluid) or by refined FSI approaches (see Sect. 5.3).

■ **Valves**, which ensure directed flow of blood through the heart chambers, present a particular challenge in cardiac flow modeling. Their pronounced movements, intruding structures, and contact with intraventricular surfaces mean that they have significant impact on intracardiac blood flow patterns [348, 436]. At the same time, their complex structure and fine motion make them equally difficult to accurately characterize and define in a modeling setup. A range of modeling approaches have been presented to account for the impact of valve motion on cardiac hemodynamics. A basic approach is to model the valve as a binary orifice opening (either completely opened, or completely closed), without any physical valve leaflets [354, 414]. Advancing from this, time-dependent orifice openings have also been used [236, 424], creating more realistic transition stages between fully opened and fully closed configurations, while still neglecting the impact of the actual leaflets on the resulting flow fields. To include leaflets while still avoiding complete anatomical segmentation, simplified descriptors of intracardiac leaflets have been used in 3D models (e.g., extruding thin-walled tubes into the intracardiac space to resemble an open-valve configuration [259, 415]). Recently, complete valve structures have been included into modeling setups making use of high-resolution images to capture anatomical motion and morphology over the cardiac cycle [35, 504]. Appreciating the complex interplay between the thin valvular leaflets and the generated high-velocity flows, simulations using fluid–structure interaction (FSI) have also been used in the field of 3D cardiac flow modeling, where high-fidelity image data is used to define appropriate anatomical features [87, 176, 261, 356].

5.2.3 Quantifying Model Output

Regardless of the modeling approach, completed simulations are typically followed by a data analysis step, where output is quantified into comparable metrics either describing the assessed cardiovascular phenomena, or being coupled to other clinically assessed data points (Fig. 11b). The heart also exhibits distinct hemodynamic behavior that can be studied in detail using cardiac modeling: intracardiac post-valvular vortices helping to redirect blood from filling to ejection [383], intraventricular pressure gradients driving both systolic [381] and diastolic flow [80], and regional modifications in residence time or intracardiac force patterns indicative of pathology [16, 313].

For reduced-order models, output is generated as scalar outputs representing flow or pressure over time and either through compartments or along 1D vessel segments [443]. These entities may then be further processed to describe other relevant clinical measures, including aortic pulse wave velocities (PWV) [315], coronary fractional flow reserves (FFR) [44], or global wall shear stresses (WSS) [63]. For higher-order modeling, simulation output is typically generated as 3D velocity components and a scalar field at discrete spatial points. Quantification is typically performed by extracting regional mean values [340], estimating global maxima [149], or characterizing regions with velocity or pressure values above a given threshold, all in order to reduce simulated output to comprehensible and easily appreciated information. Generated velocity fields can also be utilized to visualize streamline paths [53, 349], including the differentiation of intracardiac flow components [88] and areas of flow stagnation [96], again attempting to condense output to scalar descriptors. Importantly, modification of streamline directions, intracardiac flow components, or excessive stagnation can all be linked to pathological manifestations [117, 136, 470], underlining the importance of capturing these entities through flow modeling studies.

Coupled clinical metrics can also be derived from 3D output including FFR [472] or relative pressure [305]. 3D simulations also allow for quantification of vorticity (curl of the velocity field), or other surrogate measures of vortex formation (e.g., Lagrangian coherent structures [474] or Lambda2 (λ_2) identification [112]). For structural coupling, quantification of WSS or the oscillatory shear index has also become central in many cardiac flow studies [244, 254, 335]. Regardless, the output of any flow simulation is most effectively used when correlated or coupled to conjunctive clinical phenomena.

5.3 Fluid–Structure Interaction in the Heart

Many physiological processes cannot be described through either flow or solid mechanics alone. Instead, instances exist where the interplay between the tissue and hemodynamic states is fundamental to the functioning heart. Examples include aortic stenosis (where LV outflow is impeded by structural changes in the aortic valve), obstructive hypertrophic cardiomyopathy, and implantable devices (e.g., valves, assist devices). Here, fluid–structure interaction (FSI) modeling circumvents the limitations of standalone models by coupling the two problems. Different FSI methods achieve this by imposing various kinematic and kinetic constraints. In practice, the challenges that these methods typically have to overcome are related to the differences in constitutive behavior between blood and tissue.

For cardiac FSI, research has initially focused on the study of flow around the valves, pioneered by McQueen and Peskin [322, 387]. These models, based on the immersed boundary approach [386], modeled the tissue as a collection of 1D fibers embedded in a fluid domain. Continued interest and technological advances have led to a series of related studies focusing on both the mitral [62, 467] and tri-leaflet valves [262]. To examine the importance of FSI in the characterization of valve dynamics, Lau et al. [260] compared a purely structural model of the mitral valve with an FSI-based one and found significant differences in the resulting closure patterns, even in idealized rigid flow domain geometries. Other studies have also been able to incorporate prescribed LV motion in order to more accurately capture the flow patterns around the valves, for both mitral [163, 503] and aortic [459] configurations. Focusing on the aortic arch region, Hsu et al. [205] were able to show how including deformable walls into an FSI aortic valve model resulted in more realistic representations of flow, as opposed to using rigid walls. A more advanced version of this model was later used to analyze the effect of valve thickness [224], showing how thinner designs can cause leaflet flutter, a phenomena associated with blood damage and material fatigue.

A separate area of study has been that of coupling hemodynamics and myocardial biomechanics. Watanabe et al. [501, 502] developed an idealized 3D model of the left ventricle incorporating tissue, blood flow, and electrophysiology, capable of simulating both passive filling and active contraction. In this chapter, a low-order conforming approach was used to couple the fluid and solid problems, limiting the resolution and accuracy of the results. Alternatively, Cheng et al. [66] presented an approach where the solid was modeled as a hyperelastic shell, lacking the anatomical details of Watanabe's, but being better suited to resolve its constitutive components. To account for differences in constitutive behavior between the two phases, a non-conforming monolithic framework [360] was used to build patient-specific models able to simulate both active and passive mechanics [361]. Later adaptations of this model were used to study the effect of left ventricular assist devices [316–318] as well as diastolic function in hypoplastic left heart patients [89, 486]. More recently, Gao et al. [143] demonstrated the feasibility of merging the LV and mitral valve into a monolithic anatomically accurate FSI model.

6 Applications of Biomechanical Modeling in the Heart

In previous sections, methods have been presented for modeling the hyperelasticity and viscoelasticity of myocardial tissue (Sect. 3), growth and remodeling (Sect. 4), and blood flow (Sect. 5). Cardiac pathologies often result in microstructural and tissue-level changes, which impact material behavior, changing its hyper- and viscoelastic responses. Additionally, diseases can cause abnormal growth, on an organ level or localized scale, as well as reorganization of cardiomyocytes. These tissue- and organ-level changes can lead to alterations in cardiac hemodynamics, changing blood flow patterns, for example. Therefore, with appropriate data measuring the pathological changes, the previously proposed modeling methods can be used to better understand the underlying mechanisms behind different pathologies.

In this section, four cardiac pathologies will be reviewed: dilated cardiomyopathy (DCM), hypertrophic cardiomyopathy (HCM), aortic stenosis (AS), and myocardial infarction (MI). They represent diseases with significantly different pathological presentations, including increased preload (DCM), increased afterload (AS), decreased global compliance (HCM), and decreased contractility (MI). Although their primary characteristics are distinct, progression of each can lead to heart failure and adverse outcomes. A brief overview of each disease is presented in Fig. 12. These pathologies often have complex etiologies with more than one confounding factor (e.g., obesity, diabetes, hypertension) contributing to its presentation and progression. Therefore, the disease's phenotype in each patient is unique, complicating diagnosis, prediction of its development as well as treatment planning. However, data-informed models can provide mechanistic explanations to help clinicians predict disease progression as well as test possible therapies in patient-specific in silico settings.

6.1 Dilated Cardiomyopathy

Dilated cardiomyopathy (DCM) is a chronic cardiac disorder, predominantly characterized by the enlargement of the cardiac chambers along with a general deterioration in cardiac performance. Although originally considered a rare disease, recent findings that showed that DCM has a familial substrate [142] enabled the diagnosis of DCM in asymptomatic young patients, raising its incidence to 1 in 250 individuals [186]. DCM is considered a primary etiology for heart failure, while sudden cardiac death and severe arrhythmias represent possible outcomes of the disease, contributing to its high mortality rates (25% over 1 year and 50% over 5 years). Common risk factors include the male gender, advanced age as well as drug and alcohol abuse. DCM is also a common outcome resulting from many other diseases (e.g., coronary artery disease, hypertension, viral infections). Often, no specific etiology can be determined (idiopathic DCM), although recent studies have shown that up to 50% of idiopathic cases could be identified as familial DCM

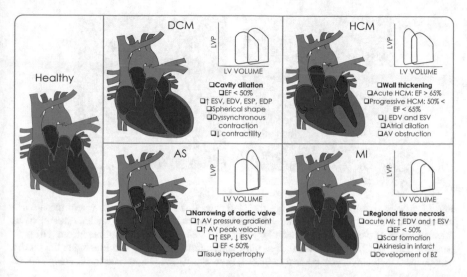

Fig. 12 Illustration of cardiac geometries (top row) and pressure–volume loops (second row) for a healthy heart as well as four diverse cardiac pathologies covered in this section: dilated cardiomyopathy (DCM), hypertrophic cardiomyopathy (HCM), aortic stenosis (AS), and myocardial infarction (MI)

[142]. A wide variety of gene mutations have been linked to DCM, the most frequent being a gene variant altering titin isoforms and impacting myocardial stiffness [186]. Among more than 30 genes implicated in the development of familial DCM are also genes encoding proteins of the sarcomere—also present in HCM cases— and proteins relevant to ion channels and z disks that could alter myocardial contraction. Due to the vast heterogeneity in genetic causes and the overall multifactorial development of the disease, identification of the underlying etiology—and thus selection of appropriate treatment—constitutes a challenging task requiring an improved understanding of DCM pathophysiology.

6.1.1 Pathological Changes

The main characteristic of DCM is the dilation of the LV, often accompanied with enlargement of the right ventricle and of the atria. LV enlargement is reflected in substantial increases in both the end-diastolic and end-systolic volumes. Changes in volume along with impaired contractility result in a reduction in stroke volume. While changes in wall thickness are moderate, a pronounced change is observed in the shape of the heart, with the LV transitioning from elliptical to spherical. These geometric changes are thought to occur progressively, as a result of sustained chronic volume overload on the myocardial wall, e.g., increased venous return as a consequence of ventricular impairment (myocardial infarction, viral inflammation, coronary heart diseases). Interestingly, DCM can also result from sustained pressure

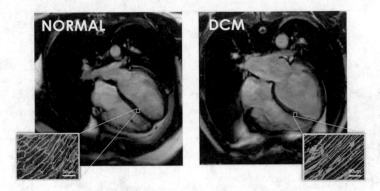

Fig. 13 Long-axis MR images of a healthy (left) and a DCM (right) heart, where the disease has caused the enlargement of both the left and right ventricles. DCM also induces alterations in microstructure, as demonstrated by histology images. (©2019, Elsevier. Reprinted from [78] with permission.) The cardiac sections, stained with wheat germ agglutinin that highlights the border of individual cardiomyocytes (shown in black), show that DCM is associated with elongation of cardiomyocytes

overload [412], as an end-stage pathology of hypertrophic remodeling [160]. Histologically, DCM is associated with interstitial fibrosis and degradation and disruption of fibrillar collagen [21], while elongation of cardiomyocytes (see Fig. 13 insets) and their serial rearrangement are considered responsible for the observed changes in LV size and shape [268].

Importantly, the anatomical remodeling observed in DCM hearts is associated with marked alterations in heart function. The heart's mechanical efficiency declines substantially, as evidenced by ejection fraction as low as 20% [351]. Systolic function is severely impaired, characterized by suppressed contractility [184], prolonged relaxation times, and ventricular conduction abnormalities [4]. Deterioration in the heart's systolic function is linked to severe disease outcomes, suggesting the potential of systolic indices to assist in risk stratification of DCM patients. Marked differences are also observed in DCM hemodynamics, with elevated end-diastolic and end-systolic pressures. Reported abnormalities in LV filling due to low intraventricular pressure gradients [530] might contribute to diastolic dysfunction, while stronger and larger diastolic vortices might act as a potential compensatory mechanism, facilitating blood transport without large pressure losses [39].

The decline in the heart's efficiency is also reflected in myocardial wall motion. Speckle tracking echocardiography [105] and tagged MRI studies revealed significant reduction in radial, circumferential, and longitudinal strains in DCM cases throughout the cardiac cycle. Notably, long-axis strain, established as a reliable indicator of global LV function, has been shown to decrease in DCM and correlate with a higher rate of cardiac events [403]. Substantial reduction was also reported in torsion and torsional shear [71], with abnormal rotational patterns observed compared to healthy individuals [435, 531]. Recent advances in DTMRI demonstrated similar trends in the changes in fiber architecture through the cardiac cycle, whereby the steepening of helix angles observed in healthy hearts during contraction was almost absent in DCM hearts [490]. A key finding in many studies was the regional variability in displacements and strains in DCM hearts, with a more severe decline in function observed in the septal LV wall compared to the lateral. Mechanical dyssynchrony is also reported in DCM patients, a potential clinical index associated with deteriorated LV systolic function [342].

6.1.2 Diagnosis and Therapies

DCM is commonly clinically diagnosed using imaging modalities such as echocardiography, cardiac MRI, or CT, as ECG is often remarkably normal in DCM cases. Diagnostic criteria for DCM include an ejection fraction <0.45 and LV end-diastolic volume or dimension larger than two standard deviations from normal values, corrected for age and body surface area.

Numerous treatments have been suggested for DCM patients aimed at limiting the development of the disease and treating heart failure symptoms. DCM interventions start from lifestyle changes and extend to pharmacological therapies and surgical procedures. DCM patients are typically administered angiotensin-converting enzyme (ACE) inhibitors and beta-blockers, which have been shown to significantly increase ejection fraction, lower end-systolic pressures, and reduce chamber size in DCM patients [110]. Improvement in overall function was also observed with administration of vasolidator drugs such as nitroprusside, by enabling lower end-systolic and end-diastolic pressures [183]. Furthermore, positive inotropic agents contribute to the prevention of cardiac dilation and severe cardiac events, by increasing calcium sensitivity of myofilaments and enhancing cardiac contractility [104]. Importantly, studies have highlighted the possibility for reverse LV remodeling through tailored pharmacological treatments with up to one-third of patients showing substantial improvement in terms of chamber size and ejection fraction [327].

In DCM cases where pharmacological options might not suffice, implantable devices are often considered. Specifically, cardiac resynchronization therapy (CRT), with pacing devices applied on either one or both ventricles, has drawn particular attention as a means of synchronizing cardiac contraction and improving systolic function. Large clinical trials with heart failure patients demonstrated the ability of CRT to reduce mortality rates and hospital admissions, when used solely or in combination with implantable cardioverter defibrillators (ICD) [429]. Studies focusing on DCM patients have reported enhancement in cardiac contractile function and amelioration of electrical dyssynchrony, especially with single-site pacing in regions with the greatest conduction delay [229]. Comparisons of CRT with pharmacological treatments involving positive inotropic agents suggest that CRT can improve DCM prognosis at a low cardiac energy cost [351], whereas drug therapies tend to increase energy expenditure in an already failing heart. Finally, implantation of left ventricular assist devices (LVAD) may be suggested for end-stage DCM patients not responding to the aforementioned treatments or awaiting heart transplantation [476].

6.1.3 Modeling Approaches

As DCM is intrinsically linked to the structural integrity of the heart muscle, numerous modeling efforts focused on examining changes in tissue mechanics, including material properties and stresses. Early works relied on bulk measures such as pressure–volume loops to infer changes in chamber compliance and reductions in ventricular stroke work in DCM [8]. Rough estimates of wall stresses (e.g., thick-walled ellipsoid formulas) were employed to demonstrate elevated wall stress in DCM hearts [8, 531], along with increased myocardial stiffness [48] and reductions in regional work. Combined with heterogeneities in regional myocardial work [521], modeling studies have identified cardiac mechanoenergetics as a major determinant of the disease, with DCM hearts failing to satisfy the increased metabolic demand [249]. Similar conclusions were also drawn by finite element models that incorporated physiologically relevant material descriptions [370] and personalized geometries from animal data [79]. Elevated stiffness in DCM patients was also suggested by in vivo models focusing on reliable parameterization, which utilized comprehensive kinematic data for stiffness quantification [173]. By incorporating patient-specific anatomies, loading, and boundary conditions, such models were also able to infer that DCM hearts need to generate higher active tension in order to drive the cardiac cycle [18].

Mathematical models can also provide insights into the mechanisms underpinning structural and functional changes accompanying DCM. For instance, models have been used to assess dilation and changes in fiber architecture as potential mechanisms for common DCM characteristics. By combining finite element modeling with longitudinal imaging and cavity pressure data, the role of dilation as a compensatory mechanism was investigated, revealing that initial dilation might improve systolic function by altering tissue stiffness and chamber compliance [79]. Additionally, reduction in ejection fraction and abnormalities in ventricular twist were found to be predominantly caused by damage in the myofibrils, rather than by ventricular dilation [164]. Similar conclusions were drawn by a biomechanical model utilizing DTMRI data, which inferred that the experimentally observed steeper helix angle contributes to reduced ventricular torsion [490]. The study used computational simulations to verify the data-derived conclusion of a minimal change in helix angle during contraction in DCM, and to assess LV size and shape as potential mechanisms for abnormalities in fiber orientation, reporting minimal effects.

6.1.4 Challenges and Future Directions

Challenges in modeling and treatment of DCM remain manifold as its pathophysiology and related disorders are yet to be fully understood, be they associated to the heart's electrical activity [5], myosin gene mutations [141, 523], or deleterious neurohormonal activation patterns [180]. Biomechanical models that have the ability to reliably predict DCM disease progression are still in an early stage. While there are numerous models that characterize the shape and volume changes of the typical eccentrically enlarged left ventricle [151, 159, 233], the primary stimuli for DCM development as well as its long-term progression still need further attention. For example, multiscale continuum models, which examine the relationship between the cavity volume and myocyte lengthening through sarcomerogenesis [78], could deliver further insights into the link between micro- and macroscopic changes.

Modeling CRT treatment for heart failure patients by use of electrophysiological, electromechanical, or hemodynamic models [264] has the potential to aid therapy planning, yet still suffers from the difficulty of reliably estimating model parameters, often requiring extensive amounts of patient data. Models that quantify the outcome of an experimental left ventricle volume reduction surgery have confirmed the potential hazards associated with this type of DCM treatment [500] but, in the future, could help to design new surgical treatments that minimize risk.

Furthermore, ventricular assist devices (VADs), which are used in patients with end-stage heart failure as a consequence of advanced DCM, are still a topic of ongoing research. While the primary scope of these devices is bridge-to-transplant solutions [400], improvements in device technology with reduction in adverse event rates qualify VADs for destination therapy [115, 419, 433, 515]. The success of novel biventricular assist devices, which can be applied minimally invasively, has been shown in both animal experiments [220] as well as computational modeling studies [193]. Patient-specific modeling, with personalized model parameters, which combines electrophysiology, biomechanics, and hemodynamics [230], pose a promising approach for designing therapies and tracking an individual's disease progression.

6.2 Hypertrophic Cardiomyopathy

Hypertrophic cardiomyopathy (HCM) is a chronic cardiac disease characterized by an increase in wall thickness, myofiber disarray, and tissue fibrosis. HCM is a primary etiology caused by an inherited autosomal-dominant genetic mutation. Although previously thought to occur in every 1 out of 500 people, recent genetic testing estimates that the genetic mutation is present in nearly 1 in 200 people [434]. Many individuals with the mutation have no (genotype positive/phenotype negative) or only mild complications. However, in phenotype positive individuals, tissue hypertrophy and remodeling can lead to severe arrhythmias, heart failure, and, in some cases, sudden cardiac death (SCD) [307]. Additionally, in some patients, severe asymmetrical hypertrophy in the septum along with systolic anterior motion (SAM) of the mitral valve leaflet can lead to aortic outflow tract obstruction [309], treated by septal myectomy or alcohol septal ablation along with mitral valve repair. Due to the diverse presentation of the disease along with the potential severity of outcomes, a better understanding of the disease progression is needed in order to identify patients at a higher risk of severe outcomes [367], predict pathological changes, and plan surgical treatments [520].

6.2.1 Pathological Changes

HCM is caused by genetic mutations that affect the assembly of sarcomeres, the force-generating unit within cardiomyocytes. Concentric growth occurs, either regionally or globally, as sarcomeres are deposited in parallel. Due to its progressive nature, HCM can be classified into four stages. *Non-hypertrophic HCM* represents the state when no hypertrophy can yet be observed. However, stage I shows existing abnormalities, such as impaired relaxation in the left ventricle (LV) and a mild degree of dilation in the right ventricle (RV). Additionally, in this phase, elevated amounts of type I collagen have been observed [366]. Despite these modifications, the cardiac function in stage I is within a normal range.

With increasing mass deposition in the LV, the disease transitions from stage I to stage II. An increase in wall thickness can occur in any region but is most common in the interventricular septum [302]. Diffuse hypertrophy can also occur in the RV [310] and is correlated with RV dysfunction. Additionally, the myocyte orientation changes and the parallel alignment of the fibers degrades over time [147] (Fig. 14

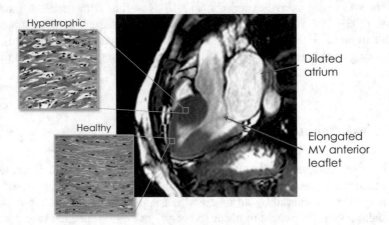

Fig. 14 Long-axis MR image of an HCM heart showing a hypertrophic septum, an elongated anterior mitral valve (MV) leaflet, tending toward obstruction of the outflow tract, and a dilated atrium. Insets show histology images of healthy and hypertrophic myocardial tissue (adapted with permission from [516], Copyright (2005) National Academy of Sciences, U.S.A.) illustrating the disarray observed in HCM hypertrophic myocardium with H&E staining. Blue: nucleus, red: cytoplasma, varying red: collagen fibers and ECM

insets). The *Classic Phenotype* defines stage II when wall thickening and fiber disarray in the hypertrophic region are prevalent. In these regions, the heart tissue exhibits hypercontraction in the systolic phase and shows impaired relaxation during diastole [127]. Therefore, the LV ejection fraction is often >65% in the absence of enhanced fibrotic tissue. Three quarters of HCM patients stabilize at this stage.

During *Adverse Remodeling* (stage III), the tissue undergoes structural modifications that lead to increased LV fibrosis, resulting in a deterioration of LV function (50% < LVEF <65%). Through a combination of cardiac magnetic resonance (CMR), late gadolinium enhancement (LGE), and post-contrast T1 imaging, it is possible to link diffuse myocardial fibrosis in the hypertrophic region of the LV to abnormal diastolic function [113]. Fibrosis in the LV myocardium can also result in non-sustained LV tachycardia [445]. Besides remodeling in the LV, the LA may also become dilated. Caused by the dysfunction and reduced EF, a part of this subgroup continues into stage IV, or *Overt Dysfunction*. Here, the LV exhibits a severe functional deterioration, which subsequently ends in heart failure. This stage is associated with extreme degrees of fibrosis and pathologically reduced LVEF <50% [302].

Up to two-thirds of all HCM patients experience obstruction of the aortic outflow tract [290] due to (a) septal hypertrophy, which protrudes into the ventricular cavity, (b) elongation of the anterior mitral valve leaflet, and (c) augmented anterior motion of the papillary muscle attached to the elongated leaflet. Together, these pathophysiological changes not only result in obstruction of the outflow tract, but also mitral valve regurgitation, due to improper closure of the mitral valve leaflets [440] (Fig. 14). Obstruction of flow through the aortic outflow tract is marked by an increased pressure gradient across the valve. This increases the cardiac work needed to maintain the stroke volume, in turn, leading to further remodeling and progressive heart failure. Due to MV regurgitation, the left atrium becomes dilated and exhibits high left-atrial pressures [366]. Therefore, atrial dilation is a common pathological change in patients with later stages of HCM. Alongside MV regurgitation, LV outflow tract obstruction (LVOTO), and atrial dilation, atrial fibrillation becomes more common, characterized by dyssynchronous contraction of the atria [365].

6.2.2 Diagnosis and Therapies

Clinical diagnosis of HCM commonly involves an ECG, with different ECG abnormalities suggesting different pathological changes observed in HCM. Echocardiography is also commonly performed to identify myocardial sections with abnormal thickness (>13 mm), while Doppler echocardiography is used to investigate the presence of obstruction or SAM. Cardiac CT or MRI with LGE might also be advised, to detect the extent of scar tissue and fibrosis.

Most first-line therapies for HCM patients include medications such as beta or calcium channel blockers [162] that address hemodynamic changes. However, for patients with interventricular hypertrophy along with severe outflow tract obstruction, hypertrophied tissue is either treated through alcohol septal ablation or surgical excision of excess tissue in the septum through a myectomy. Alcohol septal ablation is a nonsurgical treatment in which tissue is ablated in order to induce necrosis. Scar tissue, thinner than the previously hypertrophic tissue, replaces the dead tissue in the hypertrophic region [196]. Scar tissue in the ventricle causes a reorganization of the electrical conductance, which can result in ventricular tachycardias [321] or arrhythmias [362]. In the case of a septal myectomy, the chest cavity is opened and hypertrophic tissue is excised from the affected region. In patients with mitral valve regurgitation or SAM, additional surgical treatments (e.g., papillary muscle relocation) are often needed, making concomitant myectomy an appropriate option [153]. Adequate pre-surgical planning is necessary to ensure that the procedure successfully removes an adequate amount of tissue to restore systolic ejection [520]. In patients who are deemed to be at high risk of SCD due to chronic tachycardia or arrhythmia, implantable cardioverter defibrillators (ICD) are implanted. However, it has been shown that, in HCM patients, risk is often overestimated and ICDs are implanted unnecessarily [368].

6.2.3 Modeling Approaches

Modeling studies have been used to provide a mechanistic understanding of the wide range of pathological changes observed in HCM. Models have focused on studying not only the changes in regional morphology of HCM ventricles, but also the local myofiber disarray, and alterations in blood flow patterns and electrical propagation.

Increased wall thickness, myofiber disarray as well as shortened sarcomeres were shown to each be necessary in order to accurately model altered strain patterns seen experimentally in mouse hearts with septal hypertrophy [480]. Computational modeling has also been applied to study blood flow patterns in the HCM LV, particularly with outflow tract obstruction. Using CFD simulations, Su et al. [466] demonstrated that, during both systolic contraction and diastolic filling, flow patterns in an HCM heart were considerably different than in a healthy individual. For example, multiple small vortices developed in the constricted HCM LV rather than two large vortices, as seen in the healthy case. Additionally, studies of flow patterns pre- and post-myectomy using FSI showed a reduction in pressure gradients across the aortic valve as well as lower shear stresses exerted on the anterior mitral valve leaflet after surgery [95, 206]. A parametric study that varied the degree of systolic anterior motion of the mitral valve leaflet in a computational fluid dynamics model similarly associated higher pressure gradients and wall shear stress as well as greater vorticity in the ascending aorta with severity of obstruction [137]. Simulations of blood flow in the constricted HCM cavity, represented as flow through a narrow tube, illustrated that increased pressure gradients alone may not be synonymous with obstructive HCM [380]. Therefore, wall shear stress and downstream vorticity may provide better metrics by which to choose suitable candidates for myectomy, plan which regions of tissue to remove, and measure success of the following surgery.

Due to the development of both ventricular tachycardia and arrhythmia, which can lead to SCD, abnormal electrophysiological patterns pose a critical problem for HCM patients [308]. However, the mechanisms that contribute to pathological conductance patterns are not well understood. An early computational modeling study [505] proposed that increased wall thickness alone could not account for the change in electrical conductivity but that the addition of myofiber disarray was needed to generate the abnormal pattern in ventricular depolarization, typical in HCM patients. However, more recent studies [294, 295] revealed that only alterations in coupling between Purkinje fibers and surrounding myocardium, not wall thickness, fiber disarray, or myocardial abnormalities, were able to explain the alterations. Studies were also able to use machine learning models to separate HCM patients into distinct phenotypes based on their ECG measurements alone [293]. Multiscale computational models illustrated that negative inotropic agents reduced outflow tract gradients in HCM patients with LVOTO [75].

6.2.4 Challenges and Future Directions

Numerous challenges still remain in modeling HCM, in terms of examining kinetics of the HCM heart, investigating blood flow in obstructed [380] and post-myectomy cases as well as studying tissue growth and remodeling. In contrast to typical growth modes such as eccentric and concentric growth discussed in (Sect. 4.3), growth and remodeling in HCM demonstrate a more heterogeneous growth pattern, adding complexity to the modeling problem. The main challenges for modeling growth in HCM are to identify an accurate kinematic growth tensor based on the microstructural changes as well as a sufficient growth evolution law that mimics the same growth behavior as is observed in patients. Further information is needed in order to understand the triggers for heterogeneous growth in HCM. Multiscale models, as proposed by Cansiz and colleagues [55], which provide a bridge between protein-level mutations, sarcomeric changes, and contractile function, may enable a better understanding of the course of growth and remodeling in HCM.

Morphological differences alone may be a predictive factor for LVOTO in HCM hearts. For example, mitral valve angles with respect to the long axis were significantly different between HCM patients with and without obstruction [312], suggesting that shape alone may be a predictive factor for future obstruction. Although interesting, small sample sizes limit the power of this study. However, a recent large-scale multi-center study of 2773 patients with HCM [247, 352] has provided a rich set of data including imaging (short- and long-axis cine, T1 mapping, LGE), ECG, and blood tests at multiple time points. This rich dataset will enable similar shape and functional analyses for a large cohort, providing valuable insights into cardiac morphology in HCM patients.

Further, personalized electromechanical models of HCM [2], including accurate activation properties of the myocardium, are needed in order to effectively assess the need for ICDs in individual HCM patients. Large-scale studies of patients' pre- and post-surgical intervention would aid the development of models of tissue mechanics and blood flow of the ventricular septum and mitral valve. These models could then assist clinicians in pre-surgical planning phases, when deciding how much and which tissue to remove in a surgical myectomy. Myectomy surgery has been shown to be effective in reducing mortality in patients with obstructive HCM, yet is still not widely used [520]. Evolution of surgical techniques for repair of concomitant abnormalities, such as mitral valve leaflet SAM, may also be aided by modeling studies. Finally, HCM, although often thought of as a disease of the left ventricle, impacts the entire heart as seen by RV wall thickening [310], increased right-atrial pressures [319], reduced RV systolic function [128] and left-atrial dilation [90]. Therefore, whole heart mechanical models of HCM should be developed in order to account for and predict whole heart structural and functional behavior and changes.

6.3   Aortic Stenosis

Aortic stenosis (AS) is a common cardiac disease, characterized by the narrowing of the aortic valve orifice, leading to obstruction of and increased pressure gradients across the aortic valve. AS is the most common of all valvular disorders [397], present in about 5% of the population at age 65, with prevalence increasing with age [372]. This comparatively high prevalence is complicated by a predominantly asymptomatic onset, followed by significant deterioration of cardiac function once symptoms arise [411]. Deterioration is also aggravated by significant morbidity: individuals with symptomatic AS have an annual mortality rate of around 25% [57] and an average survival of 2–6 years after symptom onset [150, 411]. Thus, there is an urgent need for improved risk stratification, and better understanding of what drives disease progression.

6.3.1   Pathological Changes

The pathogenesis of AS varies between patients. However, three representative disease categories have been recognized: calcific, congenital, and rheumatic AS. *Calcific AS* is by far the most common form of AS, with calcific deposits forming on the valve leaflets (see Fig. 15a1–b2). Being closely related to atherosclerosis [396], calcific AS originates from local inflammation and altered hemodynamic stresses [107]. Biomechanical and hemodynamic alterations also play distinct roles in calcific AS, with modified shear stresses impacting the acceleration of valvular sclerosis [107], and patterns of intravalvular stress being linked to local calcific depositions [374]. *Congenital AS* is a less common form, manifested as either narrowing of the valve orifice or LV outflow tract, or in the form of a bicuspid aortic valve [231, 405]. Congenital AS anatomies give rise to flow patterns that have been postulated as both driving disease [157], and triggering associated aortic disorders [42, 204]. *Rheumatic AS* is the least frequently occurring form in the western world, following the decreasing incidence of rheumatic fever [432]. Worldwide, however, rheumatic AS is still a distinctive form of AS, with the development being a direct consequence of infection, causing valvular damage and fusion of leaflet

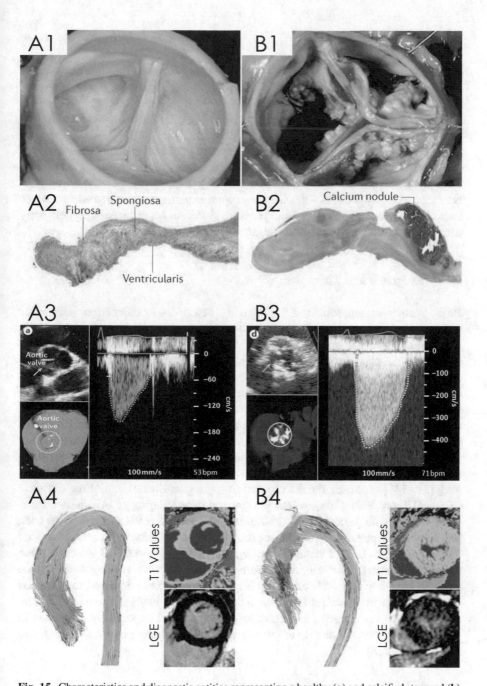

Fig. 15 Characteristics and diagnostic entities representing a healthy (**a**) and calcified stenosed (**b**) aortic valve. From the top: post-mortem tissue specimens, showing the smooth unaffected native valve (**a1**) compared to the apparent calcium deposits in calcific AS (**b1**). Histopathological data shows similar native trilaminar structure in the native valve (**a2**) compared to calcific nodules present in calcific AS (**b2**). The affect on flow is also apparent in Doppler echocardiography where native velocities (**a3**) are severely elevated in the stenosed case, and where calcium deposits are detectable on CT short-axis views (**b3**). Similar effects on flow can be observed by means of 4D Flow MRI with elevated post-stenotic velocities shown in the stenosed patient (**a4**) vs. (**b4**). Similarly, high-fidelity MRI (using T1 mapping and late gadolinium enhancement (LGE) sequences) showcases signs of myocardial fibrosis in AS compared to the native equivalent (**a4**) vs. (**b4**). (©2016, Springer Nature. Adapted from [284] with permission)

commissures [303]. Rheumatic AS is also frequently associated with conjunctive aortic regurgitation, causing complex and combined pressure and overload on the already congested left ventricle.

While pathogenesis differs between groups, the pathophysiology is similar over the entire AS population. In general, complications are directly associated with cardiac pressure overload and conjunctive hypertrophic developments. As valve area decreases, the pressure gradient over the valve increases, requiring increased cardiac work to maintain sufficient cardiac output [57]. Left ventricular hypertrophy is here seen as a compensatory mechanism, suppressing increase in myocardial stress caused by increasing intraventricular pressure [165]. As AS progresses, hypertrophy can further add to congestive heart failure symptoms [207].

6.3.2 Diagnosis and Therapies

Being mainly asymptomatic, AS is usually detected in conjunction with other diseases or during routine stethoscope auscultation (indicated by typical systolic murmurs). Echocardiography represents the main diagnostic tool with which to assess and classify obstructed flow [32], with assessment based on aortic valve area (1 cm^2 indicating severe AS), peak velocities (>4 m/s indicating severe AS), and peak pressure drop (>40 mmHg indicating severe AS) [33] (see Fig. 15a3–b3). In particular, the pressure drop provides direct quantification of the work required to maintain systemic blood flow. In standard care, pressure drops are derived by assessing peak velocities and utilizing the simplified Bernoulli equation [182]. Albeit simple and effective in instances of severe obstruction, assumptions made within its derivation, however, obstruct its usage as a general quantification tool [34, 146]. Instead, the incorporation of higher-dimensional flow imaging (e.g., 4D Flow MRI [304], ultrasound vector flow imaging [514]) allows for a more complete fluid-mechanical description, and several novel methods have been proposed for improved estimation of pressure drops in clinics [102, 109, 248, 305] (see Fig. 15a4–b4). Gold standard treatment in cases of severe AS is aortic valve replacement. Alternative surgical treatments also exist with varying performance reported [85], however, still markedly improving outcomes compared to untreated controls. Surgical valve replacement has thus moved beyond standard of care, and challenges only remain in certain sub-cohorts. Most importantly, outcomes in bicuspid valve patients remain unknown, where large clinical trial data remains to be collected [218].

6.3.3 Modeling Approaches

With AS driven by, and severely impacting, the local hemodynamic state, cardiac flow modeling has been extensively utilized to study AS. Similarly, with the highly dynamic and flexible leaflets present at the site of high-velocity jets, AS has also become a mainstay for FSI modeling within the cardiac domain.

Early attempts to model AS were performed using reduced-order descriptions, where the flow in and around the stenosis was described as a sequence of inertial and loss terms coupled in series [353], or by using larger-scale connectors to capture systemic cardiovascular behavior in the presence of AS [468]. Using these simplified models, predictions of 1D flow and pressure could be achieved as a function of stenosis severity. However, they did not address any details on how local alterations in flow state might impact disease progression. Nevertheless, instead of detailing local flow, contemporary reduced-order models have been effective in understanding how AS affects other compartments of the cardiovascular system, or how AS, in conjunction with other co-morbidities, impacts homeostatic perfusion. Reduced-order models have been used to address the impairment of coronary flow reserves in the presence of AS [145], to evaluate left ventricular workload in concomitant AS and aortic coarctation [235], and to quantify the joint predictive power of peak velocity, pressure drop, and anatomical assessment of the ventricular domains in risk stratification of AS patients [36]. In an attempt to include the effect of valve motion, tailor-made lumped parameter models have even been suggested [346] to be incorporated in subsequent AS studies [375, 376].

The flow in and around the obstructed valve has also been studied using 3D modeling. While early attempts were performed in simplified geometries [91] or in idealized vessels without leaflet-like structures [278, 441], they carry limited information regarding regional flow in AS. Using medical imaging as input, patient-specific 3D models have also been employed to study several different AS-related pathological scenarios. However, the dynamic movement of the leaflets still complicates inclusion of detailed valvular structures and, in many instances, 3D modeling of valvular disorders (including variations of AS) still utilizes statically defined leaflets [423, 532]. In more contemporary studies, refined image-based valve tracking has enabled the inclusion of anatomical valve leaflet motion into 3D flow modeling, showing how valve leaflet inclusion significantly impacts derived estimates of regional wall shear stresses, oscillatory flow, and turbulent kinetic energy [509]. Similar image-driven simulations have also been performed for bicuspid anatomies [46], again showing how leaflet inclusion significantly impacts flow. For AS, Weese et al. used a combination of CT-based segmentation and planimetry to create anatomically accurate valve models, showing how the measures of geometric and effective orifice area correlate strongly with modeling-derived pressure drops [504]. CT-based work has also been used to propose anatomical surrogate metrics to quantify underlying pressure drops [194], and likewise, leaflet-resolved 3D modeling has been used to elucidate apparent measurement differences between different imaging modalities [336].

While 3D flow modeling provides insights into the hemodynamics of AS, these models do not convey any information about structural integrity, nor do they capture any interplay between hemodynamic alterations and structural modification. For this, FSI modeling represents a viable alternative, and a variety of numerical approaches have been used to study aortic valves including immersed boundary methods [356], fictitious domain approaches [489], and overlapping domain techniques [86]. Questions relating to AS have also been addressed: De Hart et al.

showed how collagen fiber orientation optimizes intravalvular stress distribution during ejection [87]; Halevi et al. illustrated how calcific deposits generate eccentric flow jets and disturbed shear stress patterns, also implying that mechanical alterations are more dominant in driving progressive AS growth as compared to flow alterations [176]; Lavon et al. performed FSI assessment of bicuspid geometries showing how a decreased opening angle of the non-fused cusp relates to higher maximal stresses, giving evidence as to why certain AS anatomies develop symptoms earlier than others [261]. In an interesting multiscale FSI approach, Weinberg et al. investigated drivers of progressive calcific AS, differentiating outcome as a function of tricuspid versus bicuspid aortic valve patients [508]. Here, bicuspid geometries generated greater leaflet flexure following more excessive jet formations but still experienced similar cellular deformations as compared to their tricuspid counterpart following fibrotic wrinkling on a cellular level. As such, the simulation supports the hypothesis that differences in calcification patterns between tricuspid and bicuspid patients are not driven by geometrical differences, but rather by underlying vascular factors.

Lastly, AS surgical intervention has also been studied in a numerical setting, both through image-based assessment pre/post-intervention as well as using means of virtual implantation. As for primary AS, different modeling approaches have been utilized to assess surgical intervention [185, 291, 300], with the most effective descriptors possibly provided by FSI modeling [292]. These types of simulations have also been suggested as a surgical planning tool to predict paravalvular leakage [40] or valvular migration [306].

6.3.4 Challenges and Future Directions

While cardiac modeling has been used extensively to study AS, modeling challenges remain in order to disentangle the impact of flow and biomechanics on progressive disease. Additionally, the complexity of the disorder, involving large deformations under high-flow conditions, means that computational hurdles remain. First, rapid advancements in imaging and data-driven approaches should be incorporated into AS modeling, using full-field flow imaging [304] to prescribe accurate boundary conditions, or utilizing machine learning techniques to improve leaflet segmentation [281] or allow for super-resolution approaches in the problem definition [125].

Second, technical aspects remain in the area of AS modeling. More effective FSI descriptors, including enhanced parallelization schemes [195, 475], should be explored in the AS field. Similarly, with severe AS generating post-stenotic turbulence [108], incorporating turbulence modeling in the setting of AS should also be attempted. Examples of such exist [155]. However, the field could benefit from using contemporary or alternative effective turbulence descriptors [285, 457].

Other aspects of AS modeling could also be envisioned in the future. In Sect. 6.3.3, a recent multiscale approach comparing tricuspid vs. bicuspid behavior was noted, and similar approaches have also been applied to study the impact of AS on global cardiovascular output [280] and to tailor predicted behavior of implantable aortic valve devices [451]. Pushing these multiscale approaches further, incorporating how macroscopic obstruction could trigger microscopic remodeling, or how microscopic endothelial activation could be initiated by concomitant global cardiac pathologies could further help us understand underlying factors behind progressive AS. Likewise, chronic pressure overload of the left ventricle caused by the concomitant AS is also related to long-term cardiac growth and remodeling— representing a separate focus area for future modeling endeavors (see Sect. 4 for details). Lastly, including longitudinal data or even creating data-driven growth models can also help models uncover driving factors behind progressing stenosis.

6.4 Myocardial Infarction

A myocardial infarction (MI), commonly known as a heart attack, is a result of coronary heart disease (CHD). Although mortality due to CHD is decreasing in developed countries, it remains the leading cause of death in Europe [355] and gives rise to one out of every five deaths. An MI is caused by occlusion of a coronary artery, causing loss of blood supply to downstream myocardial tissue. A lack of oxygenated blood delivery to the heart muscle leads to tissue necrosis, scar formation, and, eventually, cardiac remodeling. Vessel occlusion is often caused by plaque formation in one of the major coronary arteries, caused by hypercholesterolemia, and individuals who have previously been diagnosed with hypertension are most at risk [407]. The infarct size, impacted by the location, degree, and duration of the occlusion and subsequent ischemia, can have a large effect on mortality [197]. In severe cases, the healthy and infarcted myocardium undergo significant remodeling, which can lead to heart failure [444]. Therefore, much research has been devoted to understanding, halting, and even reversing the tissue and organ-level changes that occur as a result of an MI.

6.4.1 Pathological Changes

Immediately after occlusion, the heart undergoes microstructural changes that initiate a cascade of remodeling processes [197]. The initial phase post-MI is known as the acute phase, spanning the first minutes to hours after occlusion. In this time, the lack of oxygen causes a disruption to the structural protein titin, an important constituent of the sarcomere, involved in active force generation and the region, which, without oxygen, becomes passive. During the acute phase, the infarct region can become distended during systolic contraction, resulting in dyskinesia. Both end-diastolic and end-systolic volumes are often elevated due to impaired contraction. The region of myocardium between the infarct and remote healthy tissue is known as the border zone (BZ), which is mechanically and electrically coupled to the infarct. The second phase is known as necrosis that takes place over the proceeding hours to a few days. At this time, inflammation of the affected tissue and necrosis, or cell death, occurs, causing infarcted tissue to lose its striations. Simultaneously, there is a deterioration of the collagen network. This is also the stage post-MI when there is the greatest risk of ventricular wall and chordae tendineae rupture [41].

During the subsequent fibrotic phase, spanning days to weeks after the infarction, collagen is deposited in the infarct region. New collagen is composed primarily of highly aligned type I fibers [198], oriented primarily within 30 degrees of circumferential [130]. The increase in collagen causes infarct stiffness to increase, subsequently leading to impaired filling (i.e., diastolic dysfunction). Systolic function is further impaired in this period due to the electrical and mechanical coupling between the infarct region and the healthy adjacent tissue [479]. The final phase, known as remodeling, lasts between weeks to years after infarction. During this time, there is often general improvement in cardiac function as the heart stabilizes. The infarct typically reduces in size [189], and infarct stiffness does not increase further. Healthy tissue, however, undergoes hypertrophic remodeling to maintain normal pump function and cavity dilation occurs [444] (Fig. 16).

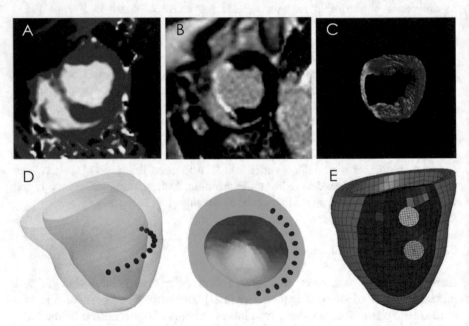

Fig. 16 The top row illustrates an infarcted porcine heart using three different imaging modalities: (**a**) contrast-enhanced T1 MR mapping, (**b**) late gadolinium-enhanced (LGE) MR imaging, and (**c**) triphenyl tetrazolium chloride (TTC) stained histology. (**a**, **b**, **c** are reproduced from [82] under Creative Commons Attribution License 4.0 from Springer Nature. ©2018.) The bottom row illustrates models used to investigate (**d**) hydrogel injection therapy (©2014, Elsevier. Adapted from [272] with permission.) and (**e**) the Coapsys restraint device (©2012, Elsevier. Adapted from [58] with permission)

6.4.2 Diagnosis and Therapies

The clinical diagnosis of the acute phase of MI is made through an ECG, whose trace is impacted by the change in electrical signal coming from the infarcted region of the heart [118]. The occurrence of an MI is also reflected in the levels of cardiac troponin, a protein that is integral to the normal contraction of cardiac muscle tissue. In this early stage, treatment of MI focuses on clearing the obstructed artery. It has been shown that prompt reperfusion post-MI leads to lower mortality rates [197]. The most common procedure to restore blood flow is through percutaneous coronary intervention in which a catheter is inserted through a peripheral artery (e.g., femoral or radial artery) into the occluded coronary artery. After which, a balloon is inflated to open the blocked vessel. A stent may also be placed in the coronary artery to prevent future re-occlusion [51]. This mechanical intervention is accompanied by pharmacological therapy to reduce the formation of blood clots and promote vasodilation in order to lower the systemic blood pressure. Post-MI, CMR imaging is used to detect and quantify fibrosis using LGE or a T1 mapping technique. Treatments that tackle the long-term remodeling post-MI have been developed in recent years. Hydrogel injection into the infarct region of the heart has been proposed as a therapy that aims to restore wall thickness and reduces the stress in the myocardium. This method has proven to reduce infarct size and preserve cardiac function [538]. Another approach is the use of mechanical devices that restrain the heart to decrease LV dilation and potentially limit the infarct zone expansion. This is a promising technique, and many approaches have appeared in the past years to optimize the use of these devices by adding mechanical anisotropy, biological or electrical functionality, and adaptability of the restraint [350, 485].

6.4.3 Modeling Approaches

Computational modeling has been used to investigate pathological growth and remodeling at various stages post-infarction. An acute-phase study of infarct tissue developed a model of wall thinning, driven by cell death and collagen degradation, which were mediated by the local oxygen concentration, defined by a diffusion problem[418]. Naim et al. [347] used a similar oxygen diffusion model coupled with a model of electrical activation to study the formation of the BZ. The study illustrated that, after MI, the BZ continues to develop and that the conductivity within the BZ is heterogeneous. The process of scar formation and remodeling post-MI was also studied by Rouillard and Holmes [413] by using a multiscale finite element model that accounted for the process of tissue fibrosis. Their results suggest the existence of a negative feedback between fibroblast alignment and the level of anisotropy [130]. Stress-driven tissue growth post-MI in the border zone and remote regions was also modeled on a patient-specific geometry, illustrating heterogeneous growth patterns [241], demonstrating the importance of geometry on predicting growth and remodeling. A patient-specific MI geometry was also employed in an electromechanical growth model, which was able to accurately depict long-term changes in PV loops for different levels of stiffness in the infarct region [271].

Numerous modeling studies have also aimed to estimate personalized passive and active parameters in the infarcted heart. Although global passive stiffness parameters may be indicative of overall cardiac function and able to differentiate infarcted from non-infarcted hearts [123], tissue properties differ between the infarcted myocardium, healthy tissue, and the border zone due to relative collagen content [60, 270, 337, 494]. In personalized models, the scarred region was found to be more than 20 times stiffer than the remote healthy tissue [494]. Similar conclusions were drawn by estimating a larger number of parameters [270], whereby an increase was reported for ground, fiber, and transverse stiffness in the infarct region when compared to healthy tissue. Balaban et al. illustrated the feasibility of estimating four spatially heterogeneous material parameters in the entire infarcted LV [26], rather than estimating homogeneous parameters for each region of interest. Another study has suggested that there may exist an optimal infarct stiffness that limits dilation but does not impede contraction of healthy tissue [491]. In addition to altered passive parameters, personalized active tension and myocardial contractility parameters, estimated in both the acute (2 days post-MI) and remodeling (6 months post-MI) phases, highlighted increased contractility required after 6 months [144]. For a comprehensive review of biomechanical modeling studies in MI, see [279].

In addition to modeling the regional stiffness of the infarcted ventricle, biomechanical models are also widely used to test potential therapies in silico. For example, hydrogel injection therapy has been shown to limit strain in the infarct region, reducing infarct expansion, and decrease stress in the healthy tissue, limiting long-term remodeling in the healthy and border zone regions [416]. Hydrogel within the tissue has been modeled in simplified form as either a bulk injection, homogeneous striations [246, 334], or spherical inclusions [237]; however, the injectate was also modeled as a realistic distribution measured directly from histology images [448]. In each of these computational studies, the addition of a non-contractile material that increases average wall thickness and material stiffness resulted in overall reduction wall strain in the infarct as well as reduction in fiber stress in the healthy region. The degree of impact was dependent on the volume, properties, and location of the injectate (e.g., [496, 510]). Computational studies into hydrogel as a potential post-infarct therapy to limit remodeling complement animal studies that have illustrated similar benefits [227]. Modeling studies have also been used to test restraint devices aimed at limiting dyskinesia and the increase in strain often seen in the infarct region [168, 379]. However, preliminary results have shown modest improvements in limiting dilation with no improvement of systolic function [485].

6.4.4 Challenges and Future Directions

Although biomechanical studies investigating subject-specific parameters in the infarcted myocardium may give a glimpse into the pathological changes occurring, they are limited in sample size and have utilized diverse material models and methods to estimate material parameters. Personalized modeling along with the use of structurally informed constitutive models such as [181, 491] could provide further insights into collagen deposition and tissue remodeling within the infarct and healthy regions. Similarly, advancements in imaging techniques, such as in vivo diffusion tensor imaging, allow for quantification of fiber orientations in the entire heart, including the infarct region [519], which can be incorporated into more accurate biomechanical models. Dempsey et al. [94] proposed a method for stratifying MI patients based on the stage post-MI by using a neural network trained on a synthetic set of stress–strain curves over time generated from a composite model of infarcted myocardium. Advancements in imaging capabilities in conjunction with novel machine learning methods can enable robust tools for clinically assessing infarct characteristics.

Computational studies of injection therapy have provided valuable information complementing animal studies. However, to date, these studies have not incorporated models of the tissue's inflammatory processes as well as the material degradation, both of which play an important role in deciding the optimal time of injection [314]. Additionally, future computational models investigating post-MI therapies could be incorporated along with growth models to predict not only acute changes in tissue stress and strain, but also ensuing growth and remodeling. Further, regeneration of viable myocardium in the infarct region through the use of stem cells is beginning to show promise [156]. Accurate computational models of cell growth in infarcted tissue could be a valuable complement to current animal studies.

7 Conclusions

The delivery of blood to the body and lungs depends on the effective mechanical contraction of the myocardium. This function relies on the hierarchical construction of the myocardium, beginning with the fundamental contractile unit (the sarcomere) and spanning all the way to the structure of the whole organ itself. In turn, pathologies of the heart often afflict the heart at multiple levels of this structural and functional hierarchy, resulting in complex, evolving disease characteristics that, in many cases, remain poorly understood. The biomechanical function of tissue, the behavior of blood, and the adaptability of the muscle tap into the rich continuum mechanics theory that enables predictive models of many aspects of heart function. Utilizing this theory through computational modeling provides an important framework for addressing challenging biomechanical questions that get at the heart of how we understand and treat disease.

References

1. Atlas of Human Cardiac Anatomy. URL www.vhlab.umn.edu/atlas
2. Adeniran, I., MacIver, D., Zhang, H.: Myocardial electrophysiological, contractile and metabolic properties of hypertrophic cardiomyopathy: insights from modelling. Computing in Cardiology **41**, 1037–1040 (2014)
3. Adolfsson, K., Enelund, M.: Fractional derivative viscoelasticity at large deformations. Nonlinear Dynamics **33**(3), 301–321 (2003)
4. Akar, F.G., Spragg, D.D., Tunin, R.S., Kass, D.A., Tomaselli, G.F.: Mechanisms underlying conduction slowing and arrhythmogenesis in nonischemic dilated cardiomyopathy. Circulation Research **95**, 717–725 (2004)
5. Akar, F.G., Tomaselli, G.F.: Electrophysiological remodeling in dilated cardiomyopathy and heart failure. In: Electrical Diseases of the Heart, pp. 290–304. Springer (2008)
6. Alberts, B., Johnson, A., Lewis, J., Raff, M., Roberts, K., Walter, P.: Molecular Biology of the Cell, Fourth Edition. Taylor & Francis Inc (2002)
7. Alford, P., Humphrey, J., Taber, L.: Growth and remodeling in a thick-walled artery model: effects of spatial variations in wall constituents. Biomechanics and Modeling in Mechanobiology **7**(4), 245–262 (2008)
8. Alter, P., Rupp, H., Rominger, M., Klose, K., Maisch, B.: A new methodological approach to assess cardiac work by pressure-volume and stress-length relations in patients with aortic valve stenosis and dilated cardiomyopathy. Pflugers Archiv: European Journal of Physiology **455**, 627–636 (2008)
9. Ambrosi, D., Ateshian, G., Arruda, E., Cowin, S., Dumais, J., Goriely, A., Holzapfel, G., Humphrey, J., Kemkemer, R., Kuhl, E., Olberding, J., Taber, L., Garikipati, K.: Perspectives on biological growth and remodeling. Journal of the Mechanics and Physics of Solids **59**(4), 863–883 (2011)
10. Arts, T., Costa, K., Covell, J., McCulloch, A.: Relating myocardial laminar architecture to shear strain and muscle fiber orientation. American Journal of Physiology: Heart and Circulatory Physiology **280**(5), H2222–H2229 (2001)
11. Arts, T., Delhaas, T., Bovendeerd, P., Verbeek, X., Prinzen, F.: Adaptation to mechanical load determines shape and properties of heart and circulation: the CircAdapt model. American Journal of Physiology: Heart and Circulatory Physiology **288**(4), H1943–H1954 (2005)
12. Arts, T., Delhaas, T., Bovendeerd, P., Verbeek, X., Prinzen, F.W.: Adaptation to mechanical load determines shape and properties of heart and circulation: the CircAdapt model. American Journal of Physiology-Heart and Circulatory Physiology **288**(4), H1943–H1954 (2005)
13. Arts, T., Prinzen, F., Snoeckx, L., Rijcken, J., Reneman, R.: Adaptation of cardiac structure by mechanical feedback in the environment of the cell: A model study. Biophysical Journal **66**(4), 953–961 (1994)
14. Arts, T., Reneman, R., Veenstra, P.: A model of the mechanics of the left ventricle. Annals of Biomedical Engineering **7**(3–4), 299–318 (1979)
15. Arumugam, J., Mojumder, J., Kassab, G., Lee, L.C.: Model of anisotropic reverse cardiac growth in mechanical dyssynchrony. Scientific Reports **9**(12670), 1–12 (2019)
16. Arvidsson, P.M., Töger, J., Pedrizzetti, G., Heiberg, E., Borgquist, R., Carlsson, M., Arheden, H.: Hemodynamic forces using four-dimensional flow MRI: an independent biomarker of cardiac function in heart failure with left ventricular dyssynchrony? American Journal of Physiology-Heart and Circulatory Physiology **315**(6), H1627–H1639 (2018)
17. Ashikaga, H., Coppola, B.A., Yamazaki, K.G., Villarreal, F.J., Omens, J.H., Covell, J.W.: Changes in regional myocardial volume during the cardiac cycle: Implications for transmural blood flow and cardiac structure. American Journal of Physiology: Heart and Circulatory Physiology **295**(2), 610–618 (2008)

18. Asner, L., Hadjicharalambous, M., Chabiniok, R., Peressutti, D., Sammut, E., Wong, J., Carr-White, G., Razavi, R., King, A., Smith, N., Lee, J., Nordsletten, D.: Patient-specific modeling for left ventricular mechanics using data-driven boundary energies. Computer Methods in Applied Mechanics and Engineering **314**, 269–295 (2017)
19. Asner, L., Hadjicharalambous, M., Chabiniok, R., Peresutti, D., Sammut, E., Wong, J., Carr-White, G., Chowienczyk, P., Lee, J., King, A., Smith, N., Razavi, R., Nordsletten, D.: Estimation of passive and active properties in the human heart using 3D tagged MRI. Biomechanics and Modeling in Mechanobiology **15**(5), 1121–1139 (2015)
20. Augenstein, K., Cowan, B., LeGrice, I., Young, A.: Estimation of cardiac hyperelastic material properties from MRI tissue tagging and diffusion tensor imaging. In: Proceedings of the International Conference on Medical Image Computing and Computer-Assisted Intervention, vol. 9, pp. 628–635 (2006)
21. Aurigemma, G.P., Zile, M.R., Gaasch, W.H.: Contractile behavior of the left ventricle in diastolic heart failure with emphasis on regional systolic function. Circulation **113**, 296–304 (2006)
22. Avazmohammadi, R., Hill, M.R., Simon, M.A., Zhang, W., Sacks, M.S.: A novel constitutive model for passive right ventricular myocardium: evidence for myofiber–collagen fiber mechanical coupling. Biomechanics and Modeling in Mechanobiology **16**(2), 561–581 (2017)
23. Avazmohammadi, R., Li, D.S., Leahy, T., Shih, E., Soares, J.S., Gorman, J.H., Gorman, R.C., Sacks, M.S.: An integrated inverse model-experimental approach to determine soft tissue three-dimensional constitutive parameters: application to post-infarcted myocardium. Biomechanics and Modeling in Mechanobiology **17**(1), 31–53 (2017)
24. Avazmohammadi, R., Soares, J.S., Li, D.S., Eperjesi, T., Pilla, J., Gorman, R.C., Sacks, M.S.: On the in vivo systolic compressibility of left ventricular free wall myocardium in the normal and infarcted heart. Journal of Biomechanics **107**, 109767 (2020)
25. Baccani, B., Domenichini, F., Pedrizzetti, G., Tonti, G.: Fluid dynamics of the left ventricular filling in dilated cardiomyopathy. Journal of Biomechanics **35**(5), 665–671 (2002)
26. Balaban, G., Finsberg, H., Funke, S., Håland, T.F., Hopp, E., Sundnes, J., Wall, S., Rognes, M.E.: In vivo estimation of elastic heterogeneity in an infarcted human heart. Biomechanics and Modeling in Mechanobiology **17**(5), 1317–1329 (2018)
27. Balakumar, P., Maung-U, K., Jagadeesh, G.: Prevalence and prevention of cardiovascular disease and diabetes mellitus. Pharmacological Research **113**, 600–609 (2016)
28. Ballyk, P., Steinman, D., Ethier, C.: Simulation of non-Newtonian blood flow in an end-to-side anastomosis. Biorheology **31**(5), 565–586 (1994)
29. Barnett, V.A.: Cellular myocytes. In: P.A. Iaizzo (ed.) Handbook of Cardiac Anatomy, Physiology, and Devices, pp. 201–214. Springer International Publishing (2015)
30. Basser, P., Mattiello, J., LeBihan, D.: MR diffusion tensor spectroscopy and imaging. Biophysical Journal **66**(1), 259–267 (1994)
31. Bastos, M.B., Burkhoff, D., Maly, J., Daemen, J., Den Uil, C.A., Ameloot, K., Lenzen, M., Mahfoud, F., Zijlstra, F., Schreuder, J.J., Van Mieghem, N.M.: Invasive left ventricle pressure-volume analysis: Overview and practical clinical implications. European Heart Journal **41**, 1286–1297 (2020)
32. Baumgartner, H., Falk, V., Bax, J.J., De Bonis, M., Hamm, C., Holm, P.J., Iung, B., Lancellotti, P., Lansac, E., Rodriguez Munoz, D., et al.: 2017 ESC/EACTS guidelines for the management of valvular heart disease. European Heart Journal **38**(36), 2739–2791 (2017)
33. Baumgartner, H., Hung, J., Bermejo, J., Chambers, J.B., Evangelista, A., Griffin, B.P., Iung, B., Otto, C.M., Pellikka, P.A., Quiñones, M.: Echocardiographic assessment of valve stenosis: EAE/ASE recommendations for clinical practice. Journal of the American Society of Echocardiography **22**(1), 1–23 (2009)
34. Baumgartner, H., Stefenelli, T., Niederberger, J., Schima, H., Maurer, G.: Overestimation of catheter gradients by Doppler ultrasound in patients with aortic stenosis: a predictable manifestation of pressure recovery. Journal of the American College of Cardiology **33**(6), 1655–1661 (1999)

35. Bavo, A., Pouch, A.M., Degroote, J., Vierendeels, J., Gorman, J.H., Gorman, R.C., Segers, P.: Patient-specific CFD models for intraventricular flow analysis from 3D ultrasound imaging: Comparison of three clinical cases. Journal of Biomechanics **50**, 144–150 (2017)

36. Ben-Assa, E., Brown, J., Keshavarz-Motamed, Z., Jose, M., Leiden, B., Olender, M., Kallel, F., Palacios, I.F., Inglessis, I., Passeri, J.J., et al.: Ventricular stroke work and vascular impedance refine the characterization of patients with aortic stenosis. Science Translational Medicine **11**(509), eaaw0181 (2019)

37. Berger, D.S., Li, J., Noordergraaf, A.: Differential effects of wave reflections and peripheral resistance on aortic blood pressure: a model-based study. American Journal of Physiology: Heart and Circulatory Physiology **266**(4), H1626–H1642 (1994)

38. Berger, D.S., Li, J.J.: Temporal relationship between left ventricular and arterial system elastances. IEEE Transactions on Biomedical Engineering **39**(4), 404–410 (1992)

39. Bermejo, J., Benito, Y., Alhama, M., Yotti, R., Martinez-Legazpi, P., del Villar, C.P., Perez-David, E., Gonzalez-Mansilla, A., Santa-Marta, C., Barrio, A., Fernandez-Aviles, F., del Alamo, J.C.: Intraventricular vortex properties in nonischemic dilated cardiomyopathy. American Journal of Physiology: Heart and Circulatory Physiology **306**(5), 718–729 (2014)

40. Bianchi, M., Marom, G., Ghosh, R.P., Rotman, O.M., Parikh, P., Gruberg, L., Bluestein, D.: Patient-specific simulation of transcatheter aortic valve replacement: Impact of deployment options on paravalvular leakage. Biomechanics and Modeling in Mechanobiology **18**(2), 435–451 (2019)

41. Birnbaum, Y., Chamoun, A.J., Anzuini, A., Lick, S.D., Ahmad, M., Uretsky, B.F.: Ventricular free wall rupture following acute myocardial infarction. Coronary Artery Disease **14**, 463–470 (2003)

42. Bissell, M.M., Hess, A.T., Biasiolli, L., Glaze, S.J., Loudon, M., Pitcher, A., Davis, A., Prendergast, B., Markl, M., Barker, A.J., et al.: Aortic dilation in bicuspid aortic valve disease: flow pattern is a major contributor and differs with valve fusion type. Circulation: Cardiovascular Imaging **6**(4), 499–507 (2013)

43. Blix, M.: Die länge und die spannung des muskels. Skandinavisches Archiv für Physiologie **3**(1), 295–318 (1892)

44. Boileau, E., Pant, S., Roobottom, C., Sazonov, I., Deng, J., Xie, X., Nithiarasu, P.: Estimating the accuracy of a reduced-order model for the calculation of fractional flow reserve (FFR). International Journal for Numerical Methods in Biomedical Engineering **34**(1), e2908 (2018)

45. Bonet, J., Wood, R.D.: Nonlinear continuum mechanics for finite element analysis. Cambridge University Press (1997)

46. Bonomi, D., Vergara, C., Faggiano, E., Stevanella, M., Conti, C., Redaelli, A., Puppini, G., Faggian, G., Formaggia, L., Luciani, G.B.: Influence of the aortic valve leaflets on the fluid-dynamics in aorta in presence of a normally functioning bicuspid valve. Biomechanics and Modeling in Mechanobiology **14**(6), 1349–1361 (2015)

47. Booz, G.W., Baker, K.M.: Molecular signalling mechanisms controlling growth and function of cardiac fibroblasts. Cardiovascular Research **30**(4), 537–543 (1995)

48. Bortone, A.S., Hess, O.M., Chiddo, A., Gaglione, A., Locuratolo, N., Caruso, G., Rizzon, P.: Functional and structural abnormalities in patients with dilated cardiomyopathy. Journal of the American College of Cardiology **14**(3), 613–23 (1989)

49. Bovendeerd, P.: Modeling of cardiac growth and remodeling of myofiber orientation. Journal of Biomechanics **45**, 872–881 (2012)

50. Braeu, F., Seitz, A., Aydin, R., Cyron, C.: Homogenized constrained mixture models for anisotropic volumetric growth and remodeling. Biomechanics and Modeling in Mechanobiology **16**, 889–906 (2017)

51. Braunwald, E.: The treatment of acute myocardial infarction: The past, the present, and the future. European Heart Journal: Acute Cardiovascular Care **1**(1), 9–12 (2012)

52. Budday, S., Ovaert, T.C., Holzapfel, G.A., Steinmann, P., Kuhl, E.: Fifty shades of brain: A review on the mechanical testing and modeling of brain tissue. Archives of Computational Methods in Engineering **27**(4), 1187–1230 (2019)

53. Buonocore, M.H.: Visualizing blood flow patterns using streamlines, arrows, and particle paths. Magnetic Resonance in Medicine **40**(2), 210–226 (1998)

54. Burattini, R., Natalucci, S.: Complex and frequency-dependent compliance of viscoelastic Windkessel resolves contradictions in elastic Windkessels. Medical Engineering & Physics **20**(7), 502–514 (1998)
55. Campbell, S.G., Mcculloch, A.D.: Multi-scale computational models of familial hypertrophic cardiomyopathy: genotype to phenotype. Journal of the Royal Society Interface **8**, 1550–1561 (2011)
56. Cansız, F.B.C., Dal, H., Kaliske, M.: An orthotropic viscoelastic material model for passive myocardium: theory and algorithmic treatment. Computer Methods in Biomechanics and Biomedical Engineering **18**(11), 1160–1172 (2015)
57. Carabello, B.A., Paulus, W.J.: Aortic stenosis. The Lancet **373**(9667), 956–966 (2009)
58. Carrick, R., Ge, L., Lee, L.C., Zhang, Z., Mishra, R., Axel, L., Guccione, J.M., Grossi, E.A., Ratcliffe, M.B.: Patient-specific finite element–based analysis of ventricular myofiber stress after Coapsys: Importance of residual stress. The Annals of Thoracic Surgery **93**(6), 1964–1971 (2012)
59. Carver, W., Nagpal, M., Nachtigal, M., Borg, T., Terracio, L.: Collagen expression in mechanically stimulated cardiac fibroblasts. Circulation Research **69**(1), 116–122 (1991)
60. Chabiniok, R., Chapelle, D., Lesault, P.F., Rahmouni, A., Deux, J.F.: Validation of a biomechanical heart model using animal data with acute myocardial infarction. In: MICCAI Workshop on Cardiovascular Interventional Imaging and Biophysical Modelling, pp. 1–9 (2009)
61. Chabiniok, R., Wang, V.Y., Hadjicharalambous, M., Asner, L., Lee, J., Sermesant, M., Kuhl, E., Young, A.A., Moireau, P., Nash, M.P., et al.: Multiphysics and multiscale modelling, data–model fusion and integration of organ physiology in the clinic: ventricular cardiac mechanics. Interface Focus **6**(2), 20150083 (2016)
62. Chandran, K.B., Kim, H.: Computational mitral valve evaluation and potential clinical applications. Annals of Biomedical Engineering **43**(6), 1348–1362 (2015)
63. Chang, G.H., Schirmer, C.M., Modarres-Sadeghi, Y.: A reduced-order model for wall shear stress in abdominal aortic aneurysms by proper orthogonal decomposition. Journal of Biomechanics **54**, 33–43 (2017)
64. Chapelle, D., Gerbeau, J.F., Sainte-Marie, J., Vignon-Clementel, I.E.: A poroelastic model valid in large strains with applications to perfusion in cardiac modeling. Computational Mechanics **46**(1), 91–101 (2010)
65. Chapelle, D., Tallec, P.L., Moireau, P., Sorine, M.: An energy-preserving muscle tissue model: formulation and compatible discretizations. International Journal for Multiscale Computational Engineering **10**(2), 189–211 (2012)
66. Cheng, Y., Oertel, H., Schenkel, T.: Fluid-structure coupled CFD simulation of the left ventricular flow during filling phase. Annals of Biomedical Engineering **33**(5), 567–576 (2005)
67. Chiastra, C., Morlacchi, S., Gallo, D., Morbiducci, U., Cárdenes, R., Larrabide, I., Migliavacca, F.: Computational fluid dynamic simulations of image-based stented coronary bifurcation models. Journal of the Royal Society Interface **10**(84), 1–13 (2013)
68. Chien, S., Usami, S., Dellenback, R.J., Gregersen, M.I.: Shear-dependent deformation of erythrocytes in rheology of human blood. American Journal of Physiology-Legacy Content **219**(1), 136–142 (1970)
69. Chien, S., Usami, S., Taylor, H.M., Lundberg, J.L., Gregersen, M.I.: Effects of hematocrit and plasma proteins on human blood rheology at low shear rates. Journal of Applied Physiology **21**(1), 81–87 (1966)
70. Cho, Y.I., Kensey, K.R.: Effects of the non-Newtonian viscosity of blood on flows in a diseased arterial vessel. Part 1: Steady flows. Biorheology **28**(3–4), 241–262 (1991)
71. Chuang, J.S., Zemljic-Harpf, A., Ross, R.S., Frank, L.R., McCulloch, A.D., Omens, J.H.: Determination of three-dimensional ventricular strain distributions in gene-targeted mice using tagged MRI. Magnetic Resonance in Medicine **64**(July), 1281–1288 (2010)
72. Cleutjens, J.P., Creemers, E.E.: Integration of concepts: cardiac extracellular matrix remodeling after myocardial infarction. Journal of cardiac failure **8**(6), S344–S348 (2002)

73. Cokelet, G.R., Meiselman, H.J.: Macro-and micro-rheological properties of blood. BIOMED-ICAL AND HEALTH RESEARCH-COMMISSION OF THE EUROPEAN COMMUNI-TIES THEN IOS PRESS **69**, 45 (2007)

74. Cokelet, G.R., Merrill, E., Gilliland, E., Shin, H., Britten, A., Wells Jr, R.: The rheology of human blood—measurement near and at zero shear rate. Transactions of the Society of Rheology **7**(1), 303–317 (1963)

75. Coppini, R., Ferrantini, C., Pioner, J.M., Santini, L., Wang, Z.J., Palandri, C., Scardigli, M., Vitale, G., Sacconi, L., Stefano, P., Flink, L., Riedy, K., Pavone, F.S., Cerbai, E., Poggesi, C., Mugelli, A., Bueno-Orovio, A., Olivotto, I., Sherrid, M.V.: Translational study of the electrophysiological and contractile effects of disopyramide in patients with obstructive hypertrophic cardiomyopathy. JACC: Basic to Translational Science **4**(7), 795–813 (2019)

76. Costa, K.D., Holmes, J.W., McCulloch, A.D.: Modelling cardiac mechanical properties in three dimensions. Philosophical Transactions of the Royal Society of London A: Mathematical, Physical and Engineering Sciences **359**(1783), 1233–1250 (2001)

77. Costa, K.D., Takayama, Y., McCulloch, A.D., Covell, J.W.: Laminar fiber architecture and three-dimensional systolic mechanics in canine ventricular myocardium. American Journal of Physiology: Heart and Circulatory Physiology **276**(2), H595–H607 (1999)

78. Costabal, F.S., Choy, J., Sack, K., Guccione, J., Kassab, G., Kuhl, E.: Multiscale characterization of heart failure. Acta Biomaterialia **86**, 66–76 (2019)

79. Costandi, P.N., Frank, L.R., McCulloch, A.D., Omens, J.H.: Role of diastolic properties in the transition to failure in a mouse model of cardiac dilatation. American Journal of Physiology: Heart and Circulatory Physiology **291**(6), H2971–H2979 (2006)

80. Courtois, M., Kovács Jr, S.J., Ludbrook, P.: Transmitral pressure-flow velocity relation. Importance of regional pressure gradients in the left ventricle during diastole. Circulation **78**(3), 661–671 (1988)

81. Cowin, S.: Continuum kinematical modeling of mass increasing biological growth. International Journal of Engineering Science **48**, 1137–1145 (2010)

82. Cui, C., Wang, S., Lu, M., Duan, X., Wang, H., Jia, L., Tang, Y., Sirajuddin, A., Prasad, S.K., Kellman, P., Arai, A.E., Zhao, S.: Detection of recent myocardial infarction using native T1 mapping in a swine model: A validation study. Scientific Reports **8**(1), 1–10 (2018)

83. Cyron, C., Aydin, R., Humphrey, J.: A homogenized constrained mixture (and mechanical analog) model for growth and remodeling of soft tissue. Biomechanics and Modeling in Mechanobiology **15**(6), 1389–1403 (2016)

84. Dao, M., Lim, C.T., Suresh, S.: Mechanics of the human red blood cell deformed by optical tweezers. Journal of the Mechanics and Physics of Solids **51**(11–12), 2259–2280 (2003)

85. Daubert, M.A., Weissman, N.J., Hahn, R.T., Pibarot, P., Parvataneni, R., Mack, M.J., Svensson, L.G., Gopal, D., Kapadia, S., Siegel, R.J., et al.: Long-term valve performance of TAVR and SAVR: a report from the partner I trial. JACC: Cardiovascular Imaging **10**(1), 15–25 (2017)

86. De Hart, J., Baaijens, F., Peters, G., Schreurs, P.: A computational fluid-structure interaction analysis of a fiber-reinforced stentless aortic valve. Journal of Biomechanics **36**(5), 699–712 (2003)

87. De Hart, J., Peters, G., Schreurs, P., Baaijens, F.: Collagen fibers reduce stresses and stabilize motion of aortic valve leaflets during systole. Journal of Biomechanics **37**(3), 303–311 (2004)

88. De Vecchi, A., Marlevi, D., Nordsletten, D.A., Ntalas, I., Leipsic, J., Bapat, V., Rajani, R., Niederer, S.A.: Left ventricular outflow obstruction predicts increase in systolic pressure gradients and blood residence time after transcatheter mitral valve replacement. Scientific Reports **8**(1), 1–11 (2018)

89. De Vecchi, A., Nordsletten, D.A., Remme, E.W., Bellsham-Revell, H., Greil, G., Simpson, J.M., Razavi, R., Smith, N.P.: Inflow typology and ventricular geometry determine efficiency of filling in the hypoplastic left heart. The Annals of Thoracic Surgery **94**(5), 1562–1569 (2012)

90. Debonnaire, P., Joyce, E., Hiemstra, Y., Mertens, B.J., Atsma, D.E., Schalij, M.J., Bax, J.J., Delgado, V., Marsan, N.A.: Left atrial size and function in hypertrophic cardiomyopathy

patients and risk of new-onset atrial fibrillation. Circulation: Arrhythmia and Electrophysiology **10**(2), e004052 (2017)

91. DeGroff, C.G., Shandas, R., Valdes-Cruz, L.: Analysis of the effect of flow rate on the Doppler continuity equation for stenotic orifice area calculations: a numerical study. Circulation **97**(16), 1597–1605 (1998)

92. Demer, L.L., Yin, F.: Passive biaxial mechanical properties of isolated canine myocardium. The Journal of physiology **339**(1), 615–630 (1983)

93. Demer, L.L., Yin, F.: Passive biaxial mechanical properties of isolated canine myocardium. The Journal of physiology **339**(1), 615–630 (1983)

94. Dempsey, S., Jafari, P., So, A., Samani, A.: A composite material based neural network for tissue mechanical properties estimation toward stage assessment of infarction. In: 2020 42nd Annual International Conference of the IEEE Engineering in Medicine & Biology Society (EMBC). IEEE (2020)

95. Deng, L., Huang, X., Zuo, H., Zheng, Y., Yang, C., Song, Y., Tang, D.: Angle of attack between blood flow and mitral valve leaflets in hypertrophic obstructive cardiomyopathy: An in vivo multi-patient CT-based FSI study. Computer Modeling in Engineering and Sciences **116**(2), 115–125 (2018)

96. Dillon-Murphy, D., Marlevi, D., Ruijsink, B., Qureshi, A., Chubb, H., Kerfoot, E., O'Neill, M., Nordsletten, D., Aslanidi, O., De Vecchi, A.: Modeling left atrial flow, energy, blood heating distribution in response to catheter ablation therapy. Frontiers in Physiology **9**, 1757 (2018)

97. Doehring, T.C., Freed, A.D., Carew, E.O., Vesely, I.: Fractional order viscoelasticity of the aortic valve cusp: An alternative to quasilinear viscoelasticity. Journal of Biomechanical Engineering **127**(4), 700–708 (2005)

98. Dokos, S., LeGrice, I.J., Smaill, B.H., Kar, J., Young, A.A.: A triaxial-measurement shear-test device for soft biological tissues. Journal of biomechanical engineering **122**(5), 471–478 (2000)

99. Dokos, S., Smaill, B., Young, A., LeGrice, I.: Shear properties of passive ventricular myocardium. American Journal of Physiology - Heart and Circulatory Physiology **283**(6), H2650–H2659 (2002)

100. Dokos, S., Smaill, B.H., Young, A.A., LeGrice, I.J.: Shear properties of passive ventricular myocardium. American Journal of Physiology: Heart and Circulatory Physiology **283**(6), H2650–H2659 (2002)

101. Domenichini, F., Pedrizzetti, G.: Intraventricular vortex flow changes in the infarcted left ventricle: numerical results in an idealised 3D shape. Computer Methods in Biomechanics and Biomedical Engineering **14**(1), 95–101 (2011)

102. Donati, F., Figueroa, C.A., Smith, N.P., Lamata, P., Nordsletten, D.A.: Non-invasive pressure difference estimation from PC-MRI using the work-energy equation. Medical Image Analysis **26**(1), 159–172 (2015)

103. Doost, S.N., Zhong, L., Su, B., Morsi, Y.S.: The numerical analysis of non-Newtonian blood flow in human patient-specific left ventricle. Computer Methods and Programs in Biomedicine **127**, 232–247 (2016)

104. Du, C.K., Morimoto, S., Nishii, K., Minakami, R., Ohta, M., Tadano, N., Lu, Q.W., Wang, Y.Y., Zhan, D.Y., Mochizuki, M., Kita, S., Miwa, Y., Takahashi-Yanaga, F., Iwamoto, T., Ohtsuki, I., Sasaguri, T.: Knock-in mouse model of dilated cardiomyopathy caused by troponin mutation. Circulation Research **101**, 185–194 (2007)

105. Duan, F., Xie, M., Wang, X., Li, Y., He, L., Jiang, L., Fu, Q.: Preliminary clinical study of left ventricular myocardial strain in patients with non-ischemic dilated cardiomyopathy by three-dimensional speckle tracking imaging. Cardiovascular Ultrasound **10**(1), 8 (2012)

106. Duncan, J., Shi, P., Constable, T., Sinusas, A.: Physical and geometrical modeling for image-based recovery of left ventricular deformation. Progress in Biophysics and Molecular Biology **69**(2–3), 333–351 (1998)

107. Dweck, M.R., Boon, N.A., Newby, D.E.: Calcific aortic stenosis: a disease of the valve and the myocardium. Journal of the American College of Cardiology **60**(19), 1854–1863 (2012)

108. Dyverfeldt, P., Hope, M.D., Tseng, E.E., Saloner, D.: Magnetic resonance measurement of turbulent kinetic energy for the estimation of irreversible pressure loss in aortic stenosis. JACC: Cardiovascular Imaging **6**(1), 64–71 (2013)

109. Ebbers, T., Farnebäck, G.: Improving computation of cardiovascular relative pressure fields from velocity MRI. Journal of Magnetic Resonance Imaging **30**(1), 54–61 (2009)

110. Eichhorn, E.J., Heesch, C.M., Barnett, J.H., Alvarez, L.G., Fass, S.M., Grayburn, P.A., Hatfield, B.A., Marcoux, L.G., Malloy, C.R.: Effect of metoprolol on myocardial function and energetics in patients with nonischemic dilated cardiomyopathy: A randomized, double-blind, placebo-controlled study. American College of Cardiology **24**(5), 1310–1320 (1994)

111. El-Hamamsy, I., Chester, A.H., Yacoub, M.H.: Cellular regulation of the structure and function of aortic valves. Journal of Advanced Research **1**(1), 5–12 (2010)

112. Elbaz, M.S., Calkoen, E.E., Westenberg, J.J., Lelieveldt, B.P., Roest, A.A., Van Der Geest, R.J.: Vortex flow during early and late left ventricular filling in normal subjects: quantitative characterization using retrospectively-gated 4d flow cardiovascular magnetic resonance and three-dimensional vortex core analysis. Journal of Cardiovascular Magnetic Resonance **16**(1), 1–12 (2014)

113. Ellims, A.H., Iles, L.M., Ling, L.H., Hare, J.L., Kaye, D.M., Taylor, A.J.: Diffuse myocardial fibrosis in hypertrophic cardiomyopathy can be identified by cardiovascular magnetic resonance, and is associated with left ventricular diastolic dysfunction. Journal of Cardiovascular Magnetic Resonance **14**(1), 76 (2012)

114. Emery, J., Omens, J., McCulloch, A.: Strain softening in rat left ventricular myocardium. Journal of Biomechanical Engineering **119**(1), 6–12 (1997)

115. Enciso, J., Adler, E., Greenberg, B.: Current status of the left ventricular assist device as a destination therapy. US Cardiology Review **10**(2) (2016)

116. Epstein, M., Maugin, G.: Thermomechanics of volumetric growth in uniform bodies. International Journal of Plasticity **16**, 951–978 (2000)

117. Eriksson, J., Carlhäll, C.J., Dyverfeldt, P., Engvall, J., Bolger, A.F., Ebbers, T.: Semi-automatic quantification of 4d left ventricular blood flow. Journal of Cardiovascular Magnetic Resonance **12**(1), 9 (2010)

118. Eskola, M.J., Nikus, K.C., Niemela, K.O., Sclarovsky, S.: How to use ECG for decision support in the catheterization laboratory: Cases with inferior ST elevation myocardial infarction. Journal of Electrocardiology **37**(4), 257–266 (2004)

119. Estrada, A.C., Yoshida, K., Saucerman, J.J., Holmes, J.W.: A multiscale model of cardiac concentric hypertrophy incorporating both mechanical and hormonal drivers of growth. Biomech Model Mechanobiol **20**(1), 293–307 (2021)

120. Fagard, R.: Athlete's heart. Heart **89**(12), 1455–1461 (2003)

121. Fåhraeus, R.: The suspension stability of the blood. Physiological Reviews **9**(2), 241–274 (1929)

122. Fahraeus, R., Lindqvist, T.: The viscosity of the blood in narrow capillary tubes. American Journal of Physiology **96**(3), 562–568 (1931)

123. Fan, L., Yao, J., Yang, C., Wu, Z., Xu, D., Tang, D.: Material stiffness parameters as potential predictors of presence of left ventricle myocardial infarction: 3d echo - based computational modeling study. Biomedical Engineering Online **15**(1), 34 (2016)

124. Fata, B., Zhang, W., Amini, R., Sacks, M.S.: Insights Into Regional Adaptations in the Growing Pulmonary Artery Using a Meso-Scale Structural Model: Effects of Ascending Aorta Impingement. Journal of Biomechanical Engineering **136**(2) (2014)

125. Ferdian, E., Suinesiaputra, A., Dubowitz, D.J., Zhao, D., Wang, A., Cowan, B., Young, A.A.: 4DFlowNet: Super-resolution 4D flow MRI using deep learning and computational fluid dynamics. Frontiers in Physics **8**, 138 (2020)

126. Ferdian, E., Suinesiaputra, A., Fung, K., Aung, N., Lukaschuk, E., Barutcu, A., Maclean, E., Paiva, J., Piechnik, S.K., Neubauer, S., Petersen, S.E., Young, A.A.: Fully automated myocardial strain estimation from cardiovascular MRI–tagged images using a deep learning framework in the UK Biobank. Radiology: Cardiothoracic Imaging **2**(1), e190032 (2020)

127. Ferreira, P.F., Kilner, P.J., Mcgill, L.A., Nielles-Vallespin, S., Scott, A.D., Ho, S.Y., Mccarthy, K.P., Haba, M.M., Ismail, T.F., Gatehouse, P.D., De Silva, R., Lyon, A.R., Prasad, S.K., Firmin, D.N., Pennell, D.J.: In vivo cardiovascular magnetic resonance diffusion tensor imaging shows evidence of abnormal myocardial laminar orientations and mobility in hypertrophic cardiomyopathy. Journal of Cardiovascular Magnetic Resonance 16(87), 87 (2014)

128. Finocchiaro, G., Knowles, J.W., Pavlovic, A., Perez, M., Magavern, E., Sinagra, G., Haddad, F., Ashley, E.A.: Prevalence and clinical correlates of right ventricular dysfunction in patients with hypertrophic cardiomyopathy. American Journal of Cardiology 113(2), 361–367 (2014)

129. Fischer, P.F., Loth, F., Lee, S.E., Lee, S.W., Smith, D.S., Bassiouny, H.S.: Simulation of high-Reynolds number vascular flows. Computer Methods in Applied Mechanics and Engineering 196(31–32), 3049–3060 (2007)

130. Fomovsky, G.M., Rouillard, A.D., Holmes, J.W.: Regional mechanics determine collagen fiber structure in healing myocardial infarcts. Journal of Molecular and Cellular Cardiology 52(5), 1083–1090 (2012)

131. Formaggia, L., Gerbeau, J.F., Nobile, F., Quarteroni, A.: On the coupling of 3D and 1D Navier–Stokes equations for flow problems in compliant vessels. Computer methods in applied mechanics and engineering 191(6–7), 561–582 (2001)

132. Formaggia, L., Nobile, F.: Stability analysis of second-order time accurate schemes for ALE–FEM. Computer methods in applied mechanics and engineering 193(39–41), 4097–4116 (2004)

133. Formaggia, L., Nobile, F., Quarteroni, A., Veneziani, A.: Multiscale modelling of the circulatory system: a preliminary analysis. Computing and Visualization in Science 2(2–3), 75–83 (1999)

134. Frangi, A., Niessen, W., Viergever, M.: Three-dimensional modeling for functional analysis of cardiac images, a review. IEEE Transactions on Medical Imaging 20(1), 2–5 (2001)

135. Fratzl, P. (ed.): Collagen Structure and Mechanics. Springer US, Boston, MA (2008)

136. Fredriksson, A.G., Zajac, J., Eriksson, J., Dyverfeldt, P., Bolger, A.F., Ebbers, T., Carlhäll, C.J.: 4-D blood flow in the human right ventricle. American Journal of Physiology-Heart and Circulatory Physiology 301(6), H2344–H2350 (2011)

137. Fumagalli, I., Fedele, M., Vergara, C., Dede', L., Ippolito, S., Nicolò, F., Antona, C., Scrofani, R., Quarteroni, A.: An image-based computational hemodynamics study of the Systolic Anterior Motion of the mitral valve. Computers in Biology and Medicine 123(July), 103922 (2020)

138. Fung, Y.: What are the residual stresses doing in our blood vessels? Annals of Biomedical Engineering 19(3), 237–249 (1991)

139. Fung, Y.C.: A first course in continuum mechanics. Prentice-Hall, Englewood Cliffs, N.J (1969)

140. Fung, Y.c.: Biomechanics: mechanical properties of living tissues. Springer Science & Business Media (2013)

141. Gacita, A., Puckelwartz, M., McNally, E.: Modeling human dilated cardiomyopathy using humans. JACC: Basic to Translational Science 3(6), 741–743 (2018)

142. Ganesh, S.K., Arnett, D.K., Assimes, T.L., Basson, C.T., Chakravarti, A., Ellinor, P.T., Engler, M.B., Goldmuntz, E., Herrington, D.M., Hershberger, R.E., et al.: Genetics and genomics for the prevention and treatment of cardiovascular disease. Circulation 128(25), 2813–2851 (2013)

143. Gao, H., Feng, L., Qi, N., Berry, C., Griffith, B.E., Luo, X.: A coupled mitral valve—left ventricle model with fluid–structure interaction. Medical Engineering & Physics 47, 128–136 (2017)

144. Gao, H., Mangion, K., Carrick, D., Husmeier, D., Luo, X., Berry, C.: Estimating prognosis in patients with acute myocardial infarction using personalized computational heart models. Scientific Reports 7(1) (2017)

145. Garcia, D., Camici, P.G., Durand, L.G., Rajappan, K., Gaillard, E., Rimoldi, O.E., Pibarot, P.: Impairment of coronary flow reserve in aortic stenosis. Journal of Applied Physiology 106(1), 113–121 (2009)

146. Garcia, D., Dumesnil, J.G., Durand, L.G., Kadem, L., Pibarot, P.: Discrepancies between catheter and Doppler estimates of valve effective orifice area can be predicted from the pressure recovery phenomenon: practical implications with regard to quantification of aortic stenosis severity. Journal of the American College of Cardiology **41**(3), 435–442 (2003)

147. Garcia-Canadilla, P., Cook, A.C., Mohun, T.J., Oji, O., Schlossarek, S., Carrier, L., McKenna, W.J., Moon, J.C., Captur, G.: Myoarchitectural disarray of hypertrophic cardiomyopathy begins pre-birth. Journal of Anatomy **235**(5), 962–976 (2019)

148. Garrido, L., Wedeen, V.J., Kwong, K.K., Spencer, U.M., Kantor, H.L.: Anisotropy of water diffusion in the myocardium of the rat. Circulation Research **74**(5), 789–793 (1994)

149. Ge, L., Leo, H.L., Sotiropoulos, F., Yoganathan, A.P.: Flow in a mechanical bileaflet heart valve at laminar and near-peak systole flow rates: CFD simulations and experiments. Journal of Biomechanical Engineering **127**(5), 782–797 (2005)

150. Généreux, P., Stone, G.W., O'Gara, P.T., Marquis-Gravel, G., Redfors, B., Giustino, G., Pibarot, P., Bax, J.J., Bonow, R.O., Leon, M.B.: Natural history, diagnostic approaches, and therapeutic strategies for patients with asymptomatic severe aortic stenosis. Journal of the American College of Cardiology **67**(19), 2263–2288 (2016)

151. Genet, M., Lee, L., Baillargeon, B., Guccione, J., Kuhl, E.: Modeling pathologies of diastolic and systolic heart failure. Ann Biomed Eng **44**(1), 112–127 (2016)

152. Gerdes, A.M., Capasso, J.M.: Structural remodeling and mechanical dysfunction of cardiac myocytes in heart failure. Journal of Molecular and Cellular Cardiology **27**(3), 849–856 (1995)

153. Gersh, B.J., Maron, B.J., Bonow, R.O., Dearani, J.A., Fifer, M.A., Link, M.S., Naidu, S.S., Nishimura, R.A., Ommen, S.R., Rakowski, H., Seidman, C.E., Towbin, J.A., Udelson, J.E., Yancy, C.W.: 2011 ACCF / AHA guideline for the diagnosis and treatment of hypertrophic cardiomyopathy. Circulation **124**(24), 2761–2796 (2011)

154. Gersh, B.J., Maron, B.J., Bonow, R.O., Dearani, J.A., Fifer, M.A., Link, M.S., Naidu, S.S., Nishimura, R.A., Ommen, S.R., Rakowski, H., et al.: 2011 ACCF/AHA guideline for the diagnosis and treatment of hypertrophic cardiomyopathy. Journal of the American College of Cardiology **58**(25), 2703–2738 (2011)

155. Ghalichi, F., Deng, X., De Champlain, A., Douville, Y., King, M., Guidoin, R.: Low Reynolds number turbulence modeling of blood flow in arterial stenoses. Biorheology **35**(4–5), 281–294 (1998)

156. Giacca, M.: Cardiac regeneration after myocardial infarction: An approachable goal. Current Cardiology Reports **22**, 122 (2020)

157. Girdauskas, E., Borger, M.A., Secknus, M.A., Girdauskas, G., Kuntze, T.: Is aortopathy in bicuspid aortic valve disease a congenital defect or a result of abnormal hemodynamics? a critical reappraisal of a one-sided argument. European Journal of Cardio-Thoracic Surgery **39**(6), 809–814 (2011)

158. Glover, B.M., Buckley, O., Ho, S.Y., Sanchez-Quintana, D., Brugada, P.: Cardiac Anatomy and Electrophysiology. In: Clinical Handbook of Cardiac Electrophysiology, pp. 1–37. Springer International Publishing (2016)

159. Göktepe, S., Abilez, O., Parker, K., Kuhl, E.: A multiscale model for eccentric and concentric cardiac growth through sarcomerogenesis. Journal of Theoretical Biology **265**(3), 433–442 (2010)

160. Goto, D., Kinugawa, S., Hamaguchi, S., Sakakibara, M., Tsuchihashi-Makaya, M., Yokota, T., Yamada, S., Yokoshiki, H., Tsutsui, H.: Clinical characteristics and outcomes of dilated phase of hypertrophic cardiomyopathy: Report from the registry data in Japan. Journal of Cardiology **61**(1), 65–70 (2013)

161. Gray, R.A., Pathmanathan, P.: Patient-specific cardiovascular computational modeling: Diversity of personalization and challenges. Journal of Cardiovascular Translational Research **11**(2), 80–88 (2018)

162. Gregor, P., Karol, C.: Medical treatment of hypertrophic cardiomyopathy – what do we know about it today? Cor et Vasa **57**(3), e219–e224 (2015)

163. Griffith, B.E., Luo, X., McQueen, D.M., Peskin, C.S.: Simulating the fluid dynamics of natural and prosthetic heart valves using the immersed boundary method. International Journal of Applied Mechanics **1**(01), 137–177 (2009)

164. Grosberg, A., Gharib, Æ.M.: Modeling the macro-structure of the heart: healthy and diseased. Medical and Biological Engineering and Computing **47**(3), 301–311 (2009)

165. Grossman, W., Jones, D., McLaurin, L., et al.: Wall stress and patterns of hypertrophy in the human left ventricle. The Journal of Clinical Investigation **56**(1), 56–64 (1975)

166. Guccione, J., Costa, K., McCulloch, A.: Finite element stress analysis of left ventricular mechanics in the beating dog heart. Journal of Biomechanics **28**(10), 1167–1177 (1995)

167. Guccione, J., McCulloch, A., Waldman, L.: Passive material properties of intact ventricular myocardium determined from a cylindrical model. Journal of Biomechanical Engineering **113**(1), 42–55 (1991)

168. Guccione, J.M., Salahieh, A., Moonly, S.M., Kortsmit, J., Wallace, A.W., Ratcliffe, M.B.: Myosplint decreases wall stress without depressing function in the failing heart: A finite element model study. Annals of Thoracic Surgery **76**, 1171–1180 (2003)

169. Gultekin, O., Sommer, G., Holzapfel, G.A.: An orthotropic viscoelastic model for the passive myocardium: continuum basis and numerical treatment. Computer Methods in Biomechanics and Biomedical Engineering **19**(15), 1647–1664 (2016)

170. Gurtin, M.: An introduction to continuum mechanics. Academic Press, New York (1981)

171. Guyton, A.C., Coleman, T.G., Granger, H.J.: Circulation: overall regulation. Annual review of physiology **34**, 13–46 (1972)

172. Gwathmey, J.K., Hajjar, R.J.: Relation between steady-state force and intracellular [Ca2+] in intact human myocardium. Index of myofibrillar responsiveness to Ca2+. Circulation **82**(4), 1266–1278 (1990)

173. Hadjicharalambous, M., Asner, L., Chabiniok, R., Sammut, E., Wong, J., Peressutti, D., Kerfoot, E., King, A., Lee, J., Razavi, R., Smith, N., Carr-White, G., Nordsletten, D.: Non-invasive model-based assessment of passive left-ventricular myocardial stiffness in healthy subjects and in patients with non-ischemic dilated cardiomyopathy. Annals of Biomedical Engineering **45**(3), 605–618 (2017)

174. Hadjicharalambous, M., Chabiniok, R., Asner, L., Sammut, E., Wong, J., Carr-White, G., Lee, J., Razavi, R., Smith, N., Nordsletten, D.: Analysis of passive cardiac constitutive laws for parameter estimation using 3D tagged MRI. Biomechanics and Modeling in Mechanobiology **14**(4), 807–828 (2014)

175. Haidekker, M.A., Tsai, A.G., Brady, T., Stevens, H.Y., Frangos, J.A., Theodorakis, E., Intaglietta, M.: A novel approach to blood plasma viscosity measurement using fluorescent molecular rotors. American Journal of Physiology-Heart and Circulatory Physiology **282**(5), H1609–H1614 (2002)

176. Halevi, R., Hamdan, A., Marom, G., Lavon, K., Ben-Zekry, S., Raanani, E., Bluestein, D., Haj-Ali, R.: Fluid–structure interaction modeling of calcific aortic valve disease using patient-specific three-dimensional calcification scans. Medical & Biological Engineering & Computing **54**(11), 1683–1694 (2016)

177. Hall, J.E., Hall, M.E.: Guyton and Hall Textbook of Medical Physiology. Elsevier, Philadelphia, PA (2016)

178. Hanna, P., Rajendran, P.S., Ajijola, O.A., Vaseghi, M., Armour, J.A., Ardell, J.L., Shivkumar, K.: Cardiac neuroanatomy-imaging nerves to define functional control. Autonomic Neuroscience **207**, 48–58 (2017)

179. Harana, J.M.: Non-invasive, MRI-based calculation of the aortic blood pressure waveform by. PhD thesis **20**, 9 (2020)

180. Hartupee, J., Mann, D.: Neurohormonal activation in heart failure with reduced ejection fraction. Nature Reviews Cardiology **14**(1), 30–38 (2017)

181. Hasaballa, A.I., Wang, V.Y., Sands, G.B., Wilson, A.J., Young, A.A., Legrice, I.J., Nash, M.P.: Microstructurally motivated constitutive modeling of heart failure mechanics. Biophysical Journal **117**(12), 2273–2286 (2019)

182. Hatle, L., Brubakk, A., Tromsdal, A., Angelsen, B.: Noninvasive assessment of pressure drop in mitral stenosis by Doppler ultrasound. Heart **40**(2), 131–140 (1978)

183. Hayashida, W., Kumada, T., Kohno, F., Noda, M., Ishikawa, N., Kambayashi, M., Kawai, C.: Left ventricular relaxation in dilated cardiomyopathy: Relation to loading conditions and regional nonuniformity. Journal of the American College of Cardiology **20**(5), 1082–1091 (1992)

184. Hayashida, W., Kumada, T., Nohara, R., Tanio, H., Kambayashi, M., Ishikawa, N., Nakamura, Y., Himura, Y., Kawai, C.: Left ventricular regional wall stress in dilated cardiomyopathy. Circulation **82**(6), 2075–2083 (1990)

185. Hellmeier, F., Nordmeyer, S., Yevtushenko, P., Bruening, J., Berger, F., Kuehne, T., Goubergrits, L., Kelm, M.: Hemodynamic evaluation of a biological and mechanical aortic valve prosthesis using patient-specific MRI-based CFD. Artificial Organs **42**(1), 49–57 (2018)

186. Hershberger, R.E., Hedges, D.J., Morales, A.: Dilated cardiomyopathy: The complexity of a diverse genetic architecture. Nature Reviews Cardiology **10**(9), 531–547 (2013)

187. Hessenthaler, A., Röhrle, O., Nordsletten, D.: Validation of a non-conforming monolithic fluid-structure interaction method using phase-contrast MRI. International Journal for Numerical Methods in Biomedical Engineering **33**(8), e2845 (2017)

188. Hill, M.R., Duan, X., Gibson, G.A., Watkins, S., Robertson, A.M.: A theoretical and nondestructive experimental approach for direct inclusion of measured collagen orientation and recruitment into mechanical models of the artery wall. Journal of Biomechanics **45**(5), 762–771 (2012)

189. Hillenbrand, H.B., Sandstede, J., Stork, S., Ramsayer, B., Hahn, D., Ertl, G., Koestler, H., Bauer, W., Ritter, C.: Remodeling of the infarct territory in the time course of infarct healing in humans. Magnetic Resonance Materials in Physics, Biology and Medicine **24**, 277–284 (2011)

190. Himpel, G., Kuhl, E., Menzel, A., Steinmann, P.: Computational modelling of isotropic multiplicative growth. Computer Modeling in Engineering and Sciences **8**(2), 119–134 (2005)

191. Hinton, R.B., Yutzey, K.E.: Heart valve structure and function in development and disease. Annual Review of Physiology **73**, 29–46 (2011)

192. Hirschvogel, M., Bassilious, M., Jagschies, L., Wildhirt, S.M., Gee, M.W.: A monolithic 3D-0D coupled closed-loop model of the heart and the vascular system: experiment-based parameter estimation for patient-specific cardiac mechanics. International Journal for Numerical Methods in Biomedical Engineering **33**(8), e2842 (2017)

193. Hirschvogel, M., Jagschies, L., Maier, A., Wildhirt, S., Gee, M.: An in-silico twin for epicardial augmentation of the failing heart. International Journal for Numerical Methods in Biomedical Engineering **35**(10) (2019)

194. Hoeijmakers, M., Soto, D.S., Waechter-Stehle, I., Kasztelnik, M., Weese, J., Hose, D., van de Vosse, F.: Estimation of valvular resistance of segmented aortic valves using computational fluid dynamics. Journal of Biomechanics **94**, 49–58 (2019)

195. Hoffman, J., Jansson, J., de Abreu, R.V., Degirmenci, N.C., Jansson, N., Müller, K., Nazarov, M., Spühler, J.H.: Unicorn: Parallel adaptive finite element simulation of turbulent flow and fluid–structure interaction for deforming domains and complex geometry. Computers & Fluids **80**, 310–319 (2013)

196. Holmes, D.R., Valeti, U.S., Nishimura, R.A.: Alcohol septal ablation for hypertrophic cardiomyopathy: Indications and technique. Catheterization and Cardiovascular Interventions **66**(3), 375–389 (2005)

197. Holmes, J., Borg, T., Covell, J.: Structure and mechanics of healing myocardial infarcts. Annual Review of Biomedical Engineering **7**, 223–253 (2005)

198. Holmes, J., Covell, J.: Collagen fiber orientation in myocardial scar tissue. Cardiovascular Pathobiology **1**, 15–22 (1996)

199. Holopainen, T., Rasanen, M., Anisimov, A., Tuomainen, T., Zheng, W., Tvorogov, D., Hulmi, J.J., Andersson, L.C., Cenni, B., Tavi, P., Mervaala, E., Kivela, R., Alitalo, K.: Endothelial bmx tyrosine kinase activity is essential for myocardial hypertrophy and remodeling. Proceedings of the National Academy of Sciences of the United States of America **112**(42), 13063–13068 (2015)

200. Holzapfel, G.A.: Nonlinear Solid Mechanics: A Continuum Approach for Engineering. John Wiley & Sons, Ltd (2000)
201. Holzapfel, G.A.: Nonlinear Solid Mechanics: A Continuum Approach for Engineering Science. 4/5. Springer Science and Business Media LLC (2002)
202. Holzapfel, G.A., Niestrawska, J.A., Ogden, R.W., Reinisch, A.J., Schriefl, A.J.: Modelling non-symmetric collagen fibre dispersion in arterial walls. Journal of the Royal Society Interface 12(106), 20150188 (2015)
203. Holzapfel, G.A., Ogden, R.W.: Constitutive modelling of passive myocardium: a structurally based framework for material characterization. Philosophical Transactions of the Royal Society A: Mathematical, Physical and Engineering Sciences 367(1902), 3445–3475 (2009)
204. Hope, M.D., Hope, T.A., Crook, S.E., Ordovas, K.G., Urbania, T.H., Alley, M.T., Higgins, C.B.: 4D flow CMR in assessment of valve-related ascending aortic disease. JACC: Cardiovascular Imaging 4(7), 781–787 (2011)
205. Hsu, M.C., Kamensky, D., Bazilevs, Y., Sacks, M.S., Hughes, T.J.: Fluid–structure interaction analysis of bioprosthetic heart valves: significance of arterial wall deformation. Computational Mechanics 54(4), 1055–1071 (2014)
206. Huang, X., Deng, L., Yang, C., Lesperance, M., Tang, D.: Comparisons of patient-specific active and passive models for left ventricle in hypertrophic obstructive cardiomyopathy. Molecular & Cellular Biomechanics 16(S2), 58–58 (2019)
207. Huber, D., Grimm, J., Koch, R., Krayenbuehl, H.P.: Determinants of ejection performance in aortic stenosis. Circulation 64(1), 126–134 (1981)
208. Humphrey, J., Rajagopal, K.: A constrained mixture model for growth and remodeling of soft tissues. Mathematical Models and Methods in Applied Sciences 12(3), 407–430 (2002)
209. Humphrey, J., Strumpf, R., Yin, F.: Determination of a constitutive relation for passive myocardium: II. Parameter estimation. Journal of Biomechanical Engineering 112, 340–346 (1990)
210. Humphrey, J., Yin, F.: On constitutive relations and finite deformations of passive cardiac tissue: I. a pseudostrain-energy function. Journal of Biomechanical Engineering 109(4), 298–304 (1987)
211. Humphrey, J., Yin, F.: Biomechanical experiments on excised myocardium: theoretical considerations. Journal of Biomechanics 22(4), 377–383 (1989)
212. Humphrey, J.D.: Cardiovascular Solid Mechanics. Springer New York (2002)
213. Hunter, P.J., McCulloch, A.D., Ter Keurs, H.: Modelling the mechanical properties of cardiac muscle. Progress in biophysics and molecular biology 69(2–3), 289–331 (1998)
214. Husse, B., Franz, W.M.: Generation of cardiac pacemaker cells by programming and differentiation. Biochimica et Biophysica Acta - Molecular Cell Research 1863(7), 1948–1952 (2016)
215. Huyghe, J., van Campen, D., Arts, T., Heethaar, R.: The constitutive behaviour of passive heart muscle tissue: a quasi-linear viscoelastic formulation. Journal of Biomechanics 24(9), 841–849 (1991)
216. Huyghe, J.M., van Campen, D.H., Arts, T., Heethaar, R.M.: The constitutive behaviour of passive heart muscle tissue: A quasi-linear viscoelastic formulation. Journal of Biomechanics 24(9), 841–849 (1991)
217. Iaizzo, P.A.: General features of the cardiovascular system. In: P.A. Iaizzo (ed.) Handbook of Cardiac Anatomy, Physiology, and Devices, pp. 3–12. Springer International Publishing (2015)
218. Ielasi, A., Latib, A., Tespili, M., Donatelli, F.: Current results and remaining challenges of trans-catheter aortic valve replacement expansion in intermediate and low risk patients. International Journal of Cardiology: Heart & Vasculature 23, 100375 (2019)
219. Imanparast, A., Fatouraee, N., Sharif, F.: The impact of valve simplifications on left ventricular hemodynamics in a three dimensional simulation based on in vivo MRI data. Journal of Biomechanics 49(9), 1482–1489 (2016)
220. Jagschies, L., Hirschvogel, M., Matallo, J., Maier, A., Mild, K., Brunner, H., Hinkel, R., Gee, M., Radermacher, P., Wildhirt, S., Hafner, S.: Individualized biventricular epicardial

augmentation technology in a drug-induced porcine failing heart model. ASAIO **64**(4), 480–488 (2018)

221. Jagschies, L., Hirschvogel, M., Matallo, J., Maier, A., Mild, K., Brunner, H., Hinkel, R., Gee, M.W., Radermacher, P., Wildhirt, S.M., et al.: Individualized biventricular epicardial augmentation technology in a drug-induced porcine failing heart model. ASAIO Journal **64**(4), 480–488 (2018)

222. Jan, K.M., Chien, S., Bigger Jr, J.T.: Observations on blood viscosity changes after acute myocardial infarction. Circulation **51**(6), 1079–1084 (1975)

223. Johnson, A.A., Tezduyar, T.E.: Mesh update strategies in parallel finite element computations of flow problems with moving boundaries and interfaces. Computer methods in applied mechanics and engineering **119**(1–2), 73–94 (1994)

224. Johnson, E.L., Wu, M.C., Xu, F., Wiese, N.M., Rajanna, M.R., Herrema, A.J., Ganapathysub-ramanian, B., Hughes, T.J., Sacks, M.S., Hsu, M.C.: Thinner biological tissues induce leaflet flutter in aortic heart valve replacements. Proceedings of the National Academy of Sciences **117**(32), 19007–19016 (2020)

225. Johnson, G.A., Tramaglini, D.M., Levine, R.E., Ohno, K., Choi, N.Y., Woo, S.L.Y.: Tensile and viscoelastic properties of human patellar tendon. Journal of Orthopaedic Research **12**(6), 796–803 (1994)

226. Johnston, B.M., Johnston, P.R., Corney, S., Kilpatrick, D.: Non-Newtonian blood flow in human right coronary arteries: steady state simulations. Journal of Biomechanics **37**(5), 709–720 (2004)

227. Kadner, K., Dobner, S., Franz, T., Bezuidenhout, D., Sirry, M.S., Zilla, P., Davies, N.H.: The beneficial effects of deferred delivery on the efficiency of hydrogel therapy post myocardial infarction. Biomaterials **33**, 2060–2066 (2012)

228. Karimi, A., Haghighatnama, M., Shojaei, A., Navidbakhsh, M., Haghi, A.M., Sadati, S.J.A.: Measurement of the viscoelastic mechanical properties of the skin tissue under uniaxial loading. Proceedings of the Institution of Mechanical Engineers, Part L: Journal of Materials: Design and Applications **230**(2), 418–425 (2015)

229. Kass, D., Chen, C., Curry, C., Talbot, M., Berger, R., Fetics, B., Nevo, E.: Improved left ventricular mechanics from acute VDD pacing in patients with dilated cardiomyopathy and ventricular conduction delay. Circulation **99**(12), 1567–1573 (1999)

230. Kayvanpour, E., Mansi, T., Sedaghat-Hamedani, F., Amr, A., Neumann, D., Georgescu, B., Seegerer, P., Kamen, A., Haas, J., Frese, K., Irawati, M., Wirsz, E., King, V., Buss, S., Mereles, D., Zitron, E., Keller, A., Katus, H., Comaniciu, D., Meder, B.: Towards personalized cardiology: Multi-scale modeling of the failing heart. PLoS ONE **10**(7), e0134869 (2015)

231. Keane, J., Driscoll, D., Gersony, W., Hayes, C., Kidd, L., O'Fallon, W., Pieroni, D., Wolfe, R., Weidman, W.: Second natural history study of congenital heart defects. Results of treatment of patients with aortic valvar stenosis. Circulation **87**(2 Suppl), I16 (1993)

232. Kerckhoffs, R., Bovendeerd, P., Kotte, J., Prinzen, F., Smits, K., Arts, T.: Homogeneity of cardiac contraction despite physiological asynchrony of depolarization: A model study. Annals of Biomedical Engineering **31**(5), 536–547 (2003)

233. Kerckhoffs, R.C., Omens, J.H., McCulloch, A.D.: A single strain-based growth law predicts concentric and eccentric cardiac growth during pressure and volume overload. Mechanics Research Communications **42**, 40–50 (2012)

234. Kerckhoffs, R.C.P., Omens, J.H., McCulloch, A.D.: Mechanical discoordination increases continuously after the onset of left bundle branch block despite constant electrical dyssyn-chrony in a computational model of cardiac electromechanics and growth. Europace **15**(Suppl 5), v65–v72 (2012)

235. Keshavarz-Motamed, Z., Garcia, J., Pibarot, P., Larose, E., Kadem, L.: Modeling the impact of concomitant aortic stenosis and coarctation of the aorta on left ventricular workload. Journal of Biomechanics **44**(16), 2817–2825 (2011)

236. Khalafvand, S., Zhong, L., Ng, E.: Three-dimensional CFD/MRI modeling reveals that ventricular surgical restoration improves ventricular function by modifying intraventricular blood flow. International Journal for Numerical Methods in Biomedical Engineering **30**(10), 1044–1056 (2014)

237. Kichula, E., Wang, H., Dorsey, S., Szczesny, S., Elliot, D., Burdick, J., Wenk, J.: Experimental and computational investigation of altered mechanical properties in myocardium after hydrogel injection. Annals of Biomedical Engineering **42**(7), 1546–1556 (2014)

238. Kim, Y.H., VandeVord, P.J., Lee, J.S.: Multiphase non-Newtonian effects on pulsatile hemodynamics in a coronary artery. International Journal for Numcrical Methods in Fluids **58**(7), 803–825 (2008)

239. Klabunde, R.: Cardiovascular physiology concepts. Lippincott Williams & Wilkins (2011)

240. Klabunde, R.E.: Cardiovascular Physiology Concepts, 2 edn. Wolters Kluwer Health (2012)

241. Klepach, D., Chuan, L., Wenk, J.F., Ratcliffe, M.B., Zohdi, T.I., Navia, J.L., Kassab, G.S., Kuhl, E., Guccione, J.M.: Growth and remodeling of the left ventricle : A case study of myocardial infarction and surgical ventricular restoration. Mechanics Research Communications **42**, 134–141 (2012)

242. Kobelev, A.V., Kobeleva, R.M., Protsenko, Y.L., Berman, I.V., Kobelev, O.A.: Viscoelastic models describing stress relaxation and creep in soft tissues. MRS Proceedings **874** (2005)

243. Kohajda, Z., Loewe, A., Toth, N., Varro, A., Nagy, N.: The cardiac pacemaker story—fundamental role of the Na+/Ca2+ exchanger in spontaneous automaticity. Frontiers in Pharmacology **11**(April), 1–25 (2020)

244. Koizumi, R., Funamoto, K., Hayase, T., Kanke, Y., Shibata, M., Shiraishi, Y., Yambe, T.: Numerical analysis of hemodynamic changes in the left atrium due to atrial fibrillation. Journal of Biomechanics **48**(3), 472–478 (2015)

245. Korakianitis, T., Shi, Y.: A concentrated parameter model for the human cardiovascular system including heart valve dynamics and atrioventricular interaction. Medical Engineering & Physics **28**(7), 613–628 (2006)

246. Kortsmit, J., Davies, N.H., Miller, R., Macadangdang, J.R., Zilla, P., Franz, T.: The effect of hydrogel injection on cardiac function and myocardial mechanics in a computational post-infarction model. Computer Methods in Biomechanics and Biomedical Engineering **16**(11), 1185–1195 (2013)

247. Kramer, C.M., Appelbaum, E., Desai, M.Y., Desvigne-Nickens, P., DiMarco, J.P., Friedrich, M.G., Geller, N., Heckler, S., Ho, C.Y., Jerosch-Herold, M., Ivey, E.A., Keleti, J., Kim, D.Y., Kolm, P., Kwong, R.Y., Maron, M.S., Schulz-Menger, J., Piechnik, S., Watkins, H., Weintraub, W.S., Wu, P., Neubauer, S.: Hypertrophic cardiomyopathy registry: The rationale and design of an international, observational study of hypertrophic cardiomyopathy. American Heart Journal **170**(2), 223–230 (2015)

248. Krittian, S.B., Lamata, P., Michler, C., Nordsletten, D.A., Bock, J., Bradley, C.P., Pitcher, A., Kilner, P.J., Markl, M., Smith, N.P.: A finite-element approach to the direct computation of relative cardiovascular pressure from time-resolved MR velocity data. Medical Image Analysis **16**(5), 1029–1037 (2012)

249. Kronenberg, M.W., Cohen, G.I., Leonen, M.F., Mladsi, T.A., Di Carli, M.F.: Myocardial oxidative metabolic supply-demand relationships in patients with nonischemic dilated cardiomyopathy. Journal of Nuclear Cardiology **13**(4), 544–553 (2006)

250. Kroon, W., Delhaas, T., Bovendeerd, P.H., Arts, T.: Computational analysis of the myocardial structure: Adaptation of cardiac myofiber orientations through deformation. Medical Image Analysis **13**(2), 346–353 (2009)

251. Kuhl, E., Garikipati, K., Arruda, E.M., Grosh, K.: Remodeling of biological tissue: Mechanically induced reorientation of a transversely isotropic chain network. Journal of the Mechanics and Physics of Solids **53**(7), 1552–1573 (2005)

252. Kuhl, E., Holzapfel, G.A.: A continuum model for remodeling in living structures. Journal of Material Science **42**, 8811–8823 (2007)

253. Kurian, T., Ambrosi, C., Hucker, W., Fedorov, V.V., Efimov, I.R.: Anatomy and electrophysiology of the human AV node. Pacing and Clinical Electrophysiology **33**(6), 754–762 (2010)

254. LaDisa, J.F., Guler, I., Olson, L.E., Hettrick, D.A., Kersten, J.R., Warltier, D.C., Pagel, P.S.: Three-dimensional computational fluid dynamics modeling of alterations in coronary wall shear stress produced by stent implantation. Annals of Biomedical Engineering **31**(8), 972–980 (2003)

255. Landes, G.: Einige untersuchungen an electrischen analogieschaltungen zum kreislaufsystem. Z. Biol **101**(418), 1942–43 (1943)
256. Lanir, Y.: A structural theory for the homogeneous biaxial stress-strain relationships in flat collagenous tissues. Journal of Biomechanics **12**, 423–436 (1979)
257. Lanir, Y.: Mechanistic micro-structural theory of soft tissues growth and remodeling: tissues with unidirectional fiber. Biomechanics and Modeling in Mechanobiology **14**(2), 245–266 (2015)
258. Lanir, Y.: Fibrous tissues growth and remodeling: Evolutionary micro-mechanical theory. Journal of the Mechanics and Physics of Solids **107**, 115–144 (2017)
259. Lantz, J., Henriksson, L., Persson, A., Karlsson, M., Ebbers, T.: Patient-specific simulation of cardiac blood flow from high-resolution computed tomography. Journal of Biomechanical Engineering **138**(12) (2016)
260. Lau, K., Diaz, V., Scambler, P., Burriesci, G.: Mitral valve dynamics in structural and fluid–structure interaction models. Medical Engineering & Physics **32**(9), 1057–1064 (2010)
261. Lavon, K., Halevi, R., Marom, G., Ben Zekry, S., Hamdan, A., Joachim Schäfers, H., Raanani, E., Haj-Ali, R.: Fluid–structure interaction models of bicuspid aortic valves: The effects of nonfused cusp angles. Journal of Biomechanical Engineering **140**(3) (2018)
262. Le, T.B., Sotiropoulos, F.: Fluid–structure interaction of an aortic heart valve prosthesis driven by an animated anatomic left ventricle. Journal of Computational Physics **244**, 41–62 (2013)
263. Leach, J.K., Alexander, R.S.: Effect of epinephrine on stress relaxation and distensibility of the isolated cat heart. American Journal of Physiology–Legacy Content **209**(5), 935–940 (1965)
264. Lee, A., Costa, C., Strocchi, M., Rinaldi, C., Niederer, S.: Computational modeling for cardiac resynchronization therapy. Journal of Cardiovascular Translational Research **11**, 92–108 (2018)
265. Lee, E.: Elastic-plastic deformation at finite strains. Journal of Applied Mechanics **36**(1), 1–6 (1969)
266. Lee, H.Y., Codella, N.C., Cham, M.D., Weinsaft, J.W., Wang, Y.: Automatic left ventricle segmentation using iterative thresholding and an active contour model with adaptation on short-axis cardiac MRI. IEEE Transactions on Biomedical Engineering **57**(4), 905–913 (2009)
267. Lee, L., Genet, M., Acevedo-Bolton, G., Ordovas, K., Guccione, J., Kuhl, E.: A computational model that predicts reverse growth in response to mechanical unloading. Biomechanics and Modeling in Mechanobiology **14**(2), 217–229 (2015)
268. Lee, L., Kassab, G., Guccione, J.: Mathematical modeling of cardiac growth and remodeling. Wiley Interdisciplinary Reviews: Systems Biology and Medicine **8**(3), 211–226 (2016)
269. Lee, L., Kassab, G., Guccione, J.: Mathematical modeling of cardiac growth and remodeling. Wiley Interdisciplinary Reviews: Systems Biology and Medicine **8**(3), 211–226 (2016)
270. Lee, L., Wenk, J., Klepach, D., Kassab, G., Guccione, J.: Structural-based models of ventricular myocardium. In: Structure-Based Mechanics of Tissues and Organs, pp. 249–263. Springer (2016)
271. Lee, L.C., Sundnes, J., Genet, M., Wenk, J.F., Wall, S.T.: An integrated electromechanical-growth heart model for simulating cardiac therapies. Biomechanics and Modeling in Mechanobiology **15**(4), 791–803 (2016)
272. Lee, L.C., Wall, S.T., Genet, M., Hinson, A., Guccione, J.M.: Bioinjection treatment: Effects of post-injection residual stress on left ventricular wall stress. Journal of Biomechanics **47**(12), 3115–3119 (2014)
273. LeGrice, I., Pope, A., Smaill, B.: The architecture of the heart: Myocyte organization and the cardiac extracellular matrix. In: Interstitial Fibrosis in Heart Failure, pp. 3–21. Springer-Verlag, New York (2005)
274. LeGrice, I., Smaill, B., Chai, L., Edgar, S., Gavin, J., Hunter, P.: Laminar structure of the heart: ventricular myocyte arrangement and connective tissue architecture in the dog. American Journal of Physiology-Heart and Circulatory Physiology **269**(2), H571–H582 (1995)
275. Lei, M., Zhang, H., Grace, A.A., Huang, C.L.: SCN5A and sinoatrial node pacemaker function. Cardiovascular Research **74**(3), 356–365 (2007)

276. Levine, G.N., Bates, E.R., Bittl, J.A., Brindis, R.G., Fihn, S.D., Fleisher, L.A., Granger, C.B., Lange, R.A., Mack, M.J., Mauri, L., et al.: 2016 ACC/AHA guideline focused update on duration of dual antiplatelet therapy in patients with coronary artery disease. Circulation **134**(10), e123–e155 (2016)

277. Li, D.S., Avazmohammadi, R., Merchant, S.S., Kawamura, T., Hsu, E.W., Gorman, J.H., Gorman, R.C., Sacks, M.S.: Insights into the passive mechanical behavior of left ventricular myocardium using a robust constitutive model based on full 3D kinematics. Journal of the Mechanical Behavior of Biomedical Materials **103**, 103508 (2020)

278. Li, M., Beech-Brandt, J., John, L., Hoskins, P., Easson, W.: Numerical analysis of pulsatile blood flow and vessel wall mechanics in different degrees of stenoses. Journal of Biomechanics **40**(16), 3715–3724 (2007)

279. Li, W.: Biomechanics of infarcted left ventricle: a review of modelling. Biomedical Engineering Letters **10**(3), 387–417 (2020)

280. Liang, F., Takagi, S., Himeno, R., Liu, H.: Multi-scale modeling of the human cardiovascular system with applications to aortic valvular and arterial stenoses. Medical & Biological Engineering & Computing **47**(7), 743–755 (2009)

281. Liang, L., Kong, F., Martin, C., Pham, T., Wang, Q., Duncan, J., Sun, W.: Machine learning–based 3-D geometry reconstruction and modeling of aortic valve deformation using 3-D computed tomography images. International Journal for Numerical Methods in Biomedical Engineering **33**(5), e2827 (2017)

282. Liao, J., Yang, L., Grashow, J., Sacks, M.: Collagen fibril kinematics in mitral valve leaflet under biaxial elongation, creep, and stress relaxation. In: Society for Heart Valve Disease Third Biennial Meeting, vol. 37 (2005)

283. Liao, J., Yang, L., Grashow, J., Sacks, M.S.: The Relation Between Collagen Fibril Kinematics and Mechanical Properties in the Mitral Valve Anterior Leaflet. Journal of Biomechanical Engineering **129**(1), 78–87 (2006)

284. Lindman, B.R., Clavel, M.A., Mathieu, P., Iung, B., Lancellotti, P., Otto, C.M., Pibarot, P.: Calcific aortic stenosis. Nature Reviews Disease Primers **2**(1), 1–28 (2016)

285. Ling, J., Kurzawski, A., Templeton, J.: Reynolds averaged turbulence modelling using deep neural networks with embedded invariance. Journal of Fluid Mechanics **807**, 155–166 (2016)

286. Litviňuková, M., Talavera-López, C., Maatz, H., Reichart, D., Worth, C.L., Lindberg, E.L., Kanda, M., Polanski, K., Heinig, M., Lee, M., et al.: Cells of the adult human heart. Nature (2020)

287. Lodish, H., Berk, A., Kaiser, C.A., Krieger, M., Bretscher, A., Ploegh, H., Amon, A., Scott, M.P.: Molecular Cell Biology, 7th edition edn. W. H. Freeman, Palgrave Macmillan (2012)

288. Loeffler, L., Sagawa, K.: A one dimensional viscoelastic model of cat heart muscle studied by small length perturbations during isometric contraction. Circulation Research **36**(4), 498–512 (1975)

289. Long, Q., Merrifield, R., Yang and, G., Xu, X., Kilner and, P., Firmin, D.: The influence of inflow boundary conditions on intra left ventricle flow predictions. Journal of Biomechanical Engineering **125**(6), 922–927 (2003)

290. Luckie, M., Khattar, R.S.: Systolic anterior motion of the mitral valve - beyond hypertrophic cardiomyopathy. Heart **94**(11), 1383–1385 (2008)

291. Luraghi, G., Migliavacca, F., García-González, A., Chiastra, C., Rossi, A., Cao, D., Stefanini, G., Matas, J.F.R.: On the modeling of patient-specific transcatheter aortic valve replacement: a fluid–structure interaction approach. Cardiovascular Engineering and Technology **10**(3), 437–455 (2019)

292. Luraghi, G., Wu, W., De Gaetano, F., Matas, J.F.R., Moggridge, G.D., Serrani, M., Stasiak, J., Costantino, M.L., Migliavacca, F.: Evaluation of an aortic valve prosthesis: Fluid-structure interaction or structural simulation? Journal of Biomechanics **58**, 45–51 (2017)

293. Lyon, A., Ariga, R., Minchole, A., Mahmod, M., Ormondroyd, E., Laguna, P., Freitas, N.D., Neubauer, S., Watkins, H., Rodriguez, B.: Distinct ECG phenotypes identified in hypertrophic cardiomyopathy using machine learning associate with arrhythmic risk markers. Frontiers in Physiology **9**, 213 (2018)

294. Lyon, A., Bueno-orovio, A., Zacur, E., Ariga, R., Grau, V., Neubauer, S., Watkins, H., Rodriguez, B., Minchole, A.: Electrocardiogram phenotypes in hypertrophic cardiomyopathy caused by distinct mechanisms: apico-basal repolarization gradients vs. Purkinje-myocardial coupling abnormalities. Europace **20**, 102–112 (2018)

295. Lyon, A., Minchole, A., Rodriguez, B.: Improving the clinical understanding of hypertrophic cardiomyopathy by combining patient data, machine learning and computer simulations: A case study. Morphologie **103**(343), 169–179 (2019)

296. Machado, J.T., Kiryakova, V., Mainardi, F.: Recent history of fractional calculus. Communications in Nonlinear Science and Numerical Simulation **16**(3), 1140–1153 (2011)

297. MacKenna, D., Vaplon, S., McCulloch, A.: Microstructural model of perimysial collagen fibers for resting myocardial mechanics during ventricular filling. American Journal of Physiology: Heart and Circulatory Physiology **273**(3), H1576–H1586 (1997)

298. Malatos, S., Raptis, A., Xenos, M.: Advances in low-dimensional mathematical modeling of the human cardiovascular system. Journal of Hypertension and Management **2**(2), 1–10 (2016)

299. Malvern, L.: Introduction to the mechanics of a continuous medium. Prentice-Hall, Englewood Cliffs, N.J (1969)

300. Mao, W., Wang, Q., Kodali, S., Sun, W.: Numerical parametric study of paravalvular leak following a transcatheter aortic valve deployment into a patient-specific aortic root. Journal of Biomechanical Engineering **140**(10) (2018)

301. Marcinkowska-Gapińska, A., Gapinski, J., Elikowski, W., Jaroszyk, F., Kubisz, L.: Comparison of three rheological models of shear flow behavior studied on blood samples from post-infarction patients. Medical & biological engineering & computing **45**(9), 837–844 (2007)

302. Marian, A.J., Braunwald, E.: Hypertrophic cardiomyopathy: Genetics, pathogenesis, clinical manifestations, diagnosis, and therapy. Circulation Research **121**(7), 749–770 (2017)

303. Marijon, E., Mirabel, M., Celermajer, D.S., Jouven, X.: Rheumatic heart disease. The Lancet **379**(9819), 953–964 (2012)

304. Markl, M., Frydrychowicz, A., Kozerke, S., Hope, M., Wieben, O.: 4D flow MRI. Journal of Magnetic Resonance Imaging **36**(5), 1015–1036 (2012)

305. Marlevi, D., Ruijsink, B., Balmus, M., Dillon-Murphy, D., Fovargue, D., Pushparajah, K., Bertoglio, C., Colarieti-Tosti, M., Larsson, M., Lamata, P., et al.: Estimation of cardiovascular relative pressure using virtual work-energy. Scientific Reports **9**(1), 1–16 (2019)

306. Marom, G., Bianchi, M., Ghosh, R.P., Bluestein, D.: Patient-specific numerical model of calcific aortic stenosis and its treatment by balloon-expandable transcatheter aortic valve: Effect of positioning on the anchorage. In: Computer Methods in Biomechanics and Biomedical Engineering, pp. 259–263. Springer (2018)

307. Maron, B., Shirani, J., Poliac, L., Mathenge, R., Roberts, W., Mueller, F.: Sudden cardiac death in young competitive athletes. Journal of the American Medical Association **276**(3), 199–204 (1996)

308. Maron, B.J.: Hypertrophic cardiomyopathy a systematic review. Journal of the American Medical Association **287**, 1308–1320 (2002)

309. Maron, B.J., Epstein, S.E.: Hypertrophic cardiomyopathy. Recent observations regarding the specificity of three hallmarks of the disease: Asymmetric septal hypertrophy, septal disorganization and systolic anterior motion of the anterior mitral leaflet. The American Journal of Cardiology **45**(1), 141–154 (1980)

310. Maron, M.S., Hauser, T.H., Dubrow, E., Horst, T.A., Kissinger, K.V., Udelson, J.E., Manning, W.J.: Right ventricular involvement in hypertrophic cardiomyopathy. American Journal of Cardiology **100**(8), 1293–1298 (2007)

311. Marsden, A.L., Esmaily-Moghadam, M.: Multiscale modeling of cardiovascular flows for clinical decision support. Applied Mechanics Reviews **67**(3), 1–11 (2015)

312. Martin, R., Lairez, O., Boudou, N., Méjean, S., Lhermusier, T., Dumonteil, N., Berry, M., Cognet, T., Massabuau, P., Elbaz, M., Rousseau, H., Galinier, M.: Relation between left ventricular outflow tract obstruction and left ventricular shape in patients with hypertrophic

cardiomyopathy: A cardiac magnetic resonance imaging study. Archives of Cardiovascular Diseases **106**(8–9), 440–447 (2013)

313. Martinez-Legazpi, P., Rossini, L., del Villar, C.P., Benito, Y., Devesa-Cordero, C., Yotti, R., Delgado-Montero, A., Gonzalez-Mansilla, A., Kahn, A.M., Fernandez-Avilés, F., et al.: Stasis mapping using ultrasound: a prospective study in acute myocardial infarction. JACC: Cardiovascular Imaging **11**(3), 514–515 (2018)

314. Matsumura, Y., Zhu, Y., Jiang, H., Amore, A.D., Luketich, S.K., Charwat, V., Yoshizumi, T., Sato, H., Yang, B., Uchibori, T., Healy, K.E., Wagner, W.R.: Intramyocardial injection of a fully synthetic hydrogel attenuates left ventricular remodeling post myocardial infarction. Biomaterials **217**, 119289 (2019)

315. Matthys, K.S., Alastruey, J., Peiró, J., Khir, A.W., Segers, P., Verdonck, P.R., Parker, K.H., Sherwin, S.J.: Pulse wave propagation in a model human arterial network: assessment of 1-D numerical simulations against in vitro measurements. Journal of Biomechanics **40**(15), 3476–3486 (2007)

316. McCormick, M., Nordsletten, D., Kay, D., Smith, N.: Modelling left ventricular function under assist device support. International Journal for Numerical Methods in Biomedical Engineering **27**(7), 1073–1095 (2011)

317. McCormick, M., Nordsletten, D., Lamata, P., Smith, N.P.: Computational analysis of the importance of flow synchrony for cardiac ventricular assist devices. Computers in Biology and Medicine **49**, 83–94 (2014)

318. McCormick, M., Nordsletten, D.A., Kay, D., Smith, N.: Simulating left ventricular fluid–solid mechanics through the cardiac cycle under LVAD support. Journal of Computational Physics **244**, 80–96 (2013)

319. McCullough, S.A., Fifer, M.A., Mohajer, P., Lowry, P.A., Reen, C.O., Baggish, A.L., Vlahakes, G.J., Shimada, Y.J.: Clinical correlates and prognostic value of elevated right atrial pressure in patients with hypertrophic cardiomyopathy. Circulation Journal **82**(5), 1405–1411 (2018)

320. McEvoy, E., Holzapfel, G.A., McGarry, P.: Compressibility and anisotropy of the ventricular myocardium: Experimental analysis and microstructural modeling. Journal of Biomechanical Engineering **140**(8) (2018)

321. McGregor, J.B., Rahman, A., Rosanio, S., Ware, D., Birnbaum, Y., Saeed, M.: Monomorphic ventricular tachycardia: A late complication of percutaneous alcohol septal ablation for hypertrophic cardiomyopathy. American Journal of the Medical Sciences **328**, 185–188 (2004)

322. McQueen, D.M., Peskin, C.S., Yellin, E.L.: Fluid dynamics of the mitral valve: physiological aspects of a mathematical model. American Journal of Physiology: Heart and Circulatory Physiology **242**(6), H1095–H1110 (1982)

323. Mekkaoui, C., Jackowski, M.P., Kostis, W.J., Stoeck, C.T., Thiagalingam, A., Reese, T.G., Reddy, V.Y., Ruskin, J.N., Kozerke, S., Sosnovik, D.E.: Myocardial scar delineation using diffusion tensor magnetic resonance tractography. Journal of the American Heart Association **7**, 1–10 (2018)

324. Menzel, A.: A fibre reorientation model for orthotropic multiplicative growth. Biomechanics and Modeling in Mechanobiology **6**(5), 303–320 (2007)

325. Menzel, A., Kuhl, E.: Frontiers in growth and remodeling. Mechanics Research Communications **1**(42), 1–14 (2012)

326. Menzel, A., Kuhl, E.: Frontiers in growth and remodeling. Mechanics Research Communications **42**, 1–14 (2012)

327. Merlo, M., Pyxaras, S.A., Pinamonti, B., Barbati, G., Di Lenarda, A., Sinagra, G.: Prevalence and prognostic significance of left ventricular reverse remodeling in dilated cardiomyopathy receiving tailored medical treatment. Journal of the American College of Cardiology **57**(13), 1468–1476 (2011)

328. Merrill, E., Benis, A., Gilliland, E., Sherwood, T., Salzman, E.: Pressure-flow relations of human blood in hollow fibers at low flow rates. Journal of Applied Physiology **20**(5), 954–967 (1965)

329. Meyers, M.A., Chawla, K.K.: Mechanical Behavior of Materials. Cambridge University Press (2008)
330. Mihalef, V., Passerini, T., Mansi, T.: Multi-scale models of the heart for patient-specific simulations. In: Artificial Intelligence for Computational Modeling of the Heart, pp. 3–42. Elsevier (2020)
331. Mihl, C., Dassen, W., Kuipers, H.: Cardiac remodelling: concentric versus eccentric hypertrophy in strength and endurance athletes. Netherlands Heart Journal **16**(4), 129–133 (2008)
332. Miller, C.E., Vanni, M.A., Keller, B.B.: Characterization of passive embryonic myocardium by quasi-linear viscoelasticity theory. Journal of Biomechanics **30**(9), 985–988 (1997)
333. Miller, C.E., Wong, C.L.: Trabeculated embryonic myocardium shows rapid stress relaxation and non-quasi-linear viscoelastic behavior. Journal of Biomechanics **33**(5), 615–622 (2000)
334. Miller, R., Davies, N., Kortsmit, J., Zilla, P., Franz, T.: Outcomes of myocardial infarction hydrogel injection therapy in the human left ventricle dependent on injectate distribution. International Journal of Numerical Methods in Biomedical Engineering **29**(8), 870–884 (2013)
335. Mitoh, A., Yano, T., Sekine, K., Mitamura, Y., Okamoto, E., Kim, D.W., Yozu, R., Kawada, S.: Computational fluid dynamics analysis of an intra-cardiac axial flow pump. Artificial Organs **27**(1), 34–40 (2003)
336. Mittal, T.K., Reichmuth, L., Bhattacharyya, S., Jain, M., Baltabaeva, A., Haley, S.R., Mirsadraee, S., Panoulas, V., Kabir, T., Nicol, E.D., et al.: Inconsistency in aortic stenosis severity between CT and echocardiography: prevalence and insights into mechanistic differences using computational fluid dynamics. Open Heart **6**(2), e001044 (2019)
337. Mojsejenko, D., McGarvey, J.R., Dorsey, S.M., Gorman, J.H., Burdick, J.A., Pilla, J.J., Gorman, R.C., Wenk, J.F.: Estimating passive mechanical properties in a myocardial infarction using MRI and finite element simulations. Biomechanics and Modeling in Mechanobiology **14**(3), 633–647 (2015)
338. Monfredi, O., Dobrzynski, H., Mondal, T., Boyett, M.R., Morris, G.M.: The anatomy and physiology of the sinoatrial node-A contemporary review. Pacing and Clinical Electrophysiology **33**(11), 1392–1406 (2010)
339. Montero, P., Flandes-Iparraguirre, M., Musquiz, S., Perez Araluce, M., Plano, D., Sanmartin, C., Orive, G., Gavira, J.J., Prosper, F., Mazo, M.M.: Cells, materials, and fabrication processes for cardiac tissue engineering. Frontiers in Bioengineering and Biotechnology **8** (2020)
340. Morris, P.D., Narracott, A., von Tengg-Kobligk, H., Soto, D.A.S., Hsiao, S., Lungu, A., Evans, P., Bressloff, N.W., Lawford, P.V., Hose, D.R., et al.: Computational fluid dynamics modelling in cardiovascular medicine. Heart **102**(1), 18–28 (2016)
341. Morris, T.A., Naik, J., Fibben, K.S., Kong, X., Kiyono, T., Yokomori, K., Grosberg, A.: Striated myocyte structural integrity: Automated analysis of sarcomeric z-discs. PLoS computational biology **16**(3), e1007676 (2020)
342. Mu, Y., Chen, L., Tang, Q., Ayoufu, G.: Real time three-dimensional echocardiographic assessment of left ventricular regional systolic function and dyssynchrony in patients with dilated cardiomyopathy. Echocardiography **27**(4), 415–420 (2010)
343. Muramoto, Y., Nagasaka, Y.: High-speed sensing of microliter-order whole-blood viscosity using laser-induced capillary wave. Journal of biorheology **25**(1–2), 43–51 (2011)
344. Murphy, J.B., Boyle, F.J.: A full-range, multi-variable, CFD-based methodology to identify abnormal near-wall hemodynamics in a stented coronary artery. Biorheology **47**(2), 117–132 (2010)
345. Muslin, A.J.: The Pathophysiology of Heart Failure, 1 edn., chap. 37, pp. 523–535. Academic Press (2012)
346. Mynard, J., Davidson, M., Penny, D., Smolich, J.: A simple, versatile valve model for use in lumped parameter and one-dimensional cardiovascular models. International Journal for Numerical Methods in Biomedical Engineering **28**(6–7), 626–641 (2012)
347. Naim, W.N., Mokhtarudin, M.J., Lim, E., Chan, B.T., Bakir, A.A., Mohamed, N.A.N.: The study of border zone formation in ischemic heart using electro-chemical coupled computational model. International Journal for Numerical Methods in Biomedical Engineering p. e3398 (2020)

348. Nakamura, M., Wada, S., Yamaguchi, T.: Influence of the opening mode of the mitral valve orifice on intraventricular hemodynamics. Annals of Biomedical Engineering **34**(6), 927–935 (2006)

349. Napel, S., Lee, D.H., Frayne, R., Rutt, B.K.: Visualizing three-dimensional flow with simulated streamlines and three-dimensional phase-contrast MR imaging. Journal of Magnetic Resonance Imaging **2**(2), 143–153 (1992)

350. Naveed, M., Mohammad, I.S., Xue, L., Khan, S., Gang, W., Cao, Y., Cheng, Y., Cui, X., DingDing, C., Feng, Y., Zhijie, W., Xiaohui, Z.: The promising future of ventricular restraint therapy for the management of end-stage heart failure. Biomedicine & Pharmacotherapy **99**, 25–32 (2018)

351. Nelson, G., Berger, R., Fetics, B., Talbot, M., Spinelli, J., Hare, J., Kass, D.: Left ventricular or biventricular pacing improves cardiac function at diminished energy cost in patients with dilated cardiomyopathy and left bundle-branch block. Circulation **102**(25), 3053–3059 (2000)

352. Neubauer, S., Weintraub, W., Appelbaum, E., Desai, M., Desvigne-Nickens, P., Dimarco, J., Dolman, S., Ho, C., Jerosch-Herold, M., Kolm, P., Kwong, R., Maron, M., Schulz-Menger, J., Watkins, H., Kramer, C.: Baseline characteristics of the hypertrophic cardiomyopathy registry. In: Health economics and policy to improve cardiovascular care and outcomes / Cardiomyopathies: diagnosis and outcome, p. 647 (2018)

353. Newman, D., Westerhof, N., Sipkema, P.: Modelling of aortic stenosis. Journal of Biomechanics **12**(3), 229–235 (1979)

354. Nguyen, V.T., Loon, C.J., Nguyen, H.H., Liang, Z., Leo, H.L.: A semi-automated method for patient-specific computational flow modelling of left ventricles. Computer Methods in Biomechanics and Biomedical engineering **18**(4), 401–413 (2015)

355. Nichols, M., Townsend, N., Scarborough, P., Rayner, M.: Cardiovascular disease in Europe 2014: epidemiological update. European Heart Journal **35**, 2950–2959 (2014)

356. Nicosia, M.A., Cochran, R.P., Einstein, D.R., Rutland, C.J., Kunzelman, K.S.: A coupled fluid-structure finite element model of the aortic valve and root. The Journal of Heart Valve Disease **12**(6), 781–789 (2003)

357. Nicoud, F., Chnafa, C., Siguenza, J., Zmijanovic, V., Mendez, S.: Large-eddy simulation of turbulence in cardiovascular flows. In: Biomedical Technology, pp. 147–167. Springer (2018)

358. Niestrawska, J., Augustin, C., Plank, G.: Computational modeling of cardiac growth and remodeling in pressure overloaded hearts – Linking microstructure to organ phenotype. Acta Biomaterialia **106**, 34–53 (2020)

359. Noll, W.: Lectures on the foundations of continuum mechanics and thermodynamics. Archive for Rational Mechanics and Analysis **52**(1), 62–92 (1973)

360. Nordsletten, D., Kay, D., Smith, N.: A non-conforming monolithic finite element method for problems of coupled mechanics. Journal of Computational Physics **229**(20), 7571–7593 (2010)

361. Nordsletten, D., McCormick, M., Kilner, P., Hunter, P., Kay, D., Smith, N.: Fluid–solid coupling for the investigation of diastolic and systolic human left ventricular function. International Journal for Numerical Methods in Biomedical Engineering **27**(7), 1017–1039 (2011)

362. Noseworthy, P.A., Rosenberg, M.A., Fifer, M.A., Palacios, I.F., Lowry, P.A., Ruskin, J.N., Sanborn, D.M., Picard, M.H., Vlahakes, G.J., Mela, T., Das, S.: Ventricular arrhythmia following alcohol septal ablation for obstructive hypertrophic cardiomyopathy. American Journal of Cardiology **104**, 128–132 (2009)

363. O'Brien, L., Remington, J.: Time course of pressure changes following quick stretch in tortoise ventricle. American Journal of Physiology–Legacy Content **211**(3), 770–776 (1966)

364. Ogden, R.W.: Non-Linear Elastic Deformations. Dover Pubn INC (1997)

365. Olivotto, I., Cecchi, F., Casey, S., Dolara, A., H, T.J., Maron, B.: Impact of atrial fibrillation on the clinical course of hypertrophic cardiomyopathy. Circulation **104**(21), 1496–1501 (2001)

366. Olivotto, I., Cecchi, F., Poggesi, C., Yacoub, M.: Patterns of disease progression in hypertrophic cardiomyopathy: An individualized approach to clinical staging. Advances in Heart Failure **5**(4), 535–546 (2012)

367. O'Mahony, C., Jichi, F., Pavlou, M., Monserrat, L., Anastasakis, A., Rapezzi, C., Biagini, E., Gimeno, J.R., Limongelli, G., McKenna, W.J., Omar, R.Z., Elliott, P.M., Ortiz-Genga, M., Fernandez, X., Vlagouli, V., Stefanadis, C., Coccolo, F., Sandoval, M.J.O., Pacileo, G., Masarone, D., Pantazis, A., Tome-Esteban, M., Dickie, S., Lambiase, P.D., Rahman, S.: A novel clinical risk prediction model for sudden cardiac death in hypertrophic cardiomyopathy. European Heart Journal **35**(30), 2010–2020 (2014)

368. O'Mahony, C., Tome-Esteban, M., Lambiase, P.D., Pantazis, A., Dickie, S., McKenna, W.J., Elliott, P.M.: A validation study of the 2003 American College of Cardiology/European Society of Cardiology and 2011 American College of Cardiology Foundation/American Heart Association risk stratification and treatment algorithms for sudden cardiac death in patients with hypertrophic cardiomyopathy. Heart **99**(8), 534–541 (2013)

369. Omens, J.H.: Stress and strain as regulators of myocardial growth. Progress in Biophysics and Molecular Biology **69**(2–3), 559–572 (1998)

370. Omens, J.H., Usyk, T.P., Li, Z., McCulloch, A.D.: Muscle LIM protein deficiency leads to alterations in passive ventricular mechanics. American Journal of Physiology: Heart and Circulatory Physiology **282**, H680–H687 (2002)

371. Ong, P.K., Lim, D., Kim, S.: Are microfluidics-based blood viscometers ready for point-of-care applications? A review. Critical Reviews™ in Biomedical Engineering **38**(2) (2010)

372. Osnabrugge, R.L., Mylotte, D., Head, S.J., Van Mieghem, N.M., Nkomo, V.T., LeReun, C.M., Bogers, A.J., Piazza, N., Kappetein, A.P.: Aortic stenosis in the elderly: disease prevalence and number of candidates for transcatheter aortic valve replacement: a meta-analysis and modeling study. Journal of the American College of Cardiology **62**(11), 1002–1012 (2013)

373. Ostadfar, A.: Biofluid mechanics: Principles and applications. Academic Press (2016)

374. Otto, C.M., Kuusisto, J., Reichenbach, D.D., Gown, A.M., O'Brien, K.D.: Characterization of the early lesion of 'degenerative' valvular aortic stenosis. histological and immunohisto-chemical studies. Circulation **90**(2), 844–853 (1994)

375. Owashi, K.P., Hubert, A., Galli, E., Donal, E., Hernández, A.I., Le Rolle, V.: Model-based estimation of left ventricular pressure and myocardial work in aortic stenosis. PLOS One **15**(3), e0229609 (2020)

376. Pagoulatou, S., Stergiopulos, N.: Evolution of aortic pressure during normal ageing: A model-based study. PLOS One **12**(7), e0182173 (2017)

377. Papademetris, X., Sinusas, A., Dione, D., Constable, R., Duncan, J.: Estimation of 3-D left ventricular deformation from medical images using biomechanical models. IEEE Transactions on Medical Imaging **21**(7), 786–800 (2002)

378. Paradis, A.N., Gay, M.S., Zhang, L.: Binucleation of cardiomyocytes: the transition from a proliferative to a terminally differentiated state. Drug discovery today **19**(5), 602–609 (2014)

379. Park, J., Choi, S., Janardhan, A.H., Lee, S.y., Raut, S., Soares, J., Shin, K., Yang, S., Lee, C., Kang, K.w., Cho, H.R., Kim, S.J., Seo, P., Hyun, W., Jung, S., Lee, H.j., Lee, N., Choi, S.H., Sacks, M., Lu, N., Josephson, M.E., Hyeon, T., Kim, D.H., Hwang, H.J.: Electromechanical cardioplasty using a wrapped elasto-conductive epicardial mesh. Science Translational Medicine **8**(344), 1–12 (2016)

380. Pasipoularides, A.: Fluid dynamic aspects of ejection in hypertrophic cardiomyopathy. Hellenic Journal of Cardiology **52**, 416–426 (2011)

381. Pasipoularides, A., Murgo, J.P., Miller, J.W., Craig, W.E.: Nonobstructive left ventricular ejection pressure gradients in man. Circulation research **61**(2), 220–227 (1987)

382. Pedrizzetti, G., La Canna, G., Alfieri, O., Tonti, G.: The vortex–an early predictor of cardiovascular outcome? Nature Reviews Cardiology **11**(9), 545–53 (2014)

383. Pedrizzetti, G., La Canna, G., Alfieri, O., Tonti, G.: The vortex–an early predictor of cardiovascular outcome? Nature Reviews Cardiology **11**(9), 545–553 (2014)

384. Peirlinck, M., Costabal, F.S., Sack, K.L., Choy, J.S., Kassab, G.S., Guccione, J.M., Beule, M.D., Segers, P., Kuhl, E.: Using machine learning to characterize heart failure across the scales. Biomech Model Mechanobiol **18**(6), 1987–2001 (2019)

385. Peirlinck, M., Costabal, F.S., Yao, J., Guccione, J.M., Tripathy, S., Wang, Y., Ozturk, D., Segars, P., Morrison, T.M., Levine, S., Kuhl, E.: Precision medicine in human heart modeling: Perspectives, challenges, and opportunities. Biomech Model Mechanobiol (2021)

386. Peskin, C.S.: Flow patterns around heart valves: a numerical method. Journal of Computational Physics **10**(2), 252–271 (1972)
387. Peskin, C.S.: The immersed boundary method. Acta Numerica **11**, 479–517 (2002)
388. Peters, C.H., Sharpe, E.J., Proenza, C.: Cardiac pacemaker activity and aging. Annual Review of Physiology **82**, 21–43 (2020)
389. Pinto, A.R., Ilinykh, A., Ivey, M.J., Kuwabara, J.T., D'Antoni, M.L., Debuque, R., Chandran, A., Wang, L., Arora, K., Rosenthal, N.A., Tallquist, M.D.: Revisiting cardiac cellular composition. Circulation Research **118**(3), 400–409 (2016)
390. Pinto, J.G., Fung, Y.: Mechanical properties of the heart muscle in the passive state. Journal of Biomechanics **6**(6), 597–616 (1973)
391. Pluijmert, M., Delhaas, T., de la Parra, A., Kroon, W., Prinzen, F., Bovendeerd, P.: Determinants of biventricular cardiac function: a mathematical model study on geometry and myofiber orientation. Biomechanics and Modeling in Mechanobiology **16**, 721–729 (2017)
392. Podlubny, I.: Fractional differential equations: an introduction to fractional derivatives, fractional differential equations, to methods of their solution and some of their applications, vol. 198. Academic press (1998)
393. Pope, A., Sands, G., Smaill, B., LeGrice, I.: Three-dimensional transmural organization of perimysial collagen in the heart. American Journal of Physiology: Heart and Circulatory Physiology **295**(3), H1243–H1252 (2008)
394. Presti, R.L., Hopps, E., Caimi, G.: Hemorheological abnormalities in human arterial hypertension. Korea-Australia Rheology Journal **26**(2), 199–204 (2014)
395. Qian, L., Todo, M., Morita, Y., Matsushita, Y., Koyano, K.: Deformation analysis of the periodontium considering the viscoelasticity of the periodontal ligament. Dental Materials **25**(10), 1285–1292 (2009)
396. Rajamannan, N.M.: Update on the pathophysiology of aortic stenosis. European Heart Journal Supplements **10**(suppl_E), E4–E10 (2008)
397. Rajamannan, N.M., Bonow, R.O., Rahimtoola, S.H.: Calcific aortic stenosis: an update. Nature Clinical Practice Cardiovascular Medicine **4**(5), 254–262 (2007)
398. Rama, R.R., Skatulla, S.: Towards real-time cardiac mechanics modelling with patient-specific heart anatomies. Computer Methods in Applied Mechanics and Engineering **328**, 47–74 (2018)
399. Reese, T.G., Weisskoff, R.M., Smith, R.N., Rosen, B.R., Dinsmore, R.E., Wedeen, V.J.: Imaging myocardial fiber architecture in vivo with magnetic resonance. Magnetic Resonance in Medicine **34**(6), 786–791 (1995)
400. Reyes, G., Fernández-Yáñez, J., Rodríguez-Abella, H., Palomo, J., Pinto, Á., Duarte, J.: Ventricular assist devices as a bridge to transplantation. Revista Espanola de Cardiologia **60**(1), 72–75 (2007)
401. Reymond, P., Merenda, F., Perren, F., Rufenacht, D., Stergiopulos, N.: Validation of a one-dimensional model of the systemic arterial tree. American Journal of Physiology-Heart and Circulatory Physiology **297**(1), H208–H222 (2009)
402. Richter, Y., Edelman, E.R.: Cardiology is flow (2006)
403. Riffel, J.H., Keller, M.G., Rost, F., Arenja, N., Andre, F., Aus Dem Siepen, F., Fritz, T., Ehlermann, P., Taeger, T., Frankenstein, L., Meder, B., Katus, H.A., Buss, S.J.: Left ventricular long axis strain: A new prognosticator in non-ischemic dilated cardiomyopathy? Journal of Cardiovascular Magnetic Resonance **18**(1), 36 (2016)
404. Rigatelli, G., Zuin, M., Fong, A.: Computational flow dynamic analysis of right and left atria in patent foramen ovale: potential links with atrial fibrillation. Journal of Atrial Fibrillation **10**(5) (2018)
405. Roberts, W.C., Ko, J.M.: Frequency by decades of unicuspid, bicuspid, and tricuspid aortic valves in adults having isolated aortic valve replacement for aortic stenosis, with or without associated aortic regurgitation. Circulation **111**(7), 920–925 (2005)
406. Rodriguez, E., Hoger, A., McCulloch, A.: Stress-dependent finite growth in soft elastic tissues. Journal of Biomechanics **27**(4), 455–467 (1994)

407. Rogers, W.J., Peterson, E.D., Frederick, P.D., French, W.J., Gibson, C.M., Jr, C.V.P., Ornato, J.P., Zalenski, R.J., Penney, J., Tiefenbrunn, A.J.: Number of coronary heart disease risk factors and mortality in patients with first myocardial infarction. Journal of American Medical Association **306**(19), 2120–2127 (2011)
408. Rohmer, D., Sitek, A., Gullberg, G.: Reconstruction and visualization of fiber and laminar structure in the normal human heart from ex vivo diffusion tensor magnetic resonance imaging (DTMRI) data. Investigative Radiology **42**(11), 777–789 (2007)
409. Rose, W.C., Shoukas, A.A.: Two-port analysis of systemic venous and arterial impedances. American Journal of Physiology: Heart and Circulatory Physiology **265**(5), H1577–H1587 (1993)
410. Ross, B.: The development of fractional calculus 1695–1900. Historia Mathematica **4**(1), 75–89 (1977)
411. Ross Jr, J., Braunwald, E.: Aortic stenosis. Circulation **38**(1s5), V–61 (1968)
412. Rossi, M., Carillo, S.: Cardiac hypertrophy due to pressure and volume overload: distinctly different biological phenomena? International Journal of Cardiology **31**(2), 133–141 (1991)
413. Rouillard, A.D., Holmes, J.W.: Coupled agent-based and finite-element models for predicting scar structure following myocardial infarction. Progress in Biophysics and Molecular Biology **115**(2–3), 235–243 (2014)
414. Saber, N.R., Wood, N.B., Gosman, A., Merrifield, R.D., Yang, G.Z., Charrier, C.L., Gatehouse, P.D., Firmin, D.N.: Progress towards patient-specific computational flow modeling of the left heart via combination of magnetic resonance imaging with computational fluid dynamics. Annals of Biomedical Engineering **31**(1), 42–52 (2003)
415. Sacco, F., Paun, B., Lehmkuhl, O., Iles, T.L., Iaizzo, P.A., Houzeaux, G., Vázquez, M., Butakoff, C., Aguado-Sierra, J.: Left ventricular trabeculations decrease the wall shear stress and increase the intra-ventricular pressure drop in CFD simulations. Frontiers in Physiology **9**, 458 (2018)
416. Sack, K.L., Davies, N.H., Guccione, J.M., Franz, T., Franz, T.: Personalised computational cardiology: Patient-specific modelling in cardiac mechanics and biomaterial injection therapies for myocardial infarction. Heart Failure Reviews **21**(6), 815–826 (2016)
417. Sacks, M.S.: Incorporation of experimentally-derived fiber orientation into a structural constitutive model for planar collagenous tissues. Journal of Biomechanical Engineering **125**(2), 280–287 (2003)
418. Saez, P., Kuhl, E.: Computational modeling of acute myocardial infarction. Computer Methods in Biomechanics and Biomedical Engineering **19**(10), 1107–1115 (2016)
419. Sajgalik, P., Grupper, A., Edwards, B., Kushwaha, S., Stulak, J., Joyce, D., Joyce, L., Daly, R., Kara, T., Schirger, J.: Current status of left ventricular assist device therapy. Mayo Clinic Proceedings **91**(7), 927–940 (2016)
420. Sasaki, N., Odajima, S.: Elongation mechanism of collagen fibrils and force-strain relations of tendon at each level of structural hierarchy. Journal of Biomechanics **29**(9), 1131–1136 (1996)
421. Sasaki, N., Odajima, S.: Stress-strain curve and Young's modulus of a collagen molecule as determined by the x-ray diffraction technique. Journal of Biomechanics **29**(5), 655–658 (1996)
422. Saucerman, J.J., Tan, P.M., Buchholz, K.S., McCulloch, A.D., Omens, J.H.: Mechanical regulation of gene expression in cardiac myocytes and fibroblasts. Nature Reviews Cardiology **16**(6), 361–378 (2019)
423. Scarsoglio, S., Saglietto, A., Gaita, F., Ridolfi, L., Anselmino, M.: Computational fluid dynamics modelling of left valvular heart diseases during atrial fibrillation. PeerJ **4**, e2240 (2016)
424. Schenkel, T., Malve, M., Reik, M., Markl, M., Jung, B., Oertel, H.: MRI-based CFD analysis of flow in a human left ventricle: methodology and application to a healthy heart. Annals of Biomedical Engineering **37**(3), 503–515 (2009)
425. Schiessel, H., Blumen, A.: Hierarchical analogues to fractional relaxation equations. Journal of Physics A: Mathematical and General **26**(19), 5057–5069 (1993)

426. Schiessel, H., Metzler, R., Blumen, A., Nonnenmacher, T.F.: Generalized viscoelastic models: their fractional equations with solutions. Journal of Physics A: Mathematical and General **28**(23), 6567–6584 (1995)

427. Schmid, H., Nash, M., Young, A., Hunter, P.: Myocardial material parameter estimation — a comparative study for simple shear. Transactions of the American Society of Mechanical Engineers **128**(5), 742–750 (2006)

428. Schmid, H., O'Callaghan, P., Nash, M., Lin, W., LeGrice, I., Smaill, B., Young, A., Hunter, P.: Myocardial material parameter estimation. Biomechanics and modeling in mechanobiology **7**(3), 161–173 (2008)

429. Schultheiss, H.P., Fairweather, D.L., Caforio, A.L., Escher, F., Hershberger, R.E., Lipshultz, S.E., Liu, P.P., Matsumori, A., Mazzanti, A., McMurray, J., Priori, S.G.: Dilated cardiomyopathy. Nature Reviews Disease Primers **5**(1) (2019)

430. Schuster, A., Hor, K.N., Kowallick, J.T., Beerbaum, P., Kutty, S.: Cardiovascular magnetic resonance myocardial feature tracking: Concepts and clinical applications. Circulation: Cardiovascular Imaging **9**(4), 1–9 (2016)

431. Scollan, D.F., Holmes, A., Winslow, R., Forder, J.: Histological validation of myocardial microstructure obtained from diffusion tensor magnetic resonance imaging. American Journal of Physiology-Heart and Circulatory Physiology **275**(6), H2308–H2318 (1998)

432. Seckeler, M.D., Hoke, T.R.: The worldwide epidemiology of acute rheumatic fever and rheumatic heart disease. Clinical Epidemiology **3**, 67 (2011)

433. Selzman, C., Oberlander, J.: The price of progress: destination left ventricular assist device therapy for terminal heart failure. North Carolina Medical Journal **67**(2), 116–117 (2006)

434. Semsarian, C., Ingles, J., Maron, M.S., Maron, B.J.: New perspectives on the prevalence of hypertrophic cardiomyopathy. Journal of the American College of Cardiology **65**(12), 1249–1254 (2015)

435. Sengupta, P.P., Krishnamoorthy, V.K., Korinek, J., Narula, J., Vannan, M.A., Lester, S.J., Tajik, J.A., Seward, J.B., Khandheria, B.K., Belohlavek, M.: Left ventricular form and function revisited: applied translational science to cardiovascular ultrasound imaging. Journal of the American Society of Echocardiography **20**(5), 539–551 (2007)

436. Seo, J.H., Vedula, V., Abraham, T., Lardo, A.C., Dawoud, F., Luo, H., Mittal, R.: Effect of the mitral valve on diastolic flow patterns. Physics of Fluids **26**(12), 121901 (2014)

437. Sermesant, M., Forest, C., Pennec, X., Delingette, H., Ayache, N.: Deformable biomechanical models: Application to 4D cardiac image analysis. Medical Image Analysis **7**(4), 475–488 (2003)

438. Sharma, K., Kass, D.A.: Heart failure with preserved ejection fraction. Circulation research **115**(1), 79–96 (2014)

439. Shen, J.J.: A structurally based viscoelastic model for passive myocardium in finite deformation. Computational Mechanics **58**(3), 491–509 (2016)

440. Sherrid, M.V., Balaram, S., Kim, B., Axel, L., Swistel, D.G.: The mitral valve in obstructive hypertrophic cardiomyopathy a test in context. Journal of the American College of Cardiology **67**, 1846–1858 (2016)

441. Sherwin, S., Blackburn, H.: Three-dimensional instabilities and transition of steady and pulsatile axisymmetric stenotic flows. Journal of Fluid Mechanics **533**, 297–327 (2005)

442. Shi, P., Sinusas, A.J., Constable, R.T., Duncan, J.S.: Volumetric deformation analysis using mechanics-based data fusion: Applications in cardiac motion recovery. International Journal of Computer Vision **35**(1), 87–107 (1999)

443. Shi, Y., Lawford, P., Hose, R.: Review of zero-D and 1-D models of blood flow in the cardiovascular system. Biomedical Engineering Online **10**(1), 33 (2011)

444. Shih, H., Lee, B., Lee, R.J., Boyle, A.J., Francisco, S.: The aging heart and post-infarction left ventricular remodeling. Journal of the American College of Cardiology **57**(1), 9–17 (2011)

445. Shiozaki, A.A., Senra, T., Arteaga, E., Martinelli Filho, M., Pita, C.G., Avila, L.F.R., Parga Filho, J.R., Mady, C., Kalil-Filho, R., Bluemke, D.A., Rochitte, C.E.: Myocardial fibrosis detected by cardiac CT predicts ventricular fibrillation/ventricular tachycardia events in patients with hypertrophic cardiomyopathy. Journal of Cardiovascular Computed Tomography **7**(3), 173–181 (2013)

446. Sigaeva, T., Sommer, G., Holzapfel, G., Martino, E.D.: Anisotropic residual stresses in arteries. Journal of the Royal Society Interface **16**(151) (2019)
447. Sigg, D.C., Hezi-Yamit, A.: Cardiac and vascular receptors and signal transduction. In: P.A. Iaizzo (ed.) Handbook of Cardiac Anatomy, Physiology, and Devices, pp. 251–277. Springer International Publishing (2015)
448. Sirry, M.S., Davies, N.H., Kadner, K., Dubuis, L., Muhammad, G., Meintjes, E.M., Spottiswoode, B.S., Zilla, P., Sirry, M.S., Davies, N.H., Kadner, K., Dubuis, L., Muhammad, G., Meintjes, E.M., Spottiswoode, B.S., Zilla, P., Franz, T.: Micro-structurally detailed model of a therapeutic hydrogel injectate in a rat biventricular cardiac geometry for computational simulations (2015)
449. Smoluk, A.T., Smoluk, L.T., Protsenko, Y.L.: Stress relaxation of heterogeneous myocardial tissue. Numerical experiments with 3D model. Biophysics **57**(6), 804–807 (2012)
450. Smoluk, L., Protsenko, Y.: Viscoelastic properties of the papillary muscle: experimental and theoretical study. Acta of Bioengineering and Biomechanics; 04/2012; ISSN 1509-409X (2012)
451. Sodhani, D., Reese, S., Moreira, R., Jockenhoevel, S., Mela, P., Stapleton, S.E.: Multi-scale modelling of textile reinforced artificial tubular aortic heart valves. Meccanica **52**(3), 677–693 (2017)
452. Sommer, G., Haspinger, D.C., Andrä, M., Sacherer, M., Viertler, C., Regitnig, P., Holzapfel, G.A.: Quantification of shear deformations and corresponding stresses in the biaxially tested human myocardium. Annals of Thoracic Surgery of biomedical engineering pp. 1–15 (2015)
453. Sommer, G., Schriefl, A.J., Andrä, M., Sacherer, M., Viertler, C., Wolinski, H., Holzapfel, G.A.: Biomechanical properties and microstructure of human ventricular myocardium. Acta biomaterialia **24**, 172–192 (2015)
454. Somphone, O., Craene, M.D., Ardon, R., Mory, B., Allain, P., Gao, H., D'hooge, J., Marchesseau, S., Sermesant, M., Delingette, H., Saloux, E.: Fast myocardial motion and strain estimation in 3D cardiac ultrasound with sparse demons. In: 2013 IEEE 10th International Symposium on Biomedical Imaging (2013)
455. Souders, C.A., Bowers, S.L., Baudino, T.A.: Cardiac fibroblast: The renaissance cell. Circulation Research **105**(12), 1164–1176 (2009)
456. Sousa, P.C., Pinho, F.T., Alves, M.A., Oliveira, M.S.: A review of hemorheology: Measuring techniques and recent advances. Korea-Australia Rheology Journal **28**(1), 1–22 (2016)
457. Spalart, P.R.: Strategies for turbulence modelling and simulations. International Journal of Heat and Fluid Flow **21**(3), 252–263 (2000)
458. Spencer, A.J.M.: Continuum mechanics. Dover Publications, Mineola, N.Y (2004)
459. Spühler, J.H., Jansson, J., Jansson, N., Hoffman, J.: 3D fluid-structure interaction simulation of aortic valves using a unified continuum ALE FEM model. Frontiers in physiology **9**, 363 (2018)
460. Standring, S. (ed.): Gray's Anatomy: The Anatomical Basis of Clinical Practice. Elsevier Limited, Philadelphia (2016)
461. Stein, P.D., Sabbah, H.N., Pitha, J.V.: Continuing disease process of calcific aortic stenosis: role of microthrombi and turbulent flow. The American Journal of Cardiology **39**(2), 159–163 (1977)
462. Stewart, J.A., Massey, E.P., Fix, C., Zhu, J., Goldsmith, E.C., Carver, W.: Temporal alterations in cardiac fibroblast function following induction of pressure overload. Cell and Tissue Research **340**(1), 117–126 (2010)
463. Stoeck, C.T., Deuster, C.V., Fleischmann, T., Lipiski, M., Cesarovic, N., Kozerke, S.: Direct comparison of in vivo versus postmortem second-order motion-compensated cardiac diffusion tensor imaging. Magnetic Resonance in Medicine **79**, 2265–2276 (2018)
464. Streeter, D., Spotnitz, H., Patel, D., Ross, J., Sonnenblick, E.: Fiber orientation in the canine left ventricle during diastole and systole. Circulation Research **24**(3), 339–347 (1969)
465. Strumpf, R., Yin, F., Humphrey, J.: Determination of a constitutive relation for passive myocardium: I. A new functional form. Journal of biomechanical engineering **112**, 333 (1990)

466. Su, B., Zhang, J.M., Tang, H.C., Wan, M., Lim, C.C.W., Su, Y., Zhao, X., San Tan, R., Zhong, L.: Patient-specific blood flows and vortex formations in patients with hypertrophic cardiomyopathy using computational fluid dynamics. In: 2014 IEEE Conference on Biomedical Engineering and Sciences, pp. 276–280. IEEE (2014)

467. Su, B., Zhong, L., Wang, X.K., Zhang, J.M., San Tan, R., Allen, J.C., Tan, S.K., Kim, S., Leo, H.L.: Numerical simulation of patient-specific left ventricular model with both mitral and aortic valves by FSI approach. Computer Methods and Programs in Biomedicine **113**(2), 474–482 (2014)

468. Sud, V., Srinivasan, R., Charles, J., Bungo, M.: Mathematical modelling of the human cardiovascular system in the presence of stenosis. Physics in Medicine & Biology **38**(3), 369 (1993)

469. Sun, W., Sacks, M.S., Sellaro, T.L., Slaughter, W.S., Scott, M.J.: Biaxial mechanical response of bioprosthetic heart valve biomaterials to high in-plane shear. Journal of Biomechanical Engineering **125**(3), 372–380 (2003)

470. Svalbring, E., Fredriksson, A., Eriksson, J., Dyverfeldt, P., Ebbers, T., Bolger, A.F., Engvall, J., Carlhäll, C.J.: Altered diastolic flow patterns and kinetic energy in subtle left ventricular remodeling and dysfunction detected by 4D flow MRI. PloS one **11**(8), e0161391 (2016)

471. Tan, L.K., McLaughlin, R.A., Lim, E., Abdul Aziz, Y.F., Liew, Y.M.: Fully automated segmentation of the left ventricle in cine cardiac MRI using neural network regression. Journal of Magnetic Resonance Imaging **48**(1), 140–152 (2018)

472. Taylor, C.A., Fonte, T.A., Min, J.K.: Computational fluid dynamics applied to cardiac computed tomography for noninvasive quantification of fractional flow reserve: scientific basis. Journal of the American College of Cardiology **61**(22), 2233–2241 (2013)

473. Terracciano, C., Guymer, S. (eds.): Heart of the Matter: Key concepts in cardiovascular science. Learning Materials in Biosciences. Springer International Publishing (2019)

474. Töger, J., Kanski, M., Carlsson, M., Kovács, S.J., Söderlind, G., Arheden, H., Heiberg, E.: Vortex ring formation in the left ventricle of the heart: analysis by 4D flow MRI and Lagrangian coherent structures. Annals of Biomedical Engineering **40**(12), 2652–2662 (2012)

475. Toma, M., Oshima, M., Takagi, S.: Decomposition and parallelization of strongly coupled fluid–structure interaction linear subsystems based on the q1/p0 discretization. Computers & Structures **173**, 84–94 (2016)

476. Topilsky, Y., Pereira, N.L., Shah, D.K., Boilson, B., Schirger, J.A., Kushwaha, S.S., Joyce, L.D., Park, S.J.: Left ventricular assist device therapy in patients with restrictive and hypertrophic cardiomyopathy. Circulation: Heart Failure **4**(3), 266–275 (2011)

477. Tsanas, A., Goulermas, J.Y., Vartela, V., Tsiapras, D., Theodorakis, G., Fisher, A.C., Sfirakis, P.: The Windkessel model revisited: a qualitative analysis of the circulatory system. Medical Engineering & Physics **31**(5), 581–588 (2009)

478. Tustison, N., Davila-Roman, V., Amini, A.: Myocardial kinematics from tagged MRI based on a 4-D B-spline model. IEEE Transactions on Biomedical Engineering **50**(8), 1038–1040 (2003)

479. Ursell, P.C., Gardner, P.I., Albala, A., Fenoglio, J.J., Wit, A.L.: Structural and electrophysiological changes in the epicardial border zone of canine myocardial infarcts during infarct healing. Circulation Research **56**(3), 436–451 (1985)

480. Usyk, T.P., Omens, J.H., McCulloch, A.D.: Regional septal dysfunction in a three-dimensional computational model of focal myofiber disarray. American Journal of Physiology - Heart and Circulatory Physiology **13**, 733–745 (2001)

481. Vadakkumpadan, F., Gurev, V., Constantino, J., Arevalo, H., Trayanova, N.: Modeling of whole-heart electrophysiology and mechanics: Toward patient-specific simulations. In: R.C. Kerckhoffs (ed.) Patient-Specific Modeling of the Cardiovascular System: Technology-Driven Personalized Medicine, pp. 145–165. Springer New York, New York, NY (2010)

482. Vahl, C., Timek, T., Bonz, A., Fuchs, H., Dillman, R., Hagl, S.: Length dependence of calcium-and force-transients in normal and failing human myocardium. Journal of molecular and cellular cardiology **30**(5), 957–966 (1998)

483. Valdez-Jasso, D., Simon, M.A., Champion, H.C., Sacks, M.S.: A murine experimental model for the mechanical behaviour of viable right-ventricular myocardium. The Journal of Physiology **590**(18), 4571–4584 (2012)

484. Valentín, A., Holzapfel, G.: Constrained mixture models as tools for testing competing hypotheses in arterial biomechanics: A brief survey. Mechanics Research Communications **42**, 126–133 (2012)

485. Varela, C.E., Fan, Y., Roche, E.T.: Optimizing epicardial restraint and reinforcement following myocardial infarction: Moving towards localized, biomimetic, and multitherapeutic options. Biomimetics **4**(1), 7 (2019)

486. de Vecchi, A., Nordsletten, D.A., Razavi, R., Greil, G., Smith, N.: Patient specific fluid–structure ventricular modelling for integrated cardiac care. Medical & Biological Engineering & Computing **51**(11), 1261–1270 (2013)

487. Vedula, V., Seo, J.H., Lardo, A.C., Mittal, R.: Effect of trabeculae and papillary muscles on the hemodynamics of the left ventricle. Theoretical and Computational Fluid Dynamics **30**(1–2), 3–21 (2016)

488. van der Velden, J., de Jong, J.W., Owen, V., Burton, P., Stienen, G.: Effect of protein kinase A on calcium sensitivity of force and its sarcomere length dependence in human cardiomyocytes. Cardiovascular research **46**(3), 487–495 (2000)

489. Verkaik, A., Hulsen, M., Bogaerds, A., van de Vosse, F.: An overlapping domain technique coupling spectral and finite elements for fluid–structure interaction. Computers & Fluids **123**, 235–245 (2015)

490. Von Deuster, C., Sammut, E., Asner, L., Nordsletten, D., Lamata, P., Stoeck, C.T., Kozerke, S., Razavi, R.: Studying dynamic myofiber aggregate reorientation in dilated cardiomyopathy using in vivo magnetic resonance diffusion tensor imaging. Circulation: Cardiovascular Imaging **9**(10), e005018 (2016)

491. Voorhees, A.P., Han, H.C.: A model to determine the effect of collagen fiber alignment on heart function post myocardial infarction. Theoretical Biology and Medical Modelling **11**(1) (2014)

492. Van de Vosse, F.N., Stergiopulos, N.: Pulse wave propagation in the arterial tree. Annual Review of Fluid Mechanics **43**, 467–499 (2011)

493. Walburn, F.J., Schneck, D.J.: A constitutive equation for whole human blood. Biorheology **13**(3), 201–210 (1976)

494. Walker, J., Ratcliffe, M., Zhang, P., Wallace, A., Fata, B., Hsu, E., Saloner, D., Guccione, J.: MRI-based finite-element analysis of left ventricular aneurysm. American Journal of Physiology-Heart and Circulatory Physiology **289**(2), H692–H700 (2005)

495. Walker, S.M.: Potentiation and hysteresis induced by stretch and subsequent release of papillary muscle of the dog. American Journal of Physiology–Legacy Content **198**(3), 519–522 (1960)

496. Wall, S.T., Walker, J.C., Healy, K.E., Ratcliffe, M.B., Guccione, J.M.: Theoretical impact of the injection of material into the myocardium: A finite element model simulation. Circulation **114**, 2627–2635 (2006)

497. Wang, V., Lam, H., Ennis, D., Cowan, B., Young, A., Nash, M.: Modelling passive diastolic mechanics with quantitative MRI of cardiac structure and function. Medical Image Analysis **13**(5), 773–784 (2009)

498. Wang, X., Schoen, J.A., Rentschler, M.E.: A quantitative comparison of soft tissue compressive viscoelastic model accuracy. Journal of the Mechanical Behavior of Biomedical Materials **20**, 126–136 (2013)

499. Wang, Z.J., Wang, V.Y., Bradley, C.P., Nash, M.P., Young, A.A., Cao, J.J.: Left ventricular diastolic myocardial stiffness and end-diastolic myofibre stress in human heart failure using personalised biomechanical analysis. Journal of Cardiovascular Translational Research **11**(4), 346–356 (2018)

500. Warwick, R., Pullan, M., Poullis, M.: Mathematical modelling to identify patients who should not undergo left ventricle remodelling surgery. Interactive CardioVascular and Thoracic Surgery **10**, 661–665 (2010)

501. Watanabe, H., Sugiura, S., Hisada, T.: The looped heart does not save energy by maintaining the momentum of blood flowing in the ventricle. American Journal of Physiology: Heart and Circulatory Physiology 294(5), H2191–H2196 (2008)
502. Watanabe, H., Sugiura, S., Kafuku, H., Hisada, T.: Multiphysics simulation of left ventricular filling dynamics using fluid-structure interaction finite element method. Biophysical Journal 87(3), 2074–2085 (2004)
503. Watton, P., Luo, X., Yin, M., Bernacca, G., Wheatley, D.: Effect of ventricle motion on the dynamic behaviour of chorded mitral valves. Journal of Fluids and Structures 24(1), 58–74 (2008)
504. Weese, J., Lungu, A., Peters, J., Weber, F.M., Waechter-Stehle, I., Hose, D.R.: CFD- and Bernoulli-based pressure drop estimates: A comparison using patient anatomies from heart and aortic valve segmentation of CT images. Medical Physics 44(6), 2281–2292 (2017)
505. Wei, D., Miyamoto, N., Mashima, S.: A computer model of myocardial disarray in simulating ECG features of hypertrophic cardiomyopathy. Japanese Heart Journal 40, 819–826 (1999)
506. Wei, Z.A., Sonntag, S.J., Toma, M., Singh-Gryzbon, S., Sun, W.: Computational fluid dynamics assessment associated with transcatheter heart valve prostheses: a position paper of the iso working group. Cardiovascular Engineering and Technology 9(3), 289–299 (2018)
507. Wei-Te, L., A, R.R.: Patient specific physics-based model for interactive visualization of cardiac dynamics. Studies in Health Technology and Informatics 70, 182–188 (2000)
508. Weinberg, E.J., Mofrad, M.R.K.: A multiscale computational comparison of the bicuspid and tricuspid aortic valves in relation to calcific aortic stenosis. Journal of Biomechanics 41(16), 3482–3487 (2008)
509. Wendell, D.C., Samyn, M.M., Cava, J.R., Ellwein, L.M., Krolikowski, M.M., Gandy, K.L., Pelech, A.N., Shadden, S.C., LaDisa Jr, J.F.: Including aortic valve morphology in computational fluid dynamics simulations: initial findings and application to aortic coarctation. Medical Engineering & Physics 35(6), 723–735 (2013)
510. Wenk, J.F., Eslami, P., Zhang, Z., Xu, C., Kuhl, E., Gorman, J.H., Robb, J.D., Ratcliffe, M.B., Gorman, R.C., Guccione, J.M.: A novel method for quantifying the in-vivo mechanical effect of material injected into a myocardial infarction. Annals of Thoracic Surgery 92(3), 935–941 (2011)
511. Westerhof, N., Elzinga, G., Sipkema, P.: An artificial arterial system for pumping hearts. Journal of Applied Physiology 31(5), 776–781 (1971)
512. Whitmore, R.: The flow behaviour of blood in the circulation. Nature 215(5097), 123–126 (1967)
513. Wiegner, A.W., Bing, O.H.: Mechanics of myocardial relaxation: Application of a model to isometric and isotonic relaxation of rat myocardium. Journal of Biomechanics 15(11), 831–840 (1982)
514. Wigen, M.S., Fadnes, S., Rodriguez-Molares, A., Bjåstad, T., Eriksen, M., Stensæth, K.H., Støylen, A., Lovstakken, L.: 4-D intracardiac ultrasound vector flow imaging–feasibility and comparison to phase-contrast MRI. IEEE Transactions on Medical Imaging 37(12), 2619–2629 (2018)
515. Wilson, S., Givertz, M., Stewart, G., Mudge Jr., G.: Ventricular assist devices: the challenges of outpatient management. Journal of the American College of Cardiology 54(18), 1647–1659 (2009)
516. Wolf, C.M., Moskowitz, I.P.G., Arno, S., Branco, D.M., Semsarian, C., Bernstein, S.A., Peterson, M., Maida, M., Morley, G.E., Fishman, G., Berul, C.I., Seidman, C.E., Seidman, J.G.: Somatic events modify hypertrophic cardiomyopathy pathology and link hypertrophy to arrhythmia. Proceedings of the National Academy of Sciences of the United States of America 102(50), 18123–18128 (2005)
517. Wong, K.C., Tee, M., Chen, M., Bluemke, D.A., Summers, R.M., Yao, J.: Regional infarction identification from cardiac CT images: a computer-aided biomechanical approach. International Journal of Computer Assisted Radiology and Surgery 11(9), 1573–1583 (2016)

518. Woods, R.H.: A few applications of a physical theorem to membranes in the human body in a state of tension. Transactions of the Royal Academy of Medicine in Ireland **10**(1), 417–427 (1892)

519. Wu, M., Tseng, W., Su, M., Liu, C., Chiou, K., Wedeen, V., Reese, T., Yang, C.: Diffusion tensor magnetic resonance imaging mapping the fiber architecture remodeling in human myocardium after infarction: correlation with viability and wall motion. Circulation **114**(10), 1036–1045 (2006)

520. Yacoub, M.H., Afifi, A., Saad, H., Aguib, H., Elguindy, A.: Current state of the art and future of myectomy. Annals of Cardiothoracic Surgery **6**(4), 307 (2017)

521. Yanagida, R., Sugawara, M., Kawai, A., Koyanagi, H.: Regional differences in myocardial work of the left ventricle in patients with idiopathic dilated cardiomyopathy: Implications for the surgical technique used for left ventriculoplasty. Journal of Thoracic and Cardiovascular Surgery **122**(September), 600–607 (2001)

522. Yang, B., Lesicko, J., Sharma, M., Hill, M., Sacks, M.S., Tunnell, J.W.: Polarized light spatial frequency domain imaging for non-destructive quantification of soft tissue fibrous structures. Biomedical Optics Express **6**(4), 1520 (2015). https://doi.org/10.1364/boe.6.001520

523. Yang, K., Breitbart, A., Lange, W.D., Hofsteen, P., Futakuchi-Tsuchida, A., Xu, J., Schopf, C., Razumova, M., Jiao, A., Boucek, R., Pabon, L., Reinecke, H., Kim, D., Ralphe, J., Regnier, M., Murry, C.: Novel adult-onset systolic cardiomyopathy due to MYH7 E848G mutation in patient-derived induced pluripotent stem cells. JACC: Basic to Translational Science **3**(6), 728–740 (2018)

524. Yang, M., Taber, L.A.: The possible role of poroelasticity in the apparent viscoelastic behavior of passive cardiac muscle. Journal of Biomechanics **24**(7), 587–597 (1991)

525. Yilmaz, F., Kutlar, A.I., Gundogdu, M.Y.: Analysis of drag effects on pulsatile blood flow in a right coronary artery by using Eulerian multiphase model. Korea-Australia Rheology Journal **23**(2), 89 (2011)

526. Yin, F.: Ventricular wall stress. Circulation Research **49**(4), 829–842 (1981)

527. Yin, F.C., Strumpf, R.K., Chew, P.H., Zeger, S.L.: Quantification of the mechanical properties of noncontracting canine myocardium under simultaneous biaxial loading. Journal of Biomechanics **20**(6), 577–589 (1987)

528. Yin, M., Yazdani, A., Karniadakis, G.E.: One-dimensional modeling of fractional flow reserve in coronary artery disease: Uncertainty quantification and Bayesian optimization. Computer Methods in Applied Mechanics and Engineering **353**, 66–85 (2019)

529. Yoo, L., Kim, H., Gupta, V., Demer, J.L.: Quasilinear viscoelastic behavior of bovine extraocular muscle tissue. Investigative Opthalmology & Visual Science **50**(8), 3721 (2009)

530. Yotti, R., Bermejo, J., Antoranz, J.C., Desco, M.M., Cortina, C., Rojo-Alvarez, J.L., Allue, C., Martin, L., Moreno, M., Serrano, J.A., Munoz, R., Garcia-Fernandez, M.A.: A noninvasive method for assessing impaired diastolic suction in patients with dilated cardiomyopathy. Circulation **112**(19), 2921–2929 (2005)

531. Young, A.A., Dokos, S., Powell, K.A., Sturm, B., McCulloch, A.D., Starling, R.C., McCarthy, P.M., White, R.D.: Regional heterogeneity of function in nonischemic dilated cardiomyopathy. Cardiovascular Research **49**, 308–318 (2001)

532. Youssefi, P., Gomez, A., He, T., Anderson, L., Bunce, N., Sharma, R., Figueroa, C.A., Jahangiri, M.: Patient-specific computational fluid dynamics—assessment of aortic hemodynamics in a spectrum of aortic valve pathologies. Journal of Thoracic and Cardiovascular Surgery **153**(1), 8–20 (2017)

533. Zhang, W., Ayoub, S., Liao, J., Sacks, M.S.: A meso-scale layer-specific structural constitutive model of the mitral heart valve leaflets. Acta Biomaterialia **32**, 238–255 (2016)

534. Zhang, W., Capilnasiu, A., Nordsletten, D.A.: Comparative analysis of nonlinear viscoelastic models across common biomechanical experiments. Journal of Elasticity p. In press (2021)

535. Zhang, W., Capilnasiu, A., Sommer, G., Holzapfel, G.A., Nordsletten, D.A.: An efficient and accurate method for modeling nonlinear fractional viscoelastic biomaterials. Computer Methods in Applied Mechanics and Engineering **362**, 112834 (2020)

536. Zhang, W., Zakerzadeh, R., Zhang, W., Sacks, M.S.: A material modeling approach for the effective response of planar soft tissues for efficient computational simulations. Journal of the Mechanical Behavior of Biomedical Materials **89**, 168–198 (2019)
537. Zhong, L., Zhang, J.M., Su, B., Tan, R.S., Allen, J.C., Kassab, G.S.: Application of patient-specific computational fluid dynamics in coronary and intra-cardiac flow simulations: Challenges and opportunities. Frontiers in Physiology **9** (2018)
538. Zhu, Y., Matsumura, Y., Wagner, W.R.: Biomaterials ventricular wall biomaterial injection therapy after myocardial infarction: Advances in material design, mechanistic insight and early clinical experiences. Biomaterials **129**, 37–53 (2017)
539. Zile, M.R., Baicu, C.F., Gaasch, W.H.: Diastolic heart failure – abnormalities in active relaxation and passive stiffness of the left ventricle. New England Journal of Medicine **350**(19), 1953–1959 (2004)

Translational Cardiovascular Modeling: Tetralogy of Fallot and Modeling of Diseases

Radomír Chabiniok, Kateřina Škardová, Radek Galabov, Pavel Eichler,
Maria Gusseva, Jan Janoušek, Radek Fučík, Jaroslav Tintěra,
Tomáš Oberhuber, and Tarique Hussain

Abstract Translational cardiovascular modeling (TCM) combines clinical data with physiologically and biophysically based models of the heart, vessels, or circulation while aiming to contribute to diagnosis or optimal clinical management. Models of heart mechanics and electromechanical models are applicable when assessing ventricular function, contributing to the planning of optimal intervention. During a perioperative period or acute exacerbation of heart failure, close to real-time running models can be coupled with signals monitoring cardiovascular physiology. Blood flow assessed by combining phase-contrast magnetic resonance imaging with

R. Chabiniok (✉)
UT Southwestern Medical Center, Dallas, TX, USA

Inria France, Ecole Polytechnique,CNRS, Institut Polytechnique de Paris, Palaiseau, France

St. Thomas' Hospital, King's College, London, UK

Faculty of Nuclear Sciences and Physical Engineering, Czech Technical University in Prague, Prague, Czech Republic
e-mail: radomir.chabiniok@utsouthwestern.edu

K. Škardová · P. Eichler · R. Fučík · T. Oberhuber
Faculty of Nuclear Sciences and Physical Engineering, Czech Technical University in Prague, Prague, Czech Republic

R. Galabov
Institute for Clinical and Experimental Medicine, Prague, Czech Republic

Faculty of Nuclear Sciences and Physical Engineering, Czech Technical University in Prague, Prague, Czech Republic

M. Gusseva
Inria France, Ecole Polytechnique, CNRS, Institut Polytechnique de Paris, Palaiseau, France

J. Janoušek
Children's Heart Center, University Hospital Motol, Charles University, Prague, Czech Republic

J. Tintěra
Institute for Clinical and Experimental Medicine, Prague, Czech Republic

T. Hussain
UT Southwestern Medical Center, Dallas, TX, USA

flow models can contribute to the decision about a possible intervention, e.g., on heart valves or large vessels. Furthermore, advanced imaging and image processing constrained by biophysical models allows for the study of distinct patterns, which could contribute to early detection or mapping a disease progress. In this chapter, we demonstrate the applicability of some TCM methods on tetralogy of Fallot (TOF)—the most common cyanotic congenital heart disease. A number of already existing modeling techniques can be applied on the cohort of TOF. Likewise, some novel techniques developed specifically for the group of TOF patients could serve in some other pathologies. This whole approach leads to an acronym TOFMOD, standing for Tetralogy of Fallot and Modeling of Diseases.

1   Introduction

Translational cardiovascular modeling (TCM) plays a crucial role in extending high-fidelity engineering models into real-life clinical applications. It represents a novel multidisciplinary approach, which brings the opportunity to address clinical problems not sufficiently solved by current techniques and contributes into advancing healthcare of pathologies involving cardiovascular system. As reviewed in [1], the components of TCM are: (1) physiologically and biophysically based models of the heart, flow in vessels or circulation; (2) acquisition and processing pipelines for clinical data; (3) means of creation of personalized models (i.e., model–data fusion); and (4) translation of patient-specific modeling into patient care, contributing to diagnosis or optimal clinical management.

We will present TCM through the application into one particular disease named tetralogy of Fallot (TOF). TOF is a disease, for which several TCM techniques can be combined: for instance, assessment of myocardial functional state can be performed by employing models of heart mechanics; planning of cardiac resynchronization therapy can be assisted by using electromechanical heart models; advanced imaging and image processing constrained by biophysical models allow to study distinct patterns in TOF patients; and assessment of narrowing of a pulmonary artery branch and possible indication for an intervention (e.g., stenting) can be facilitated by using large vessel flow models. This chapter presents some of these techniques.

1.1   Tetralogy of Fallot

Tetralogy of Fallot is a common congenital heart disease. It accounts for 7–10% of all congenital heart diseases. If untreated, it would lead to death in infancy. During the embryologic phase, the aorticopulmonary septum (dividing the cardiac outflow channel into aortic and pulmonary outflows [2]) is deviated anteriorly causing the two outflow arteries—pulmonary artery (PA) and aorta (from the right and left ventricle, respectively)—to be of different size (narrower PA and wider

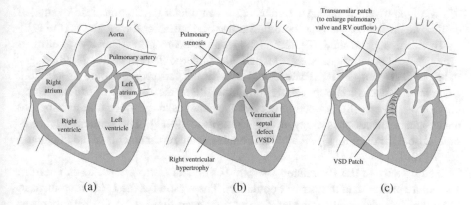

Fig. 1 Schematics of normal anatomy of heart and large vessels and of tetralogy of Fallot. (**a**) Normal heart. (**b**) Tetralogy of Fallot (TOF). (**c**) Complete surgical repair (rTOF)

proximal aorta). As depicted in Fig. 1b, this causes (1) narrowing of the right ventricular outflow tract (RVOT) and PA; (2) the aorta overriding both left and right ventricular outflow tracts (therefore, blood from both ventricles can enter the aorta); (3) ventricular septal defect (VSD); and (4) hypertrophy of the right ventricle (RV), which compensates the RV working against high afterload (due to the stenosed RVOT). Additionally, TOF patients often suffer from other associated defects, out of which stenosis of a branch of the PA (left or right pulmonary artery, LPA, RPA) is common. Mixing of oxygenated and deoxygenated blood entering the aorta from the LV and RV, respectively, causes a typical bluish color of the skin, the so-called cyanosis, and TOF represents the most common cyanotic congenital heart disease. The four classical symptoms of this *blue baby syndrome* (*la maladie bleue*) were described by the French physician Etienne-Louis Arthur Fallot in 1888 [3].

Complete surgical repair of TOF (rTOF) is usually performed during the first year of life. It consists of repairing the VSD and increasing the size of the RVOT and PA, as depicted in Fig. 1c. The correction of the PA stenosis (causing the initial pressure RV overload and RV hypertrophy), often made by using a transannular patch, typically disrupts the function of the pulmonary valve and causes a pulmonary regurgitation (PR), i.e., "leaking pulmonary valve." RV is then chronically volume overloaded. Moreover, even after the repair of tetralogy of Fallot, some rTOF patients suffer from a residual stenosis of the RV outflow tract (RVOT). This eventually leads to a combined RV volume and pressure overload. The chronic volume overload causes RV remodeling in the sense of RV dilatation, and the chronic pressure overload leads to the hypertrophy of the RV myocardium [4]. The hearts of rTOF patients often have electrical activation abnormalities either as a consequence of the surgical repair or developed in the remodeled heart. The right bundle branch block (RBBB)—i.e., a block in the main electrical conduction pathway, which normally allows a rapid transmission of electrical activation throughout RV [5]— is common. It causes a dyssynchrony in the activation and contraction (a delayed activation and contraction of the RV free wall), which decreases cardiac efficiency

[6]. Malignant ventricular arrhythmias (e.g., ventricular fibrillation) lead to sudden cardiac death and are the most common cause of death after the repair of TOF [7].

To avoid uncontrolled RV remodeling and even to allow the RV to reverse remodel back to normal size, the patients undergo later in life a pulmonary valve replacement (PVR). PVR would ideally be performed as late as possible (intervention needs to be repeated throughout life as the lifespan of the implanted valve is 5–10 years [8]) but before irreversible changes of the myocardium or before a significant electrophysiology event takes place [9–11]. The optimal timing is crucial both for the initial PVR and for the subsequent replacements of the implanted valve [12, 13].

Additionally to the RV-related issues, rTOF patients are known to suffer from LV failure earlier than in normal population. The reason for the LV involvement in TOF, which is classically considered as a "right heart disease," is not entirely clear. Both mechanical (e.g., dyssynchrony in contraction) and humoral factors (which can lead, e.g., to an increased tissue stiffness of both ventricles [14]) may be involved, and modeling has the potential to shed light on some of them.

While the complete surgical repair increases the life expectancy from a few years to many decades up to normal life expectancy [15], the rTOF patients must undergo regular follow-up exams and their clinical management is still facing limits, for example, in:

1. Optimal timing of PVR
2. Optimal indication and timing of stenting of a stenosed branch pulmonary artery
3. Optimal management of electrophysiological (EP) disorders, including atrial arrhythmias, ventricular arrhythmias, and prevention of sudden cardiac death
4. Preventing left- and right-sided heart failure

These items represent some of the challenges of current clinical care. In the next sections, we demonstrate some paths in which TCM can contribute to clinical management of rTOF patients. In Sect. 2 we show how can TCM be applied during assessing ventricular function either at regular follow-up exams when detailed clinical data are available (Sects. 2.1 and 2.2) or in a perioperative period or acute exacerbation of heart failure while coupling fast running models with continuously acquired signals monitoring cardiovascular physiology at operation theatre or intensive care unit (ICU; Sect. 2.3). Section 2.4 demonstrates some paths of including the models in the assessment of a long-term cardiac performance. Section 3 presents the pulmonary right ventricular resynchronization in congenital heart diseases, as for rTOF. Physiological and physical knowledge, the so-called model a priori, can be incorporated into advanced imaging and image processing techniques. Section 4 presents some examples in which model-constrained image processing allows to study some distinct patterns of TOF. Those could represent initial signs of early-stage heart failure or be associated with a disease progress. Section 5 shows the steps toward coupling flow models with clinical data to overcome some physical limits of the imaging techniques. Section 6 further discusses various aspects of the clinical modeling for rTOF patients and shows some connections between this particular congenital heart disease and models for general population and acquired heart diseases. Section 7 concludes this chapter.

2 Ventricular Mechanics and Its Biomechanical Modeling

Coupling image data with biophysical models allows to estimate the biophysical properties of the heart and vascular system. Such quantities are not directly measurable from the image and pressure data but are accessible via constitutive relations and equilibrium equations (see reviews [1, 16] and references therein). Moreover, the physical character predisposes the biophysical models to predictivity. In this section, we provide a more detailed look on some of these techniques, which can be applied in rTOF patients.

2.1 Assessment of Right Ventricular Mechanics in rTOF Patients

In Sect. 1.1 we explained that rTOF patients suffer from a chronic PR or RVOT obstruction, leading to a volume or pressure overload of their RV. The patients undergo PVR, and the procedure needs to be repeated approximately every 10 years. It is, therefore, of utmost interest to predict when the pathological remodeling of RV is approaching an irreversible state and suggest to perform the intervention immediately, or defer it and suggest a suitable timing for the next follow-up exam.

While the current works and clinical guidelines focus on the direct measures taken from image data (typically ventricular volumes and transvalvular flow, performed by imaging techniques), the underlying mechanism of the chronically overloaded tissue undergoing pathological remodeling is unknown.

Gusseva et al. hypothesized in [17] that accessing directly mechanical properties of the heart in rTOF patients will provide additional metrics, which may better stratify the PVR responders versus non-responders. A biomechanical model of single heart cavity developed by Caruel et al. [18] was used. The preload was imposed from the measured pressure data in diastole. The heart model was connected to a Windkessel model of circulatory system (lumped-parameter model representing the circulation by a combination of resistive and capacitive elements in the analogy to electrical circuit [19]). The schematics of the model is shown in Fig. 2a. The geometry and kinematics of the ventricle are reduced to a sphere with the inner radius R and wall thickness d. The constitutive mechanical laws were preserved as in the full 3D heart model described in [20] and [21]. The model was adjusted into a patient-specific regime. This allowed to plot RV pressure–volume (PV) loops for every patient pre-/post-PVR (see example in Fig. 2b). In addition, the patient-specific models allowed to access the parameter of RV contractility, i.e., the active stress developed in each sarcomere unit.

The myocardial contractilities prior to and after PVR for 20 patients included in study [17] are plotted in Fig. 3a and the estimate of RV stroke work (an area encompassed within the PV loop) in Fig. 3b. Thanks to the valvular component (a system of diodes with forward and backward resistance, $R_{\text{for}}^{\text{RVOT}}$ and $R_{\text{back}}^{\text{RVOT}}$,

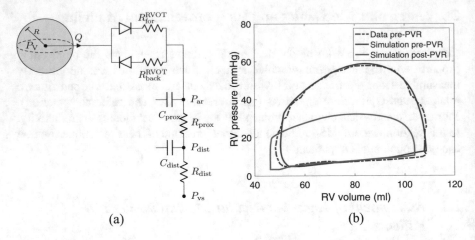

(a)   (b)

Fig. 2 Schematics of the heart model coupled to the circulation system (**a**) and an example of model-derived pressure–volume loops prior to and after pulmonary valve replacement (PVR) (**b**). Q, outflow from right ventricle (RV); R_{for}^{RVOT} and R_{back}^{RVOT}, forward and backward resistance of RV outflow tract; R_{prox}, R_{dist}, resistance of proximal, distal part of pulmonary circulation; P_V, P_{ar}, P_{dist}, P_{vs}, pressures in right ventricle, pulmonary artery, distal pulmonary circulation, and pulmonary veins. (**a**) Model schematics. (**b**) Model-derived pressure–volume loops

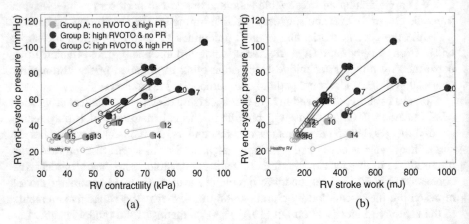

(a)   (b)

Fig. 3 Study estimating contractility and stroke work of right ventricles in 20 patients with repaired tetralogy of Fallot prior to (colored circles) and after pulmonary valve replacement (empty circles). PR: pulmonary regurgitation; RV: right ventricle; RVOTO: RV outflow tract obstruction. (**a**) Model-accessed contractility of RV. (**b**) RV stroke work

respectively, see Fig. 2a), the model allowed to study an independent effect of PR and RVOT obstruction on ventricular mechanics. Light blue and purple circles in Fig. 3 demonstrate pre- and post-PVR contractility (stroke work) for patients with predominantly high PR and RVOT obstruction, respectively. Figure 3 shows a systematic trend of a decrease of RV contractility and stroke work observed immediately after deploying the valve. The lower inotropic (contractility) level would be more favorable with respect to the energy need of the myocardium, while the heart would be still providing an adequate cardiac output into the circulation (in fact, even 20–50% higher than during the chronic PR, which is in line with experimental data by Lurz et al. [22]). Moreover, the pre-PVR models revealed lower levels of contractility (and stroke work) in patients with predominant PR, suggesting that ventricles are likely to adapt better to a long-term exposure to valvular regurgitation than to RVOT obstruction, which is also discussed in the study [17].

This example shows that biomechanical modeling has the potential to contribute to personalized intervention by enhancing the ability to predict RV performance after PVR. This represents a way to baseline and track the RV performance over time and allows for clinicians to make more informed decisions about PVR in patients with rTOF.

2.2 Early-Stage Heart Failure Assessment

Heart failure (HF) is the syndrome in which the heart is not able to pump a sufficient amount of blood to cover the metabolic needs of tissues. The HF patients suffer from symptoms such as shortness of breath, cough, palpitations, or dizziness. Early-stage HF is characterized by HF symptoms present only under more than ordinary physical activity (class I–II according to the New York Heart Association [NYHA]; see Table 1). The assessment of early-stage HF, therefore, requests performing a clinical exam under physical exercise.

A decrease in exercise tolerance in rTOF patients is frequent. There are a number of possible causes: extracardiac (such as an increased RV afterload, e.g., due to the RVOT obstruction, branch pulmonary artery stenosis, pulmonary pathology, etc.), or intracardiac (e.g., limitation of myocardial perfusion or of chronotropic response) [23]. The extra- and intracardiac causes can be combined, as, e.g., in the chronically increased afterload, requiring the ventricle to develop a higher active stress, i.e., to work chronically on a higher inotropic (contractility) level [5]. As presented in Sect. 2.1, accessing the tissue contractility requests employing a biophysical model of heart. Quantifying the contractility at rest and during physical exercise (which can be replaced by a pharmacological stress test) allows to assess the so-called contractile reserve (i.e., the level to which the contractility can increase in a given patient).

Table 1 New York Heart Association classification for the heart failure (NYHA)

Class	Objective assessment
I	Patient with cardiac disease but without any limitations in physical activity
II	Comfortable at rest, ordinary physical activity results in fatigue, palpitations, dyspnea, or anginal pain
III	Comfortable at rest, but less than ordinary physical activity results in fatigue, palpitations, dyspnea, or anginal pain
IV	Symptoms of HF may be present even at rest

In [24], Ruijsink et al. performed a quantification of pharmacological stress exams in a group of patients born with a single ventricle. Their ventricles were surgically adjusted to serve as a pump for systemic and pulmonary circulations connected in series, by redirecting the main systemic veins (superior and inferior venae cavae) directly into the pulmonary arteries (the so-called total cavo-pulmonary connection known as the Fontan circulation [25]). Exercise intolerance is a relatively common early symptom of the failure of the single ventricle, and conventional methods used in clinics such as echocardiography or magnetic resonance imaging (MRI) are often unable to define the actual causes of the onset of the failure. Simultaneous cardiac catheterization with MRI at rest and under pharmacological stress (simulating the exercise) has great potential to distinguish the actual cause of the intolerance. However, the complex physiology of the Fontan circulation and large inter-individual variations make it difficult to obtain direct conclusions from these investigations, even though having such rich datasets.

Employing a combined cardiac and circulatory model (similarly to Sect. 2.1) allowed for an enhanced diagnostic assessment of pathophysiological exercise responses in patients with Fontan circulation. Figure 4 reveals that the myocardial contractility values in Fontan patients were normal at rest. There was a good contractile reserve in most patients. The increase of contractility in three patients (marked as FP 7, 8, 9 and CC 1, we recall Fig. 4) was, however, limited. The low contractile reserve in these patients could be the cause of their early-stage HF. Good correlations of model-derived myocardial contractilities and periphery vascular resistance with their measured clinical surrogates depicted in Fig. 5 validate the modeling framework.

The work by Ruijsink et al. [24] demonstrates that modeling can provide valuable additional information to the current diagnostic assessment of Fontan patients. This work paves the way for the translation of the model-accessed myocardial contractility and contractile reserve into other cohorts of patients, such as rTOF.

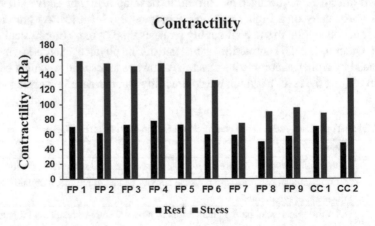

Fig. 4 Estimated contractilities at rest and under Dobutamine test in 9 Fontan patients (FP) and 2 biventricular control cases (CC)

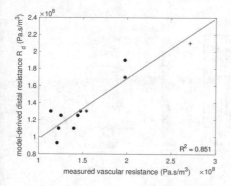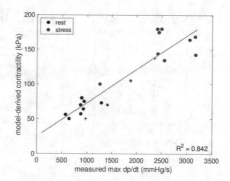

Fig. 5 Model estimated quantities versus measurements. Left: Model-derived distal Windkessel resistance (R_d) versus measured vascular resistance at rest. Right: Model estimated contractility versus measured ventricular max dp/dt (i.e., the usual clinical surrogate of myocardial contractility). Figure used from [24] under the Creative Commons Attribution (CC BY) license

2.3 Model-Augmented Monitoring During the Perioperative Period or in the ICU

Real-time (or close to real-time) biomechanical models allow for coupling with monitoring data during general anesthesia (GA) or in the ICU. Direct analysis of the arterial pressure or cardiac output may be challenging in complex clinical situations. Patient-specific biomechanical modeling allows to simulate ventricular PV loops and obtain functional indicators of the cardiovascular system, e.g., resistance of the circulatory system or myocardial contractility, evolving throughout GA or during the stay at ICU.

In [26], Le Gall et al. described a proof-of-concept study of applying biomechanical models to augment hemodynamic physiological monitoring. Patient-specific biomechanical models of heart and vasculature were created for 45 patients undergoing GA. While aortic pressure and flow were used to set up the patient-specific models, those then allowed for a quantitative assessment of heart function in each individual patient. The validation was conducted on a group of 4 patients, who underwent a combined MRI and pressure catheter procedure. Figure 6a compares the PV loop generated by the model with the measured ventricular pressure and volume and demonstrates validation of the approach. Figure 6b shows the model-derived PV loop of a patient under GA during hypotension and how the PV loop changed after administering the vasoactive drug norepinephrine, and Fig. 7 shows the quantitative changes of physiology accessed thanks to modeling.

Work [26] by Le Gall et al. demonstrates the application of biophysical models built for individual patients to estimate PV loops and cardiovascular functional indicators using data, which are available during GA. Moreover, the predictive capability of the biophysical models representing the cardiovascular system of individual patients builds the foundations for a pro-active use in optimal clinical

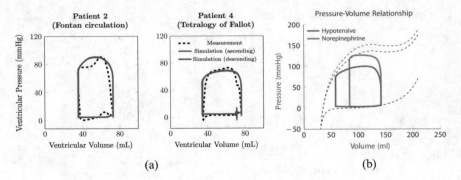

(a)   (b)

Fig. 6 Cardiac modeling coupled with monitoring signals. Figure used from [26] under the Creative Commons Attribution (CC BY) license. (**a**) Validation PV loops. (**b**) PV loop at hypotension vs. after administering vasoactive drug norepinephrine

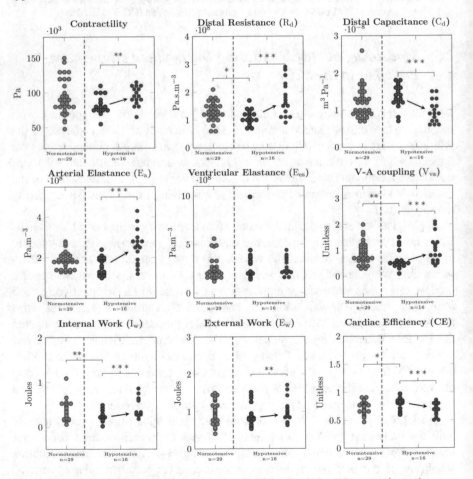

Fig. 7 Boxplots of model parameters and the results of the simulation. Normotensive patients are in green; hypotensive patients before and after administering norepinephrine are in blue and red color, respectively. $*$ $p < 0.05$; $**$ $p < 0.01$; $***$ $p < 0.001$. Figure used from [26] under the Creative Commons Attribution (CC BY) license

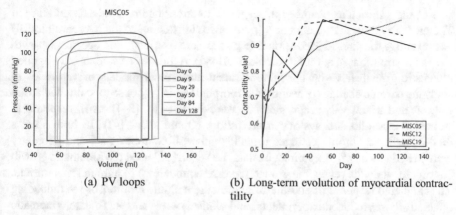

(a) PV loops

(b) Long-term evolution of myocardial contractility

Fig. 8 (**a**) Example of left ventricular PV loops in a selected patient with multisystem inflammatory syndrome in children (MIS-C) at acute stage and during the recovery. (**b**) Progressive recovery of myocardial contractility after MIS-C

handling of complex situations (e.g., assistance in clinical management of septic shock).

Both increasing the insight in pathophysiology and proposing a possible therapeutic intervention could be helpful in newly discovered diseases. For instance, there is currently a limited knowledge about Multisystem Inflammatory Syndrome in Children (MIS-C) known also as Pediatric Inflammatory Multisystem Syndrome-temporally associated with SARS-CoV-2 (PIMS-TS). It is a serious condition, which is associated to the SARS-Cov-2 virus and appearing typically 2–4 weeks after the exposure to the coronavirus, while the COVID-19 disease may pass only with mild symptoms [27, 28]. Model-augmented cardiovascular monitoring could be of a direct benefit in this life-threatening condition during the acute stage at ICU and also during the long-term clinical follow-up of patients. Both were demonstrated in a study by Waugh et al. [29]; see Fig. 8. The patient-specific models of 8 patients allowed to access the cardiovascular physiology of MIS-C patients at ICU to distinguish between the cardiac and vascular components of their cardiovascular shock state. In addition, the study demonstrated cardiac recovery 2–4 weeks after discharge from ICU, supporting the optimistic scenario that the heart is fully recovering within a month.

2.4 Assessment of a Long-Term Cardiac Performance

An assessment of a long-term cardiac performance has been studied within a kinematic growth theory framework as originally proposed by Rodriguez et al. [30], and later employed in the heart by Kroon et al. [31], and others [4, 32–34]. The growth and remodeling of the cardiac tissue is modeled as alterations in shape and size of the ventricular chamber in response to stress and/or strain-based stimuli. The model hypothesizes that concentric growth (thickening of the myocardial wall) occurs due to the stress-based stimuli (e.g., pressure overload) and eccentric growth due to the strain-based stimuli (volume overload). A number of studies have applied kinematic growth laws on the finite element anatomical models of ventricles to study the chronic changes of ventricular morphology in response to simulated aortic stenosis and mitral valve regurgitation (Kerckhoffs et al. [34]), or in response to simulated diastolic and systolic heart failure (Genet et al. [4]). In both studies model-predicted chronic changes in wall thickness, chamber size, and geometry were in agreement with experimental data. Lee et al. [35] used the kinematic growth concept to study the reversible growth (reverse remodeling) in human left ventricle. The simulated response of end-diastolic pressure–volume relationship to unloading qualitatively reproduced experimental and clinical observations. To date, kinematic growth theory is capable of predicting the consequence of growth on organ morphology while is limited in predicting tissue-level mechanical properties (e.g., myocardial contractility or stiffness) and microstructural changes (e.g., apposition of collagen fibers). Hybrid modeling approaches linking tissue-level turnover events with an evolving organ morphology could be a way to better understand the mechanics of growth and remodeling in physiological and pathological conditions. For example, microstructurally oriented models such as constrained mixture theory originally designed to capture mass turnover events in individual constituents of the mixture (e.g., solid tissue) [36, 37], if applied to the heart, could potentially describe, e.g., deposition or apposition of myofilaments or collagen in hypertensive ventricles. In the field of vascular modeling a number of attempts to combine a constrained mixture theory with kinematic growth have shown promising results in modeling the effect of elastin and collagen degradation on the development of thick-walled aortic aneurysms [38, 39]. Patient-specific modeling approach presented in Sects. 2.1–2.3 could evaluate current state tissue-level mechanical properties, and if combined with kinematic growth framework (or with a hybrid of kinematic growth and mixture theory), it has a potential to become an invaluable therapeutic tool in predicting long-term effects of therapies, e.g., response to PVR, and in predicting the optimal timing for interventions, including for complex surgeries.

3 Pulmonary Right Ventricular Resynchronization in Congenital Heart Disease

RV dysfunction and occasional failure are associated with a number of congenital heart lesions, most predominantly but not limited to rTOF. They are attributed to several factors including myocardial fibrosis due to preoperative hypoxemia and pressure overload, surgical scar, and long-term post-repair volume overload caused by pulmonary regurgitation and frequently associated with RBBB [40]. RBBB is also the most frequent cause of electromechanical dyssynchrony in congenital heart disease. In rTOF patients, RV electromechanical dyssynchrony resembles the classic dyssynchrony pattern incorporating an early septal contraction with RV free wall pre-stretch followed by late free wall contraction and septal rebound stretch. It has been associated with a decreased RV ejection fraction, mechanical inefficiency, and impaired exercise capacity [41, 42]. The dyssynchrony is hypothesized to play a significant role in the progress of RV dysfunction. A recently published study using computer modeling showed that the extent of negative influence of RV dyssynchrony may be greater than previously thought and may even overweight the pulmonary regurgitation during exercise [43]. A relief of RV volume overload by PVR is thought to reverse the pathological RV remodeling occurring due to the pulmonary regurgitation in rTOF patients. However, a decreased probability of reverse remodeling has been reported in patients with high RV end-diastolic and end-systolic volumes, low RV ejection fraction, and those with a wide QRS complex (≥ 160 ms) [44]. Thus, PVR alone may not lead to the normalization of RV performance.

Cardiac resynchronization therapy (CRT) has so far been used to treat dyssynchrony in the failing systemic (left) ventricle. Robust clinical data demonstrate a favorable CRT effect on acute hemodynamics, reverse cardiac remodeling, functional capacity, and patient survival. Limited evidence suggests a similar CRT effect in children and patients with congenital heart disease [45]. Clinical reports on subpulmonary RV resynchronization (RV-CRT) are scarce, however. In this section we briefly review the current experience with RV-CRT in patients with congenital heart disease.

3.1 Temporary RV-CRT

Published data show that acute RV-CRT through either RV or biventricular pacing improves short-term hemodynamics both in children and adults with congenital heart disease and RBBB. RV-CRT has been successfully used to treat RV dysfunction in acute postoperative setting [46]. Pacing is applied by temporary epicardial leads placed by a surgeon on the right atrium and the late activated free wall of the right ventricle. RV resynchronization is achieved by atrial-triggered RV free wall pacing in complete fusion with spontaneous ventricular activation to obtain a maximal QRS duration shortening. Specific attention is paid to a complete abolition of the broad S wave in standard lead I. The aim is to resynchronize electrically as well as mechanically the septum and RV free wall. A commercially available external dual-chamber pacemaker may be used. Temporary RV-CRT performed in this way carried a significant short-term improvement of hemodynamics in children early after surgery for TOF and may be a useful non-pharmacologic adjunct to the management of hemodynamically compromised patients. The resynchronization effect was maximized when pacing from the area of the latest RV activation. In another study, the temporary RV-CRT brought multiple positive effects on RV mechanics, synchrony, and contraction efficiency [47].

3.2   Permanent RV-CRT

The use of permanent RV-CRT to treat RV dysfunction associated with RBBB seems to be the next logical step. Unfortunately, published evidence is limited. In the first well-documented report on permanent RV-CRT in a patient late after the repair of TOF [48], both right and left ventricular function improved significantly within 6 months accompanied by an improvement in the NYHA class as well as exercise capacity. Data on a small series of congenital heart disease patients undergoing RV-CRT has been presented at the Heart Rhythm conference in 2018 [49], confirming multiple positive mid-term resynchronization effects on the RV systolic function, electromechanical synchrony, and contraction efficiency.

3.3   Toward Electromechanical Modeling for RV-CRT

Cardiac modeling was employed in several projects predicting the effect of CRT on the failing LV [50, 51]. Pilot modeling study [52] demonstrates that pacing has the potential to significantly decrease valvular regurgitation also in other pathologies— e.g., severe tricuspid regurgitation. Similar models are likely to be used in rTOF patients [43].

RV-CRT seems to be a valuable method to treat both acute and chronic subpulmonary right ventricular dysfunction and failure associated with RBBB in patients with congenital heart disease. Further studies are needed to better define its role in long-term RV failure management. Cardiac electromechanical models represent a great potential in optimization of this treatment.

4   Model-Constrained Image Processing

One of the key image processing tasks in the field of medical imaging is image registration. It is defined as the process of aligning two or more images of the same scene. It is used to deal with misalignment caused by different time of acquisition, different scaling, or different acquisition techniques. Aligning series of images taken at different times is often the first step in any subsequent analysis. However, the analysis of the detected misalignment itself might bear useful information. This applies specifically in cardiac applications, where both aspects of image registration are utilized. First, aligning a given series of the cardiac images allows for the estimation of local tissue specification that is encoded in the image series. Second, it is used to extract the motion of the moving organ from the image series. Features of interest, such as time evolution of volume and torsion, are obtained by the subsequent analysis of the detected motion.

Most image registration methods consist of defining a measure of misalignment and its minimization. The registration methods can be divided into several groups based on what is taken into account in the evaluation of misalignment [53]. These groups are: landmark-based methods, segmentation-based methods, and voxel-based methods. In the landmark-based registration methods the alignment of two images is evaluated based on the alignment of a finite set of voxels, the so-called landmarks, detected on both images. The landmarks can be detected automatically or manually, as in [54], where a landmark-based registration method is used for brain MR images. In the segmentation-based methods the alignment of images is evaluated based on the alignment of objects segmented in both images. In [55], B-spline method was used to register segmented left ventricle contours. Finally, in the voxel-based methods, the alignment is evaluated based on all voxels of the images. An optical flow method is a representative of this category. In [56], the optical flow method incorporating the image intensity constancy and gradient orientation was used for registration of multimodal images.

Mathematically speaking, image registration is an ill-posed problem and is, therefore, often formulated as a variational problem, which typically contains a data similarity term and a regularization term. The appropriate choice of the data similarity term is especially important when registering images obtained by different modalities or if the voxel intensity evolves in time. In such cases, the similarity cannot be evaluated based on the image intensity and more complex similarity terms are needed. Similarity terms based on correlation [57] or terms utilizing mutual information [58] are often employed. The regularization term deals with the ill-posedness of the problem. Moreover, it can be used to prevent non-physiological motion of the object. In advanced methods, the regularization term can incorporate physiological and physical knowledge of the processed object in the form of "model a priori knowledge."

In this section, we show some examples of both aspects of image registration: the extraction of motion indicators and estimation of the tissue characteristics indicators. In Sect. 4.1, we discuss the data similarity term used for registration of MOLLI MRI data (Modified Look-Locker inversion recovery, [59]), which are used to estimate the tissue T1 relaxation time (tissue characterization obtained by magnetic resonance related, e.g., to assess myocardial fibrosis). In Sect. 4.2 we show the incorporation of a mechanical model into the regularization term. Another approach of constructing the data similarity term using directly a model of imaging technique is discussed in Sect. 4.3. The final parts represent the crucial steps in patient-specific modeling: In Sect. 4.4 we briefly describe the problem of the fusion of clinical data and a complex biophysical model and in Sect. 4.5 we show some directions of employing machine learning and artificial intelligence.

4.1 Assessment of T1 Relaxation Time from MOLLI MRI Sequence

In this section we show the approach proposed in [60] for the registration of MOLLI MR image series. The MOLLI sequence is used for pixel-wise estimation of the tissue characteristics of T1 relaxation time. In the case of suboptimal breath-holding, image registration is needed in order to compute reliable T1 relaxation times. This holds especially for the points adjacent to the epicardial and endocardial surfaces of the myocardium, where the misalignment may significantly corrupt the computation of the T1 relaxation time.

The evolution of image intensity and large changes in image contrast are the factors that make the registration of the image series acquired by MOLLI challenging. In order to deal with this characteristic of the sequence, a special data similarity term was proposed by Škardová et al. in [60], based on evaluating signed distance of image voxels from segmented surfaces. The method requires the segmentation of epicardial and endocardial surfaces of the myocardium. The data similarity term is then based on the assumption:

$$\phi(\mathbf{x}(t), t) = const \quad \forall t \in (0, T),$$

where $\phi(\mathbf{x}(t), t)$ represents the signed distance of point $\mathbf{x}$ from the segmented contours of the myocardium. That is, the distance of each point from the myocardium contour was assumed to remain the same over time. This formulation allows for the use of the voxel-based registration methods and excludes the effects of non-constant image intensity. Since the displacement caused by suboptimal breath-holding is independent of the frame ordering, all source images can be registered with the target image independently. Three images from the MOLLI series with segmented myocardium are shown in Fig. 9. The source images S_1 and S_2 are shown in Figs. 9a,b, while the target image T in Fig. 9c. The result of registration using this method is depicted in Fig. 10.

4.2 Motion Extraction from Image Data Using Mechanical Model Constraints

A detailed extraction of motion pattern from functional cine cardiovascular MR images (see Fig. 11) has been a challenging problem since longtime. In MRI, a special magnetic saturation pulse can be applied. The tissue can be saturated in a regular pattern (typically in the form of lines or grids), with which the material points of the myocardium can be followed. The technique is known as SPAMM (SPAtial Modulation of Magnetization [61]) or as tagged MRI. The periodic character of the tag lines allows to use the so-called Harmonic Phase (HARP) [62] or Sine-Wave Modeling (SinMod) [63] analyses. Recently, an improved Harmonic

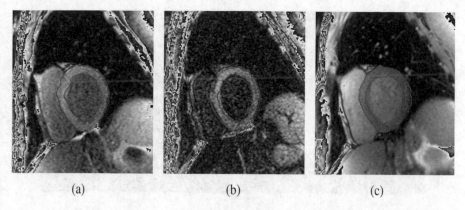

(a)   (b)   (c)

Fig. 9 Segmented epicardial and endocardial contours of myocardium of left ventricle. (**a**) Source image S_1. (**b**) Source image S_2. (**c**) Target image T

Fig. 10 Source images of Fig. 9a,b registered to target image in Fig. 9c. (**a**) Transformed source image S_1. (**b**) Transformed source image S_2

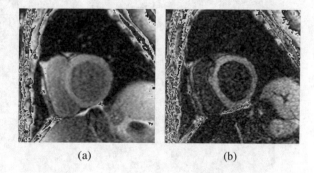

(a)   (b)

Phase analysis method was introduced by Mella et al. under the name HARP-I [64]. HARP-I has a superior efficiency to the standard HARP or SinMod methods, as was demonstrated in [64]. However, it relies on acquiring the tagged MR images, which is practically difficult to do in every patient due to lengthening of the MR exam.

Genet et al. [65] described a novel motion tracking based on image registration while imposing mechanical constraint of equilibrated gap regularization—the so-called equilibrated warping.[1] It has been shown not only to successfully track the motion in the tagged MR images but also to extract global torsion in standard cine MR images [66]. The LV torsion in rTOF patients was previously studied; however, poor reproducibility was observed when using standard clinical commercial software [67]. In [68], Castellanos et al. successfully employed the equilibrated warping in rTOF patients. In this work, the authors aimed to demonstrate that assessing the LV torsion in a robust and reproducible way has the potential to contribute to an earlier detection of heart failure.

---

[1] The implementation is available at https://gitlab.inria.fr/mgenet/dolfin_dic.

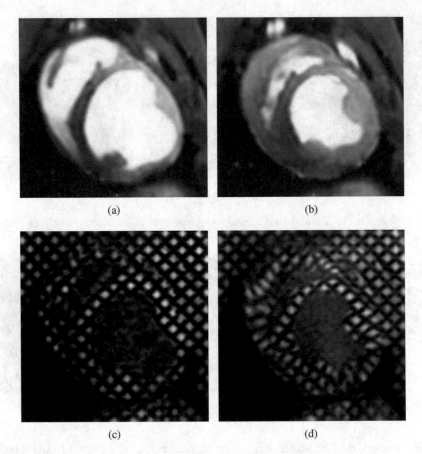

(a)   (b)

(c)   (d)

Fig. 11 Short axis views of cine MRI and tagged MRI. (**a**) Cine MRI: end-diastolic image. (**b**) Cine MRI: end-systolic image. (**c**) Tagged MRI: end-diastolic image. (**d**) Tagged MRI: end-systolic image

4.3 Motion Extraction from Image Data Using Model of Imaging Modality

In this section we discuss extracting motion from medical image series by using a model of the imaging modality. In [69], such an approach is employed to register 2D tagged MR images. This method uses a geometrical model of the segmented object that is to be registered and also the imaging model. For each setting of the geometrical model, the imaging model can be used to generate synthetic images. We look for the mapping that best matches the synthetic and acquired images.

The imaging model used by Škardová et al. in [69] includes the 2D tagging and artifacts caused by a limited image spatial resolution. The image registration is regularized by the equilibrated warping, as described in [65, 66]. We remark that the proposed method could be applied also for other medical imaging modalities.

More effects could be considered, including complex artifacts given by the imaging method, e.g., if a very high acceleration of data acquisition is used. An efficient processing of such data has the potential to reduce the overall scanning time. This could substantially assist in some challenging MR acquisition such as cardiovascular MRI exam during physiological exercise [70].

4.4   Model–Data Fusion

Advanced image processing constrained by a model gives an opportunity for a fast and reliable extraction of motion from image data. This is important in stratifying patients (e.g., as presented in Sect. 4.2). Moreover, the acquired and processed data can be incorporated into truly biophysical models in the sense of the third component of TCM (as listed in the introductory Sect. 1)—data–model fusion. The topic of assimilating clinical data into a complex biophysical model was reviewed by Chapelle et al. in [71]. The correction of model by data leads to increasing the patient-specificity of the models, as presented, e.g., in [72, 73]. The biophysical character then allows to estimate some clinically relevant parameters, contributing to a detailed description of physiology state. The parameters obtained, thanks to fusing a model with data, are represented, for instance, by passive tissue stiffness [74–76] or contractility [77–79] for the heart tissue. For large vessels, Young's modulus or relative pressures were estimated in [80–82], to give some examples. Such model-accessed parameters are expected to be included in clinical studies of a larger scale in the near future.

4.5   Data-Driven Strategies for Image Analysis

Recently, with an increasing availability of large amounts of medical image data, data-driven approaches have become more frequently used in multiple classes of medical image analysis. As examples can serve pattern recognition, segmentation, and registration. The data-driven methods learn patterns directly from image data; therefore, in principle, no mathematical model is needed. In practice, data-driven methods can be used alone or in combination with a mathematical model, if a model with an appropriate level of detail and expressiveness is available.

However, with the purely data-driven methods, problems with robustness and interpretability (which are key qualities in many medical applications) might arise. This could occur mainly for deep learning methods—neural networks with tens or even over hundreds of layers. Both approaches—model-based and data-driven—are compared in [83] in terms of robustness, accuracy, interpretability, and explainability.

As was mentioned in the beginning of Sect. 4, the model-based image registration methods usually incorporate a data similarity term and regularization terms.

Recently, data-driven methods for learning transformation parameters by minimizing traditional cost functions using neural networks have been presented. These methods can be divided into two groups: supervised—for which a set of registered images with known transformation parameters is available—and unsupervised—for which transformation parameters are learned directly from input image data. The supervised methods include, for example, a Spatial Pyramid Network (SPyNet) [84] and a convolutional neural network (CNN)-based method for optical flow estimation. Similarly to traditional optical flow method, SPyNet incorporates spatial resolution pyramid to bypass estimating large displacements. In [85] CNN-based method for multimodal registration of T1- and T2-weighted MR images of brain was presented. The unsupervised methods based on CNN include [86], used for registration of brain MR images or [87] used for cardiac MR images.

Similarly, in image segmentation the data-driven methods learn the information about the shape and intensity distribution of a segmented object from the image data. Even early approaches using simple voxel-wise intensity features, for example, the support vector machine (SVM)-based method [88] and k-nearest neighbor classifier [89], were successfully used for brain segmentation. In classical machine learning approaches, the most effective methods are based on decision forests [90], which allow to incorporate larger feature vectors, as they perform feature selection and learning at the same time. Today, most of the successful segmentation methods are based on neural networks, specifically CNNs [91].

5 Large Vessel Flow Modeling

Phase-contrast magnetic resonance imaging (PC-MRI) [92] is used to assess the level of valvular stenosis or regurgitation [93]. PC-MRI is based on the fact that the phase shift of the MR signal after applying a bipolar gradient is linearly proportional to the velocity of the moving spins in the image. The linearity of the phase shift is valid only under the assumption of laminar flow, and the measured flow is underestimated if the level of turbulence is above a certain level [94]. There exist research MR sequences specifically designed for turbulent flow (see, e.g., [95–97]). However, they are currently not a part of standard clinical MR packages and cannot be, therefore, routinely used. Furthermore, they would substantially lengthen the MR exam, which would complicate their clinical deployment.

To reduce the limitations of standard PC-MRI, the imaging method can be coupled with a biophysical model of flow, in the sense of Sect. 4.4. The model can lead to a correction of data, e.g., of the underestimation of velocity in turbulent flow mentioned above. While some parts of the vessel geometry may be inaccessible to imaging, due to, e.g., the presence of vascular stents as illustrated in Fig. 12, the computational flow dynamics (CFD) model constrained by the data measured before and after the stent has the potential to contribute to detecting the possible re-stenosis of the stented vessel segment.

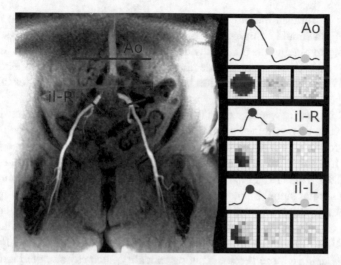

Fig. 12 Illustration of clinical PC-MRI examination of the bifurcation of abdominal aorta. Due to the presence of stents, the aortic bifurcation cannot be directly assessed. A CFD simulation has the potential to provide crucial information missing in the clinical assessment. Ao: Aorta; il-R: right common iliac artery; il-L: left common iliac artery

5.1   Phantom Experiment

Prior to clinical usage of a numerical model of flow in vessels, it should be validated on representative problems using *in vivo* or *in vitro* data [98]. *In vivo* data often suffer from beat-to-beat flow changes, limited signal-to-noise ratio, and motion artifacts [99]. *In vitro* experiments, however, can be controlled in several ways. They provide valuable information firstly for the validation of a CFD model and secondly for a successful coupling of the model with PC-MRI data. The CFD and experimental data can be compared using common statistical methods applied on quantities of interest, e.g., mean flow perpendicular to a centerline at various locations along the vessel, or flow ratio between two or more branches of the vessel tree [100]. Prior to comparison, experimental data may be used to fine-tune the CFD model parameters, such as fluid density and viscosity [100]. It can also facilitate in setting a boundary condition, e.g., whether a simple mean flow is sufficient as an input and output condition or whether a more elaborate flow profile is needed. Or, similarly, whether a zero outlet pressure provides sufficient results as opposed to a more precise estimate, such as 1D *in silico* pressure model [100, 101]. In the following paragraphs, we provide some practical aspects of *in vitro* experimenting.

First, a flow rate close to the physiological regime should be achieved. A special care must be taken to utilize a stable pump that does not experience a decrease in power when subjected to an increasing resistance. The sources of resistance are hydrostatic pressure (caused by the vertical arrangement of the pump and MRI table) and the resistance of the circulatory system (causing a pressure loss). To minimize the friction of fluid in the circulation, a hose of a large diameter and a minimum possible length should be used. This is limited by the amount of fluid available and the necessity of placing the pump outside the magnetic field of the MR scanner. Any connecting parts of the system or devices, such as a flow meter, also tend to add a resistance. Inevitable internal friction in the fluid itself is another factor influencing the pump performance.

PC-MR images may suffer from low velocity-to-noise ratio. This statistical error of velocity can be minimized by prolonged scan times. Such an approach would be often non-feasible *in vivo*. Not only precision but also the accuracy is a challenge of *in vitro* measurement especially with smaller vessels, as they are prone to partial volume effect. This artifact is reduced by selecting the material of phantom which provides a similar signal intensity as the fluid [102].

Another advantage of *in vitro* experiments is the precise knowledge of geometry. Phantoms can be created from a 3D printed model using suitable silicon-based material [103]. The geometry of the model can then be described analytically or obtained by using a more precise imaging modality, such as computed tomography.

Initial validation phases of a flow model can be conducted by using simplified *in vitro* experiments. For instance, it may be sufficient to consider constant flow through a rigid pipe with a valve-shaped orifice (narrowing) while using clean water as a medium. Such example is illustrated in Fig. 13 and was used in the study by Fučík et al. [104]. The shapes of phantom valves represent three degrees of valvular stenosis (mild, moderate, and severe). Hessenthaler et al. presented in [105]

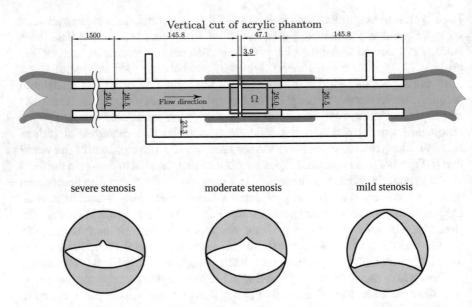

Fig. 13 Illustration of vertical cross section through experimental phantom and acrylic plates mimicking the mild (Mild-VS), moderate (Mod-VS), and severe (Sev-VS) stenoses. All dimensions are in mm. The red region Ω represents the domain of the computational simulations in 3D

a phantom with non-rigid walls suitable for assessing the computational models including fluid–structure interaction. Finally, various numerical benchmarks (e.g., [106]) can be used to evaluate the implementation issues of models.

For validation of more complex flow models, the initial experiment can be further modified. For instance, a pulsating flow pump can be used or the water medium can be replaced with some blood-like fluid. There exist commercially available blood-mimicking fluids suitable for the use in MRI. Their price may, however, be an obstacle. Several substitutes were proposed. In [107] Campo-Deaño et al. explore the properties of xanthan gum to mimic the shear-thinning behavior and of glycerin and sucrose to mimic the elasticity. These alternatives are cheap and easy to prepare. However, they are less convenient to work with, as they are difficult to rinse off in the case of leak and they also deteriorate with time.

5.2 Computational Fluid Dynamic Model

Applications of CFD in cardiovascular medicine can be found, e.g., in the treatment and diagnosis of aortic or cerebral aneurysm [108], aortic coarctation [109], vessel or valvular stenosis [110, 111], valvular regurgitation [112], or a description of the aortic arch flow [113]. The CFD is usually coupled with PC-MRI [114], where the measurement serves as the source of initial and boundary conditions for the CFD. Classical CFD methods (such as the finite volume or finite element methods) are widely used methods and were applied in the cardiovascular problems using, for example, commercial ANSYS engineering software [115] or open-source software OpenFOAM [116]. The main disadvantage of these classical CFD methods is their substantial computational cost (for reasonable spatial and temporal resolutions), which is a key obstacle for their application during medical examinations.

The lattice Boltzmann method (LBM) [117] can be considered as a modern alternative to the classical numerical methods. The main advantage of LBM is the accuracy and local nature, which allows the massive parallel implementation on computational clusters, including those with modern graphic accelerator units (GPUs), and clusters of GPUs. LBM has been already used in a wide range of computational problems. In [118], Fučík et al. showed the capability of LBM to simulate fluid flow in complex fluid domains such as aortic vessels with stenosis.

In [104], Fučík et al. address the issue of underestimating the backflow by PC-MRI. The 3D geometry was given by the geometry of the phantom with three levels of stenosis; see Fig. 13. The inlet boundary condition was given by the flow measured prior to the stenosed region. By comparing the PC-MRI measured backflow with the simulation, excellent agreement between LBM and PC-MRI results of both forward and backward flow for mild and moderate stenosis was observed, as depicted in Fig. 14. However, a large discrepancy occurred in the case of the severe stenosis, as can be seen in Fig. 15. This observation can be explained by the decrease in signal caused by the turbulence-induced accumulation of spins with different encoded phases in one voxel during the readout period. The

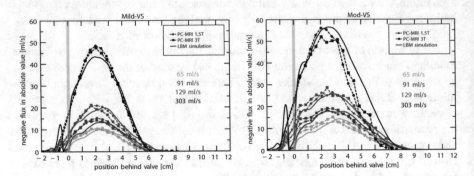

Fig. 14 Simulated backward flow for mild (left) and moderate stenosis (1.5T and 3T MRI systems) confronted with the simulation by lattice Boltzmann method. The solid vertical line represents the valve position. Each color corresponds to the flow rate on the inlet given by the measurement ranging from physiological at rest (green) to accelerated flow (black) representing, e.g., high exercise stress

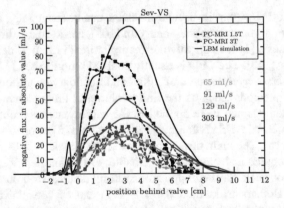

Fig. 15 Simulated backward flow for severe stenosis (1.5T and 3T MRI systems) confronted with the simulation by lattice Boltzmann method. The solid vertical line represents the valve position. Each color corresponds to the flow rate on the inlet given by the measurement ranging from physiological at rest (green) to accelerated flow (black) representing, e.g., high exercise stress

presented work is an essential initial step toward combining measured data by PC-MRI with CFD model suitable for the use in clinical workflow, allowing to enhance the interpretation of measured data by a biophysical model.

The appropriate CFD model should reflect the correct physical properties of the fluid in the vessels. Thus, Škardová et al. investigated in [119] three fluid models for the blood flow through a stenotic vessel. Two models approximated blood as a Newtonian fluid with different fluid kinematic viscosity. The third one used the Carreau–Yasuda model [120], which is designed for capturing the non-Newtonian behavior of blood.

The simulations based on the non-Newtonian model in three different aorta geometries with gradually increasing severities of stenosis showed that the severity of stenosis has a non-negligible impact on fluid viscosity. Nevertheless, the fluctuation of viscosity was small in all simulations. Thus, based on the numerical simulations, the Newtonian approach can be applied to the fluid simulations in the geometries of large vessels, however, with a careful choice of the Newtonian fluid viscosity. Phantom experiments (e.g., as described in Sect. 5.1) with various types of fluid will be the next step to shed light into the most appropriate constitutive model of blood in vessels.

6 Discussion

This chapter primarily dealt with TCM applied in congenital heart diseases. It focused on tetralogy of Fallot after complete surgical repair (rTOF). TOF may be a relatively rare disease; however, the regular follow-up exams of patients represent a significant cost for the healthcare system. Furthermore, the acquired data throughout the evolution of cardiovascular remodeling provide an opportunity to address some important clinical questions of TOF using cardiovascular models.

Sections 2–5 addressed some common issues in rTOF patients. Each topic is typically advanced by one or a few research groups. However, in a given individual patient, the optimal management may require several of these approaches (as well as some additional ones) to be put together—one of the challenges of successful clinical application of cardiovascular modeling nowadays.

While TOF represents one particular disease, a number of already existing modeling techniques can be applied to the cohort of TOF. On the contrary, the novel techniques developed specifically in the group of TOF patients could be useful for other pathologies as well. Indeed, each challenge solved in the TOF cohort has its counterparts in a much larger group of acquired heart diseases. This is in line with the TOFMOD acronym—Tetralogy of Fallot and Modeling of Diseases—as is depicted in the schematics in Fig. 16. For instance:

1. Knowledge obtained from the optimization of PVR could possibly be translated into a much larger cohort of the left-heart valve (aortic and mitral) pathologies [121].
2. Optimal timing of intervention on pulmonary artery could contribute in optimizing the management of pathologies affecting the aorta, head and neck aortic branches, or, e.g., in the interventions for ischemic disease of lower extremities.
3. Biomechanical modeling has shown to be feasible in the planning and optimization of cardiac resynchronization therapy (CRT) for failing LV in acquired heart disease [50, 51]. This can be a starting point for the translation of CRT modeling into the TOF cohort, for whom the therapy is focused on the right ventricle (RV-CRT) [47].
4. Thanks to regular follow-up of rTOF patients in cardiology clinic, the failing LV or RV might be diagnosed and treated in its early stage. The diagnosis and

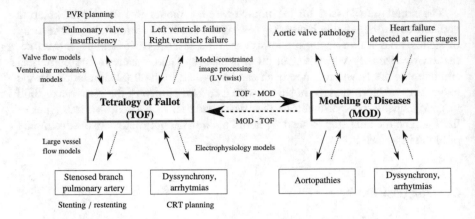

Fig. 16 Tetralogy of Fallot and Modeling of Diseases (TOFMOD) scheme

management of an early-stage HF could then serve also for the cohort of the acquired heart diseases, in order to shift the management of their heart failure into earlier stages. This has the potential to improve the current 5-year survival rate of HF being around 45% [122].

This chapter demonstrated the potential of the biomechanical modeling approaches to assist in data handling and inform about a disease's progress. To ensure reliable predictions, the rigorous assessment of the credibility of the model must be included in the workflow. Namely, the quantification of the model's ability to explain the underlying mechanism must be performed in terms of verification of the numerical method, validation of the biophysical assumptions of the model (a degree to which the model is capable to represent the reality, usually performed with a measured dataset), and uncertainty quantification [123]—a subject of the recent research within the cardiac modeling community. When personalizing the models (the so-called data–model coupling), the clinical data serving as input for the model are subject to measurement errors. Therefore, in order to correctly interpret the model-derived predictions, it is of paramount importance to understand and quantify the sensitivity of the model outputs to the noisy input data signals. For example, the study of Eck et al. [124] demonstrated the efficiency of stochastic mathematical solutions, namely the polynomial chaos method (for details, see [125]), to quantify the uncertainty of the parameters of one-dimensional patient-specific blood flow model.

The uttermost goal of precision medicine is to minimize uncertainties in the patient management workflow. The creation of "digital twins" or "avatars" of the patients [126] and their subsequent holistic translation into clinical practice could be optimized via synergistic approach of combining physics-based and statistical models. These two approaches could counterbalance limitations present in either of them. For example, a physics-based model can be used when a solid understanding of the biophysical system and the validation data are available (e.g., prediction of immediate adaptation of tissue contractility in response to PVR [17]). However, should the underlying biophysical mechanism be poorly understood (e.g., long-term tissue remodeling) or being too complex to be described by the model, a statistical approach may still be capable of identifying some predictive relations (e.g., statistical shape modeling to predict RV remodeling [127] or characterizing RV–LV interactions [128] in rTOF patients). The synergistic approach was used in the recent work by Regazzoni et al. [129] to explain an unknown long-term dynamics of the aortic flow in hypertensive conditions. A mechanical model of large vessel flow (capturing the short-term phenomena) was combined with data-driven modeling approach (machine learning) to explain the long-term phenomena.

7 Conclusion

Applying modeling methods in clinic is a very recent research topic with a number of associated challenges. First, a relevant clinical question, which can be addressed by modeling, must be defined. Second, the model including suitable components and of an adequate complexity must be selected. Third, data have to be obtainable, allowing for model personalization. When turning a model into the patient-specific setup, the physical and physiological character of the models allows them to estimate quantities that are hard (or even impossible) to measure *in vivo* while maintaining clinical relevance. In addition, the predictive capabilities of such models allow to assess the action of certain therapeutic effects *in silico*. Both have the potential to contribute to the optimal clinical management of patients. Clinical problems requesting to employ novel techniques typically represent rather large challenges. Combining various multidisciplinary approaches in a collaborative way has the potential to advance the clinical management and can be helpful for patients.

Acknowledgments The authors acknowledge the support of the Associated Team TOFMOD, created between the Inria France and UT Southwestern Medical Center Dallas; the Ministry of Health of the Czech Republic (project no. NV19-08-00071); the Ministry of Education, Youth and Sports of the Czech Republic (under the OP RDE grant number CZ.02.1.01/0.0/0.0/16_019/0000765:Research Center for Informatics); and the Grant Agency of the Czech Technical University in Prague (grant no. SGS20/184/OHK4/3T/14).

The authors would like to thank their coworkers in the research projects described in this chapter, in particular Dominique Chapelle, Martin Genet, Philippe Moireau (research team MƎDISIM, Inria, France); Arthur Le Gall, Etienne Gayat, Fabrice Vallée (Dept. of Anesthesia and Critical Care, Lariboisière hospital Paris, France); Camille Hancock Friesen, Joshua Greer, Gerald Greil, Animesh Tandon, Rebecca Waugh (Dept. of Pediatrics, UT Southwestern Medical Center, Dallas, TX); Daniel Castellanos (Children's Boston Hospital, MA); Kuberan Pushparajah, Bram Ruijsink, Konrad Zugaj (St. Thomas' Hospital, King's College London, UK); Jakub Klinkovský, Petr Pauš (Faculty of Nuclear Sciences and Physical Engineering, Czech Technical University in Prague, Czech Republic), and Robert Straka (AGH University of Science and Technology, Krakow, Poland). In addition, the authors would like to thank Federica Caforio (Gottfried Schatz Research Center, Division of Biophysics, Medical University of Graz, Graz, Austria) for her valuable comments.

References

1. R. Chabiniok, V. Y. Wang, M. Hadjicharalambous, L. Asner, J. Lee, M. Sermesant, E. Kuhl, A. A. Young, P. Moireau, M. P. Nash, *et al.*, "Multiphysics and multiscale modelling, data–model fusion and integration of organ physiology in the clinic: ventricular cardiac mechanics," *Interface focus*, vol. 6, no. 2, p. 20150083, 2016.

2. R. H. Anderson, S. Webb, N. A. Brown, W. Lamers, and A. Moorman, "Development of the heart:(3) formation of the ventricular outflow tracts, arterial valves, and intrapericardial arterial trunks," *Heart*, vol. 89, no. 9, pp. 1110–1118, 2003.

3. E. Fallot, "Contribution a l'anatomie pathologique de la maladie bleue (cyanotic cardiaque)," *Marseille méd*, vol. 25, pp. 77–138, 1888.

4. M. Genet, L. C. Lee, B. Baillargeon, J. M. Guccione, and E. Kuhl, "Modeling pathologies of diastolic and systolic heart failure," *Annals of biomedical engineering*, vol. 44, no. 1, pp. 112–127, 2016.

5. J. E. Hall and M. E. Hall, *Guyton and Hall textbook of medical physiology e-Book*. Elsevier Health Sciences, 2020.

6. L. Johnson, M. A. Simon, M. R. Pinsky, and S. G. Shroff, "Insights into the effects of contraction dyssynchrony on global left ventricular mechano-energetic function," *Pacing and Clinical Electrophysiology*, vol. 32, no. 2, pp. 224–233, 2009.

7. G. D. Nollert, S. H. Däbritz, M. Schmoeckel, C. Vicol, and B. Reichart, "Risk factors for sudden death after repair of tetralogy of Fallot," *The Annals of thoracic surgery*, vol. 76, no. 6, pp. 1901–1905, 2003.

8. T. Oosterhof, F. J. Meijboom, H. W. Vliegen, M. G. Hazekamp, A. H. Zwinderman, B. J. Bouma, A. P. van Dijk, and B. J. Mulder, "Long-term follow-up of homograft function after pulmonary valve replacement in patients with tetralogy of Fallot," *European heart journal*, vol. 27, no. 12, pp. 1478–1484, 2006.

9. M. M. Cheung, I. E. Konstantinov, and A. N. Redington, "Late complications of repair of tetralogy of Fallot and indications for pulmonary valve replacement," in *Seminars in thoracic and cardiovascular surgery*, vol. 17, pp. 155–159, Elsevier, 2005.

10. M. A. Gatzoulis, J. A. Till, J. Somerville, and A. N. Redington, "Mechanoelectrical interaction in tetralogy of Fallot: QRS prolongation relates to right ventricular size and predicts malignant ventricular arrhythmias and sudden death," *Circulation*, vol. 92, no. 2, pp. 231–237, 1995.

11. A. Sabate Rotes, H. M. Connolly, C. A. Warnes, N. M. Ammash, S. D. Phillips, J. A. Dearani, H. V. Schaff, H. M. Burkhart, D. O. Hodge, S. J. Asirvatham, *et al.*, "Ventricular arrhythmia risk stratification in patients with tetralogy of Fallot at the time of pulmonary valve replacement," *Circulation: Arrhythmia and Electrophysiology*, vol. 8, no. 1, pp. 110–116, 2015.

12. J. P. Bokma, T. Geva, L. A. Sleeper, S. V. B. Narayan, R. Wald, K. Hickey, K. Jansen, R. Wassall, M. Lu, M. A. Gatzoulis, *et al.*, "A propensity score-adjusted analysis of clinical outcomes after pulmonary valve replacement in tetralogy of Fallot," *Heart*, vol. 104, no. 9, pp. 738–744, 2018.

13. T. Geva, B. Mulder, K. Gauvreau, S. V. Babu-Narayan, R. M. Wald, K. Hickey, A. J. Powell, M. A. Gatzoulis, and A. M. Valente, "Preoperative predictors of death and sustained ventricular tachycardia after pulmonary valve replacement in patients with repaired tetralogy of Fallot enrolled in the INDICATOR cohort," *Circulation*, vol. 138, no. 19, pp. 2106–2115, 2018.

14. E. A. Nielsen, M. Sun, O. Honjo, V. E. Hjortdal, A. N. Redington, and M. K. Friedberg, "Dual Endothelin Receptor Blockade Abrogates Right Ventricular Remodeling and Biventricular Fibrosis in Isolated Elevated Right Ventricular Afterload," *PLoS ONE*, vol. 11, no. 1, p. e0146767, 2016.

15. G. Nollert, T. Fischlein, S. Bouterwek, C. Böhmer, W. Klinner, and B. Reichart, "Long-term survival in patients with repair of tetralogy of Fallot: 36-year follow-up of 490 survivors of the first year after surgical repair," *Journal of the American College of Cardiology*, vol. 30, no. 5, pp. 1374–1383, 1997.

16. V. Wang, P. Nielsen, and M. Nash, "Image-based predictive modeling of heart mechanics," *Annual review of biomedical engineering*, vol. 17, pp. 351–383, 2015.

17. M. Gusseva, T. Hussain, C. Hancock Friesen, P. Moireau, A. Tandon, G. Greil, K. Hasbani, D. Chapelle, and R. Chabiniok, "Biomechanical modeling to inform pulmonary valve replacement in tetralogy of Fallot patients after complete repair," *Canadian Journal of Cardiology*, 2021.

18. M. Caruel, R. Chabiniok, P. Moireau, Y. Lecarpentier, and D. Chapelle, "Dimensional reductions of a cardiac model for effective validation and calibration," *Biomechanics and Modeling in Mechanobiology*, vol. 13, pp. 897–914, Aug. 2014.

19. N. Stergiopulos, B. E. Westerhof, and N. Westerhof, "Total arterial inertance as the fourth element of the Windkessel model," *American Journal of Physiology-Heart and Circulatory Physiology*, vol. 276, pp. H81–H88, Jan. 1999.

20. J. Sainte-Marie, D. Chapelle, R. Cimrman, and M. Sorine, "Modeling and estimation of the cardiac electromechanical activity," *Computers & Structures*, vol. 84, pp. 1743–1759, Nov. 2006.

21. D. Chapelle, P. Le Tallec, P. Moireau, and M. Sorine, "An energy-preserving muscle tissue model: formulation and compatible discretizations," *International Journal for Multiscale Computational Engineering*, vol. 10, no. 2, pp. 189–211, 2012.

22. P. Lurz, J. Nordmeyer, V. Muthurangu, S. Khambadkone, R. Derrick, R. Yates, M. Sury, P. Bonhoeffer, and A. M. Taylor, "Comparison of Bare Metal Stenting and Percutaneous Pulmonary Valve Implantation for Treatment of Right Ventricular Outflow Tract Obstruction," *Circulation*, vol. 119, pp. 2995–3001, June 2009.

23. A. T. Yetman, K.-J. Lee, R. Hamilton, W. R. Morrow, and B. W. McCrindle, "Exercise capacity after repair of tetralogy of Fallot in infancy," *American Journal of Cardiology*, vol. 87, no. 8, pp. 1021–1023, 2001.

24. B. Ruijsink, K. Zugaj, J. Wong, K. Pushparajah, T. Hussain, P. Moireau, R. Razavi, D. Chapelle, and R. Chabiniok, "Dobutamine stress testing in patients with Fontan circulation augmented by biomechanical modeling," *PLoS one*, vol. 15, no. 2, p. e0229015, 2020.

25. P. Khairy, N. Poirier, and L.-A. Mercier, "Univentricular heart," *Circulation*, vol. 115, no. 6, pp. 800–812, 2007.

26. A. Le Gall, F. Vallée, K. Pushparajah, T. Hussain, A. Mebazaa, D. Chapelle, E. Gayat, and R. Chabiniok, "Monitoring of cardiovascular physiology augmented by a patient-specific biomechanical model during general anesthesia. A proof of concept study," *PLoS one*, vol. 15, no. 5, p. e0232830, 2020.

27. M. Ahmed, S. Advani, A. Moreira, S. Zoretic, J. Martinez, K. Chorath, S. Acosta, R. Naqvi, F. Burmeister-Morton, F. Burmeister, *et al.*, "Multisystem inflammatory syndrome in children: A systematic review," *EClinicalMedicine*, vol. 26, p. 100527, 2020.

28. Z. Belhadjer, M. Méot, F. Bajolle, D. Khraiche, A. Legendre, S. Abakka, J. Auriau, M. Grimaud, M. Oualha, M. Beghetti, *et al.*, "Acute heart failure in multisystem inflammatory syndrome in children (MIS-C) in the context of global SARS-CoV-2 pandemic," *Circulation*, 2020.

29. R. Waugh, M. Abdelghafar Hussein, J. Weller, K. Sharma, G. Greil, J. Kahn, T. Hussain, and R. Chabiniok, "Cardiac modeling for Multisystem Inflammatory Syndrome in Children (MIS-C, PIMS-TS)," in *International Conference on Functional Imaging and Modeling of the Heart (FIMH)*, pp. 435–446, Springer, 2021.

30. E. K. Rodriguez, A. Hoger, and A. D. McCulloch, "Stress-dependent finite growth in soft elastic tissues.," *Journal of biomechanics*, vol. 27, pp. 455–467, Apr. 1994.

31. W. Kroon, T. Delhaas, T. Arts, and P. Bovendeerd, "Computational modeling of volumetric soft tissue growth: application to the cardiac left ventricle," *Biomechanics and Modeling in Mechanobiology*, vol. 8, pp. 301–309, Aug. 2009.

32. S. Göktepe, O. J. Abilez, and E. Kuhl, "A generic approach towards finite growth with examples of athlete's heart, cardiac dilation, and cardiac wall thickening," *Journal of the mechanics and physics of solids*, vol. 58, pp. 1661–1680, Oct. 2010.

33. S. Göktepe, O. J. Abilez, K. K. Parker, and E. Kuhl, "A multiscale model for eccentric and concentric cardiac growth through sarcomerogenesis," *Journal of Theoretical Biology*, vol. 265, no. 3, pp. 433–442, 2010.

34. R. C. P. Kerckhoffs, J. Omens, and A. D. McCulloch, "A single strain-based growth law predicts concentric and eccentric cardiac growth during pressure and volume overload.," *Mechanics research communications*, vol. 42, pp. 40–50, June 2012.

35. L. C. Lee, M. Genet, G. Acevedo-Bolton, K. Ordovas, J. M. Guccione, and E. Kuhl, "A computational model that predicts reverse growth in response to mechanical unloading," *Biomechanics and Modeling in Mechanobiology*, vol. 14, no. 2, pp. 217–229, 2015.

36. J. A. Niestrawska, C. M. Augustin, and G. Plank, "Computational modeling of cardiac growth and remodeling in pressure overloaded hearts—Linking microstructure to organ phenotype," *Acta biomaterialia*, vol. 106, pp. 34–53, 2020.

37. G. A. Ateshian and J. D. Humphrey, "Continuum mixture models of biological growth and remodeling: past successes and future opportunities," *Annual review of biomedical engineering*, vol. 14, pp. 97–111, 2012.

38. P. W. Alford, J. D. Humphrey, and L. A. Taber, "Growth and remodeling in a thick-walled artery model: effects of spatial variations in wall constituents," *Biomechanics and Modeling in Mechanobiology*, vol. 7, pp. 245–262, Aug. 2008.

39. W. J. Lin, M. D. Iafrati, R. A. Peattie, and L. Dorfmann, "Growth and remodeling with application to abdominal aortic aneurysms," *Journal of Engineering Mathematics*, vol. 109, pp. 113–137, Apr. 2018.

40. Task Force on the Management of Grown Up Congenital Heart Disease of the European Society of Cardiology, J. Deanfield, E. Thaulow, C. Warnes, G. Webb, F. Kolbel, A. Hoffman, K. Sorenson, H. Kaemmerer, U. Thilen, *et al.*, "Management of grown up congenital heart disease," *European heart journal*, vol. 24, no. 11, pp. 1035–1084, 2003.

41. W. Hui, C. Slorach, A. Dragulescu, L. Mertens, B. Bijnens, and M. K. Friedberg, "Mechanisms of right ventricular electromechanical dyssynchrony and mechanical inefficiency in children after repair of tetralogy of Fallot," *Circulation: Cardiovascular Imaging*, vol. 7, no. 4, pp. 610–618, 2014.

42. D. Yim, W. Hui, G. Larios, A. Dragulescu, L. Grosse-Wortmann, B. Bijnens, L. Mertens, and M. K. Friedberg, "Quantification of right ventricular electromechanical dyssynchrony in relation to right ventricular function and clinical outcomes in children with repaired tetralogy of Fallot," *Journal of the American Society of Echocardiography*, vol. 31, no. 7, pp. 822–830, 2018.

43. J. Lumens, C.-P. S. Fan, J. Walmsley, D. Yim, C. Manlhiot, A. Dragulescu, L. Grosse-Wortmann, L. Mertens, F. W. Prinzen, T. Delhaas, and M. K. Friedberg, "Relative impact of right ventricular electromechanical dyssynchrony versus pulmonary regurgitation on right ventricular dysfunction and exercise intolerance in patients after repair of tetralogy of Fallot," *Journal of the American Heart Association*, vol. 8, no. 2, p. e010903, 2019.

44. T. Geva, K. Gauvreau, A. J. Powell, F. Cecchin, J. Rhodes, J. Geva, and P. del Nido, "Randomized trial of pulmonary valve replacement with and without right ventricular remodeling surgery," *Circulation*, vol. 122, no. 11_suppl_1, pp. S201–S208, 2010.

45. K. S. Motonaga and A. M. Dubin, "Cardiac resynchronization therapy for pediatric patients with heart failure and congenital heart disease: a reappraisal of results," *Circulation*, vol. 129, no. 18, pp. 1879–1891, 2014.

46. P. Vojtovič, F. Kučera, P. Kubuš, R. Gebauer, T. Matějka, T. Tláskal, M. Ložek, J. Kovanda, and J. Janoušek, "Acute right ventricular resynchronization improves haemodynamics in children after surgical repair of tetralogy of Fallot," *EP Europace*, vol. 20, no. 2, pp. 323–328, 2018.

47. J. Janoušek, J. Kovanda, M. Ložek, V. Tomek, P. Vojtovič, R. Gebauer, P. Kubuš, M. Krejčíř, J. Lumens, T. Delhaas, *et al.*, "Pulmonary right ventricular resynchronization in congenital heart disease: acute improvement in right ventricular mechanics and contraction efficiency," *Circulation: Cardiovascular Imaging*, vol. 10, no. 9, p. e006424, 2017.

48. P. Kubuš, O. Materna, P. Tax, V. Tomek, and J. Janoušek, "Successful permanent resynchronization for failing right ventricle after repair of tetralogy of Fallot," *Circulation*, vol. 130, no. 22, pp. e186–e190, 2014.

49. J. Janoušek, J. Kovanda, M. Ložek, V. Tomek, R. Gebauer, and P. Kubuš, "Cardiac resynchronization therapy for treatment of chronic pulmonary right ventricular dysfunction in congenital heart disease," in *Heart Rhythm*, vol. 15(suppl), p. S270, 2018. Abstract.

50. S. A. Niederer, G. Plank, P. Chinchapatnam, M. Ginks, P. Lamata, K. S. Rhode, C. A. Rinaldi, R. Razavi, and N. P. Smith, "Length-dependent tension in the failing heart and the efficacy of cardiac resynchronization therapy," *Cardiovascular research*, vol. 89, no. 2, pp. 336–343, 2010.

51. M. Sermesant, R. Chabiniok, P. Chinchapatnam, T. Mansi, F. Billet, P. Moireau, J.-M. Peyrat, K. Wong, J. Relan, K. Rhode, *et al.*, "Patient-specific electromechanical models of the heart for the prediction of pacing acute effects in CRT: a preliminary clinical validation," *Medical image analysis*, vol. 16, no. 1, pp. 201–215, 2012.

52. R. Chabiniok, P. Moireau, C. Kiesewetter, T. Hussain, R. Razavi, and D. Chapelle, "Assessment of atrioventricular valve regurgitation using biomechanical cardiac modeling," in *International Conference on Functional Imaging and Modeling of the Heart (FIMH)*, pp. 401–411, Springer, 2017.

53. J. A. Maintz and M. A. Viergever, "A survey of medical image registration," *Medical image analysis*, vol. 2, no. 1, pp. 1–36, 1998.

54. C. Hogea, C. Davatzikos, and G. Biros, "Brain–tumor interaction biophysical models for medical image registration," *SIAM Journal on Scientific Computing*, vol. 30, no. 6, pp. 3050–3072, 2008.

55. X. Deng and T. S. Denney, "Three-dimensional myocardial strain reconstruction from tagged MRI using a cylindrical B-spline model," *IEEE Transactions on Medical Imaging*, vol. 23, no. 7, pp. 861–867, 2004.

56. M. Heinrich, J. Schnabel, F. Gleeson, F. Brady, and M. Jenkinson, "Non-rigid multimodal medical image registration using optical flow and gradient orientation," *Proc. Medical Image Analysis and Understanding*, pp. 141–145, 2010.

57. A. Roche, G. Malandain, X. Pennec, and N. Ayache, "The correlation ratio as a new similarity measure for multimodal image registration," in *International Conference on Medical Image Computing and Computer-Assisted Intervention*, pp. 1115–1124, Springer, 1998.

58. X. Lu, S. Zhang, H. Su, and Y. Chen, "Mutual information-based multimodal image registration using a novel joint histogram estimation," *Computerized Medical Imaging and Graphics*, vol. 32, no. 3, pp. 202–209, 2008.

59. D. R. Messroghli, A. Radjenovic, S. Kozerke, D. M. Higgins, M. U. Sivananthan, and J. P. Ridgway, "Modified Look-Locker inversion recovery (MOLLI) for high-resolution T1 mapping of the heart," *Magnetic Resonance in Medicine*, vol. 52, no. 1, pp. 141–146, 2004.

60. K. Škardová, T. Oberhuber, J. Tintěra, and R. Chabiniok, "Signed-distance function based non-rigid registration of image series with varying image intensity," *Discrete & Continuous Dynamical Systems - S*, vol. 14, no. 3, pp. 1145–1160, 2021.

61. L. Axel and L. Dougherty, "MR imaging of motion with spatial modulation of magnetization," *Radiology*, vol. 171, no. 3, pp. 841–845, 1989.

62. N. F. Osman, W. S. Kerwin, E. R. McVeigh, and J. L. Prince, "Cardiac motion tracking using CINE harmonic phase (HARP) magnetic resonance imaging," *Magnetic Resonance in Medicine*, vol. 42, no. 6, pp. 1048–1060, 1999.

63. T. Arts, F. W. Prinzen, T. Delhaas, J. R. Milles, A. C. Rossi, and P. Clarysse, "Mapping displacement and deformation of the heart with local sine-wave modeling," *IEEE transactions on medical imaging*, vol. 29, no. 5, pp. 1114–1123, 2010.

64. H. Mella, J. Mura, H. Wang, M. D. Taylor, R. Chabiniok, J. Tintera, J. Sotelo, and S. Uribe, "HARP-I: A harmonic phase interpolation method for the estimation of motion from tagged MR images," *IEEE transactions on medical imaging*, vol. 40, no. 4, pp. 1240–1252, 2021.

65. M. Genet, C. T. Stoeck, C. Von Deuster, L. C. Lee, and S. Kozerke, "Equilibrated warping: finite element image registration with finite strain equilibrium gap regularization," *Medical image analysis*, vol. 50, pp. 1–22, 2018.

66. L. C. Lee and M. Genet, "Validation of equilibrated warping—image registration with mechanical regularization—on 3D ultrasound images," in *International Conference on Functional Imaging and Modeling of the Heart (FIMH)*, pp. 334–341, Springer, 2019.

67. B. E. Burkhardt, M. N. Velasco Forte, S. Durairaj, I. Rafiq, I. Valverde, A. Tandon, J. Simpson, and T. Hussain, "Timely pulmonary valve replacement may allow preservation of left ventricular circumferential strain in patients with tetralogy of Fallot," *Frontiers in Pediatrics*, vol. 5, p. 39, 2017.

68. D. A. Castellanos, K. Škardová, A. Bhattaru, G. Greil, A. Tandon, J. Dillenbeck, B. Burkhardt, T. Hussain, M. Genet, and R. Chabiniok, "Left ventricular torsion obtained using equilibrated warping in patients with repaired tetralogy of Fallot," *Pediatric Cardiology*, vol. 42, no. 6, pp.1275–1283, 2021. https://doi.org/10.1007/s,2000246-021-02608-y

69. K. Škardová, M. Rambausek, R. Chabiniok, and M. Genet, "Mechanical and imaging models-based image registration," in *ECCOMAS Thematic Conference on Computational Vision and Medical Image Processing*, pp. 77–85, Springer, 2019.

70. B. Ruijsink, E. Puyol-Antón, M. Usman, J. van Amerom, P. Duong, M. N. V. Forte, K. Pushparajah, A. Frigiola, D. A. Nordsletten, A. P. King, *et al.*, "Semi-automatic cardiac and respiratory gated MRI for cardiac assessment during exercise," in *Molecular Imaging, Reconstruction and Analysis of Moving Body Organs, and Stroke Imaging and Treatment*, pp. 86–95, Springer, 2017.

71. D. Chapelle, M. Fragu, V. Mallet, and P. Moireau, "Fundamental principles of data assimilation underlying the Verdandi library: applications to biophysical model personalization within euHeart," *Medical & biological engineering & computing*, vol. 51, no. 11, pp. 1221–1233, 2013.

72. R. Chabiniok, G. Bureau, A. Groth, J. Tintera, J. Weese, D. Chapelle, and P. Moireau, "Cardiac displacement tracking with data assimilation combining a biomechanical model and an automatic contour detection," in *International Conference on Functional Imaging and Modeling of the Heart (FIMH)*, pp. 405–414, Springer, 2019.

73. P. Moireau, C. Bertoglio, N. Xiao, C. A. Figueroa, C. Taylor, D. Chapelle, and J.-F. Gerbeau, "Sequential identification of boundary support parameters in a fluid-structure vascular model using patient image data," *Biomechanics and modeling in mechanobiology*, vol. 12, no. 3, pp. 475–496, 2013.

74. V. Y. Wang, H. Lam, D. B. Ennis, B. R. Cowan, A. A. Young, and M. P. Nash, "Modelling passive diastolic mechanics with quantitative MRI of cardiac structure and function," *Medical image analysis*, vol. 13, no. 5, pp. 773–784, 2009.

75. M. Hadjicharalambous, R. Chabiniok, L. Asner, E. Sammut, J. Wong, G. Carr-White, J. Lee, R. Razavi, N. Smith, and D. Nordsletten, "Analysis of passive cardiac constitutive laws for parameter estimation using 3D tagged MRI," *Biomechanics and modeling in mechanobiology*, vol. 14, no. 4, pp. 807–828, 2015.

76. M. Hadjicharalambous, L. Asner, R. Chabiniok, E. Sammut, J. Wong, D. Peressutti, E. Kerfoot, A. King, J. Lee, R. Razavi, *et al.*, "Non-invasive model-based assessment of passive left-ventricular myocardial stiffness in healthy subjects and in patients with non-ischemic dilated cardiomyopathy," *Annals of biomedical engineering*, vol. 45, no. 3, pp. 605–618, 2017.

77. R. Chabiniok, P. Moireau, P.-F. Lesault, A. Rahmouni, J.-F. Deux, and D. Chapelle, "Estimation of tissue contractility from cardiac cine-MRI using a biomechanical heart model," *Biomechanics and modeling in mechanobiology*, vol. 11, no. 5, pp. 609–630, 2012.

78. L. Asner, M. Hadjicharalambous, R. Chabiniok, D. Peresutti, E. Sammut, J. Wong, G. Carr-White, P. Chowienczyk, J. Lee, A. King, *et al.*, "Estimation of passive and active properties in the human heart using 3D tagged MRI," *Biomechanics and modeling in mechanobiology*, vol. 15, no. 5, pp. 1121–1139, 2016.

79. A. Imperiale, D. Chapelle, and P. Moireau, "Sequential data assimilation for mechanical systems with complex image data: application to tagged-MRI in cardiac mechanics," *Advanced Modeling and Simulation in Engineering Sciences*, vol. 8, no. 1, pp. 1–47, 2021.

80. C. Bertoglio, P. Moireau, and J.-F. Gerbeau, "Sequential parameter estimation for fluid–structure problems: Application to hemodynamics," *International Journal for Numerical Methods in Biomedical Engineering*, vol. 28, no. 4, pp. 434–455, 2012.

81. A. Caiazzo, F. Caforio, G. Montecinos, L. O. Muller, P. J. Blanco, and E. F. Toro, "Assessment of reduced-order unscented Kalman filter for parameter identification in 1-dimensional blood flow models using experimental data," *International journal for numerical methods in biomedical engineering*, vol. 33, no. 8, p. e2843, 2017.

82. F. Donati, S. Myerson, M. M. Bissell, N. P. Smith, S. Neubauer, M. J. Monaghan, D. A. Nordsletten, and P. Lamata, "Beyond Bernoulli: improving the accuracy and precision of noninvasive estimation of peak pressure drops," *Circulation: Cardiovascular Imaging*, vol. 10, no. 1, p. e005207, 2017.

83. D. Rueckert and J. A. Schnabel, "Model-based and data-driven strategies in medical image computing," *Proceedings of the IEEE*, vol. 108, no. 1, pp. 110–124, 2019.

84. A. Ranjan and M. J. Black, "Optical flow estimation using a spatial pyramid network," in *Proceedings of the IEEE conference on computer vision and pattern recognition*, pp. 4161–4170, 2017.

85. M. Simonovsky, B. Gutiérrez-Becker, D. Mateus, N. Navab, and N. Komodakis, "A deep metric for multimodal registration," in *International conference on medical image computing and computer-assisted intervention*, pp. 10–18, Springer, 2016.

86. G. Balakrishnan, A. Zhao, M. R. Sabuncu, J. Guttag, and A. V. Dalca, "An unsupervised learning model for deformable medical image registration," in *Proceedings of the IEEE conference on computer vision and pattern recognition*, pp. 9252–9260, 2018.

87. A. V. Dalca, G. Balakrishnan, J. Guttag, and M. R. Sabuncu, "Unsupervised learning for fast probabilistic diffeomorphic registration," in *International Conference on Medical Image Computing and Computer-Assisted Intervention*, pp. 729–738, Springer, 2018.

88. S. Bauer, L.-P. Nolte, and M. Reyes, "Fully automatic segmentation of brain tumor images using support vector machine classification in combination with hierarchical conditional random field regularization," in *International conference on medical image computing and computer-assisted intervention (MICCAI)*, pp. 354–361, Springer, 2011.

89. H. A. Vrooman, C. A. Cocosco, F. van der Lijn, R. Stokking, M. A. Ikram, M. W. Vernooij, M. M. Breteler, and W. J. Niessen, "Multi-spectral brain tissue segmentation using automatically trained k-nearest-neighbor classification," *NeuroImage*, vol. 37, no. 1, pp. 71–81, 2007.

90. A. Montillo, J. Shotton, J. Winn, J. E. Iglesias, D. Metaxas, and A. Criminisi, "Entangled decision forests and their application for semantic segmentation of CT images," in *Biennial International Conference on Information Processing in Medical Imaging*, pp. 184–196, Springer, 2011.

91. W. Bai, M. Sinclair, G. Tarroni, O. Oktay, M. Rajchl, G. Vaillant, A. M. Lee, N. Aung, E. Lukaschuk, M. M. Sanghvi, *et al.*, "Automated cardiovascular magnetic resonance image analysis with fully convolutional networks," *Journal of Cardiovascular Magnetic Resonance*, vol. 20, no. 1, p. 65, 2018.

92. K. S. Nayak, J.-F. Nielsen, M. A. Bernstein, M. Markl, P. D. Gatehouse, R. M. Botnar, D. Saloner, C. Lorenz, H. Wen, B. S. Hu, *et al.*, "Cardiovascular magnetic resonance phase contrast imaging," *Journal of Cardiovascular Magnetic Resonance*, vol. 17, no. 1, p. 71, 2015.

93. P. Dyverfeldt, M. Bissell, A. J. Barker, A. F. Bolger, C.-J. Carlhäll, T. Ebbers, C. J. Francios, A. Frydrychowicz, J. Geiger, D. Giese, *et al.*, "4D flow cardiovascular magnetic resonance consensus statement," *Journal of Cardiovascular Magnetic Resonance*, vol. 17, no. 1, p. 72, 2015.

94. K. R. O'Brien, B. R. Cowan, M. Jain, R. A. Stewart, A. J. Kerr, and A. A. Young, "MRI phase contrast velocity and flow errors in turbulent stenotic jets," *Journal of Magnetic Resonance Imaging*, vol. 28, no. 1, pp. 210–218, 2008.

95. A. J. Sederman, M. D. Mantle, C. Buckley, and L. F. Gladden, "MRI technique for measurement of velocity vectors, acceleration, and autocorrelation functions in turbulent flow," *Journal of Magnetic Resonance*, vol. 166, no. 2, pp. 182–189, 2004.
96. P. Dyverfeldt, R. Gårdhagen, A. Sigfridsson, M. Karlsson, and T. Ebbers, "On MRI turbulence quantification," *Magnetic resonance imaging*, vol. 27, no. 7, pp. 913–922, 2009.
97. J. Walheim, H. Dillinger, and S. Kozerke, "Multipoint 5D flow cardiovascular magnetic resonance-accelerated cardiac-and respiratory-motion resolved mapping of mean and turbulent velocities," *Journal of Cardiovascular Magnetic Resonance*, vol. 21, no. 1, pp. 1–13, 2019.
98. L. Zhong, J.-M. Zhang, B. Su, R. S. Tan, J. C. Allen, and G. S. Kassab, "Application of patient-specific computational fluid dynamics in coronary and intra-cardiac flow simulations: Challenges and opportunities," *Frontiers in Physiology*, vol. 9, p. 742, 2018.
99. J. Lotz, C. Meier, A. Leppert, and M. Galanski, "Cardiovascular flow measurement with phase-contrast MR imaging: Basic facts and implementation," *RadioGraphics*, vol. 22, no. 3, pp. 651–671, 2002. PMID: 12006694.
100. A. Roldán-Alzate, S. García-Rodríguez, P. V. Anagnostopoulos, S. Srinivasan, O. Wieben, and C. J. Fraçois, "Hemodynamic study of TCPC using in vivo and *in vitro* 4D flow MRI and numerical simulation," *Journal of Biomechanics*, vol. 48, no. 7, pp. 1325–1330, 2015.
101. P. D. Morris, A. Narracott, H. von Tengg-Kobligk, D. A. Silva Soto, S. Hsiao, A. Lungu, P. Evans, N. W. Bressloff, P. V. Lawford, D. R. Hose, and J. P. Gunn, "Computational fluid dynamics modelling in cardiovascular medicine," *Heart*, vol. 102, no. 1, pp. 18–28, 2016.
102. R. L. Wolf, R. L. Ehman, S. J. Riederer, and P. J. Rossman, "Analysis of systematic and random error in MR volumetric flow measurements," *Magnetic Resonance in Medicine*, vol. 30, no. 1, pp. 82–91, 1993.
103. P. Cao, Y. Duhamel, G. Olympe, B. Ramond, and F. Langevin, "A new production method of elastic silicone carotid phantom based on MRI acquisition using rapid prototyping technique," in *2013 35th Annual International Conference of the IEEE Engineering in Medicine and Biology Society (EMBC)*, pp. 5331–5334, IEEE, 2013.
104. R. Fučík, R. Galabov, P. Pauš, P. Eichler, J. Klinkovský, R. Straka, J. Tintěra, and R. Chabiniok, "Investigation of phase-contrast magnetic resonance imaging underestimation of turbulent flow through the aortic valve phantom: Experimental and computational study using lattice Boltzmann method," *Magnetic Resonance Materials in Physics, Biology and Medicine*, vol. 33, no. 5, pp. 649–662, 2020.
105. A. Hessenthaler, N. Gaddum, O. Holub, R. Sinkus, O. Röhrle, and D. Nordsletten, "Experiment for validation of fluid-structure interaction models and algorithms," *International journal for numerical methods in biomedical engineering*, vol. 33, no. 9, p. e2848, 2017.
106. R. Chabiniok, J. Hron, A. Jarolímová, J. Málek, K. Rajagopal, K. Rajagopal, H. Švihlová, and K. Tůma, "A benchmark problem to evaluate implementational issues for three-dimensional flows of incompressible fluids subject to slip boundary conditions," *Applications in Engineering Science*, p. 100038, 2021.
107. L. Campo-Deaño, R. P. Dullens, D. G. Aarts, F. T. Pinho, and M. S. Oliveira, "Viscoelasticity of blood and viscoelastic blood analogues for use in polydymethylsiloxane *in vitro* models of the circulatory system," *Biomicrofluidics*, vol. 7, no. 3, p. 034102, 2013.
108. J. R. Anderson, O. Diaz, R. Klucznik, Y. J. Zhang, G. W. Britz, R. G. Grossman, N. Lv, Q. Huang, and C. Karmonik, "Validation of computational fluid dynamics methods with anatomically exact, 3D printed MRI phantoms and 4D pcMRI," in *2014 36th Annual International Conference of the IEEE Engineering in Medicine and Biology Society*, pp. 6699–6701, IEEE, 2014.
109. L. Goubergrits, E. Riesenkampff, P. Yevtushenko, J. Schaller, U. Kertzscher, F. Berger, and T. Kuehne, "Is MRI-Based CFD Able to Improve Clinical Treatment of Coarctations of Aorta?," *Ann Biomed Eng*, vol. 43, no. 1, pp. 168–176, 2015.
110. H. Ha, J. Lantz, M. Ziegler, B. Casas, M. Karlsson, P. Dyverfeldt, and T. Ebbers, "Evaluation of aortic regurgitation with cardiac magnetic resonance imaging: a systematic review," *Sci Rep*, vol. 7, p. 46618, 2017.

111. J. Kweon, D. H. Yang, G. B. Kim, N. Kim, M. Paek, A. F. Stalder, A. Greiser, and Y.-H. Kim, "Four-dimensional flow MRI for evaluation of post-stenotic turbulent flow in a phantom: comparison with flowmeter and computational fluid dynamics," *Eur Radiol*, vol. 26, no. 10, pp. 3588–3597, 2016.

112. D. C. Wendell, M. M. Samyn, J. R. Cava, M. M. Krolikowski, and J. F. LaDisa, "The Impact of Cardiac Motion on Aortic Valve Flow Used in Computational Simulations of the Thoracic Aorta," *J Biomech Eng*, vol. 138, no. 9, p. 091001, 2016.

113. S. Miyazaki, K. Itatani, T. Furusawa, T. Nishino, M. Sugiyama, Y. Takehara, and S. Yasukochi, "Validation of numerical simulation methods in aortic arch using 4D Flow MRI," *Heart Vessels*, vol. 32, no. 8, pp. 1032–1044, 2017.

114. J. Sotelo, L. Dux-Santoy, A. Guala, J. Rodríguez-Palomares, A. Evangelista, C. Sing-Long, J. Urbina, J. Mura, D. E. Hurtado, and S. Uribe, "3D axial and circumferential wall shear stress from 4D flow MRI data using a finite element method and a Laplacian approach," *Magn Reson Med*, vol. 79, no. 5, pp. 2816–2823, 2018.

115. D. Hose, P. Lawford, A. Narracott, J. Penrose, and I. Jones, "Fluid-solid interaction: benchmarking of an external coupling of ANSYS with CFX for cardiovascular applications," *Journal of medical engineering & technology*, vol. 27, no. 1, pp. 23–31, 2003.

116. S. Pinto, E. Doutel, J. Campos, and J. Miranda, "Blood analog fluid flow in vessels with stenosis: Development of an OpenFOAM code to simulate pulsatile flow and elasticity of the fluid," *APCBEE Procedia*, vol. 7, pp. 73–79, 2013.

117. T. Krüger, H. Kusumaatmaja, A. Kuzmin, O. Shardt, G. Silva, and E. M. Viggen, "The lattice Boltzmann method," *Springer International Publishing*, vol. 10, no. 978-3, pp. 4–15, 2017.

118. R. Fučík, P. Eichler, R. Straka, P. Pauš, J. Klinkovský, and T. Oberhuber, "On optimal node spacing for immersed boundary–lattice Boltzmann method in 2D and 3D," *Computers & Mathematics with Applications*, vol. 77, no. 4, pp. 1144–1162, 2019.

119. K. Škardová, P. Eichler, T. Oberhuber, and R. Fučík, "Investigation of blood-like non-Newtonian fluid flow in stenotic arteries using the lattice Boltzmann method in 2D," in *Proceedings of ALGORITMY*, pp. 101–110, 2020.

120. J. Boyd, J. M. Buick, and S. Green, "Analysis of the Casson and Carreau-Yasuda non-Newtonian blood models in steady and oscillatory flows using the lattice Boltzmann method," *Physics of Fluids*, vol. 19, no. 9, p. 093103, 2007.

121. P. Faggiano, G. P. Aurigemma, C. Rusconi, and W. H. Gaasch, "Progression of valvular aortic stenosis in adults: literature review and clinical implications," *American heart journal*, vol. 132, no. 2, pp. 408–417, 1996.

122. C. J. Taylor, J. M. Ordóñez-Mena, A. K. Roalfe, S. Lay-Flurrie, N. R. Jones, T. Marshall, and F. R. Hobbs, "Trends in survival after a diagnosis of heart failure in the United Kingdom 2000–2017: population based cohort study," *BMJ*, vol. 364, 2019.

123. M. Viceconti, F. Pappalardo, B. Rodriguez, M. Horner, J. Bischoff, and F. Musuamba Tshinanu, "In silico trials: Verification, validation and uncertainty quantification of predictive models used in the regulatory evaluation of biomedical products," *Methods*, vol. 185, pp. 120–127, 2021. Methods on simulation in biomedicine.

124. V. Eck, J. Sturdy, and L. Hellevik, "Effects of arterial wall models and measurement uncertainties on cardiovascular model predictions," *Journal of Biomechanics*, vol. 50, pp. 188–194, 2017. Biofluid mechanics of multitude pathways: From cellular to organ.

125. V. G. Eck, W. P. Donders, J. Sturdy, J. Feinberg, T. Delhaas, L. R. Hellevik, and W. Huberts, "A guide to uncertainty quantification and sensitivity analysis for cardiovascular applications.," *International journal for numerical methods in biomedical engineering*, vol. 32, Aug. 2016.

126. J. Corral-Acero, F. Margara, M. Marciniak, C. Rodero, F. Loncaric, Y. Feng, A. Gilbert, J. F. Fernandes, H. A. Bukhari, A. Wajdan, *et al.*, "The 'Digital Twin' to enable the vision of precision cardiology," *European Heart Journal*, 2020.

127. T. Mansi, I. Voigt, B. Leonardi, X. Pennec, S. Durrleman, M. Sermesant, H. Delingette, A. M. Taylor, Y. Boudjemline, G. Pongiglione, *et al.*, "A statistical model for quantification and prediction of cardiac remodelling: Application to tetralogy of Fallot," *IEEE transactions on medical imaging*, vol. 30, no. 9, pp. 1605–1616, 2011.
128. C. Mauger, S. Govil, R. Chabiniok, K. Gilbert, S. Hegde, T. Hussain, A. D. McCulloch, C. J. Occleshaw, J. Omens, J. Perry, K. Pushparajah, A. Suinesiaputra, and A. A. Young, "Right-left ventricular shape variations in tetralogy of Fallot: Associations with pulmonary regurgitation," *Journal of Cardiovascular Magnetic Resonance*, 23(105), 2021. https://doi.org/10.1186/s12968-021-00780-x
129. F. Regazzoni, D. Chapelle, and P. Moireau, "Combining Data Assimilation and Machine Learning to build data-driven models for unknown long time dynamics - Applications in cardiovascular modeling," *International Journal for Numerical Methods in Biomedical Engineering*, 2021.

Printed in the United States
by Baker & Taylor Publisher Services